Food Engineering Automation with Robotics and AI

Food Engineering Automation with Robotics and AI

Abir Chakravorty

IIT Kharagpur, Kharagpur, India

WILEY

Published by John Wiley & Sons, Inc., Hoboken, New Jersey.

Published simultaneously in Canada.

Library of Congress Cataloging-in-Publication Data:

Names: Chakravorty, Abir author
Title: Food engineering automation with robotics and AI / Abir Chakravorty.
Description: Hoboken, New Jersey : John Wiley & Sons Inc, [2026] | Includes bibliographical references and index.
Identifiers: LCCN 2025026038 | ISBN 9781394316564 (hardback) | ISBN 9781394316588 (epdf) | ISBN 9781394316571 (epub) | ISBN 9781394316595 (obook)
Subjects: LCSH: Food industry and trade–Automation | Food industry and trade–Technological innovations | Artificial intelligence–Industrial applications
Classification: LCC TP372.8 .C43 2026 | DDC 664/.024–dc23/eng/20250730
LC record available at https://lccn.loc.gov/2025026038

Cover Design: Wiley
Cover Image: © onurdongel/Getty Images

Set in 9.5/12.5pts STIXTwoText by Lumina Datamatics

To my parents

Contents

About the Book

The frontier of innovation that is rapidly changing the landscape of food production is robotics and food processing. Automation, machine learning, and AI should not only be about higher productivity, but also food safety, consistency of product, sustainability, and adaptability of food systems to answer the challenges of tomorrow.

The book explains how robotics is integrated into food processing systems. It provides details of the current state of the industry, the newest technologies available, and the most important future trends that will define food production in the years to come. It goes over both theoretical knowledge and practical aspects so that it becomes a suitable read for all kinds and levels of readers, from undergraduate engineering students, researchers, up to industry practitioners.

Thus, the chapters spread widely across a gamut of topics, from the kinds of robots used in food processing to applying automation to various unit operations in food processing, and lastly, the role of AI and robotics in food safety and quality control. Each section describes case studies based on real-world examples that show what has worked and research institutions that are first off the block with these innovations.

At the gates of an age of digital transformation, synergy between robotics and food processing is no longer a possibility but a reality that will define the next generation of food systems. This book stands as a tool in navigating this exciting field, in order to learn some of the basics behind robotics and automation, or to understand state-of-the-art applications in the food sector.

We hope that the book will motivate further research, discussion, and innovation toward the achievement of more intelligent, sustainable solutions for food processing.

Preface

The food industry is being transformed by the uniting of robotics, artificial intelligence, automation techniques, and advanced sensors. Since there is more demand for food to be safer, processed more efficiently, and designed to meet specific needs and environmental standards, a large group of food processors needs to align their approaches through a unified system. In this book, we have gathered the main engineering topics, automation models used in manufacturing, AI/ML methods, and case studies that specifically match the food processing industry.

At the beginning of the book, Chapter 1 describes the use of robotics and automation in today's food processing. It gives useful knowledge about the types of robots, sorts of automation, sensors, and AI applications. It gives a thorough treatment of robotics kinematics with homogeneous matrices and using the Denavit–Hartenberg system, as well as IoT and systems for predicting food quality and safety. At the end of the chapter, there are practical case studies and exercises to help you understand the main ideas. In Chapter 2, the chapter centers on the automation of chilling and freezing because these operations are vital in the cold chain. The chapter looks into the types of controls used, any challenges encountered, applications within different sectors, and what is expected in the future. It informs us about the role of automation in changing the logistics and quality of goods stored in cold storage facilities.

In Chapter 3, the subject discusses different robots, kinematics, and artificial intelligence algorithms. At first, it looks at kinds of manipulators, then discusses both direct and inverse kinematics, various spline approaches, and basic aspects of ANN and CNN architectures. The chapter discusses ways of using A*, Dijkstra, and fuzzy logic in navigation and motion control and explains the link to food packaging and handling. In Chapter 4, it is discussed how AI/ML and robots help improve quality, promote safety, and improve tracking in the food supply. This chapter talks about sorting/grading systems, ways to handle data, ethical aspects, and economic assessments that help achieve sustainability objectives. It also studies how artificial intelligence can be used to judge food freshness and its shelf life. The fifth chapter links to robotics' use in meat, fish, and poultry processing. How to overcome certain handling difficulties, improve robotic grippers and manipulators, use CFD modeling, and use vision systems is described in this paper. Workings demonstrating robotic fish and intelligent broiler sorting give a clear example of how things have progressed.

Chapter 6 is about fuzzy-PID controllers and SCADA systems that are important in food process control. You will do simulations of pasteurization with MATLAB/Simulink and learn about SCADA and how to conduct lab experiments, mixing both theoretical and actual monitoring.

The next chapter deals with 3D food printing, self-guided robots, and controlling systems in production lines. In this chapter, 3D printing is explained along with sustainable packaging, while mobile and humanoid robots are introduced as well. Readers will also be taught about the ISA-95 hierarchy, PLCs, HMI interfaces, and ladder logic programming. Moreover, it considers modern ways of using sensors, for example, by using optical and IoT-powered devices for instant control.

Chapter 8 reviews PID strategies in detail for use in food factories. Methods of tuning, using certain control policies, and SCADA system usage are explained using examples in food heating and mixing. There are practice problems and similar examples included in the chapter.

Online spectroscopy and instrumentation are the main topics of Chapter 9 in the book. Advanced approaches such as hyperspectral imaging, X-ray detectors, and safety systems helped by AI are unveiled to the readers. The chapter further discusses how food packaging and data acquisition are used online during today's quality control and inspection processes. The final part of Chapter 10 provides information on a set of case study and simulation examples with robots. It brings together theory and practice by showing the different stages of food automation, including robotic cooking, reinforcement learning, gripper making, and cleanups in 3D printing.

This book is made for undergraduate and postgraduate students in food engineering, automation, and robotics, together with researchers, professionals, and technologists looking to improve the food industry. In every chapter, there are questions and examples worked out to help improve understanding and learning in or outside a classroom.

The goal for this book is to strengthen its readers with the skills and knowledge required to advance food processing.

Dr. Abir Chakravorty
Assistant Professor
Agricultural & Food Engineering Department
Indian Institute of Technology Kharagpur, Kharagpur, India

Acknowledgments

Thanks to my parents, Smt. Sulata Chakravorty and Shri Alok Kumar Chakravorty, I have been motivated and supported to work on this book by them.

I intend that this book helps and supports students in Food and Agricultural Engineering and Mechanical Engineering, offering useful insights and significant value during their studies and work.

About the Author

Abir Chakravorty, PhD, M.Tech, works in the field of automation, robotics, and food equipment machinery design in the Agricultural and Food Engineering Department, Indian Institute of Technology (IIT) Kharagpur. He received his PhD from the Department of Mechanical Engineering, IIT Kharagpur. He has three patents filed, one granted, fifteen and more journals, and a book published in his name.

Acronyms

AI	Artificial Intelligence
ML	Machine Learning
PLC	Programmable Logic Controller
SCADA	Supervisory Control and Data Acquisition
IoT	Internet of Things
CAD	Computer-aided Design
CAM	Computer-aided Manufacturing
M2M	Machine to Machine
HMI	Human–machine Interface
ROS	Robot Operating System
RPA	Robotic Process Automation
CNC	Computer Numerical Control
AGV	Automated Guided Vehicle
FMS	Flexible Manufacturing System
QMS	Quality Management System
OEE	Overall Equipment Efficiency
TPM	Total Productive Maintenance
PMS	Production Management System
WMS	Warehouse Management System
LIDAR	Light Detection and Ranging
MES	Manufacturing Execution System

AIoT	Artificial Intelligence of Things
CPS	Cyber-physical Systems
BMS	Building Management System
RFID	Radio Frequency Identification
UAV	Unmanned Aerial Vehicle
VMS	Visual Management System
ERP	Enterprise Resource Planning
CFD	Computational Fluid Dynamics
FPLC	Food Processing Logistics Control
DA	Data Analytics
IoS	Internet of Services
LIMS	Laboratory Information Management System
SOP	Standard Operating Procedures
PDA	Personal Digital Assistant
DCS	Distributed Control System
FAT	Factory Acceptance Testing
SAT	Site Acceptance Testing
IIoT	Industrial Internet of Things
CAD/CAM	Computer-aided Design/Computer-aided Manufacturing
SLM	Service Level Management
MOM	Manufacturing Operations Management
TQM	Total Quality Management
CMMS	Computerized Maintenance Management System
NFC	Near Field Communication
FIFO	First In, First Out
FEA	Finite Element Analysis
VFD	Variable Frequency Drive
CIM	Computer Integrated Manufacturing

Introduction

Robotics is a testimony of human ingenuity and combines into machines that are supposed to do work previously assumed to be man's alone, the complexity of engineering, artificial intelligence, and design. In the past decades, we have witnessed just how much technology in robotics has evolved from simple machines that help in the completion of repetitive tasks to self-sustaining robots with the capability of learning and adapting themselves. Robotics, integrated into industries ranging from manufacturing and healthcare to entertainment and space exploration, continues to redefine how we interact with technology.

This manuscript seeks to delve deeper into the complex world of robotics in all its multifaceted glory—the theoretical frameworks and practical applications are not excluded. The document incorporates the latest research and technologies along with countless real-life examples in a manner that will help make this overview of key concepts, challenges, and innovations comprehensive. It will touch upon some aspects of the area: the design and development of robotic systems; human-robot interaction; artificial intelligence as used in delivering robotic autonomy; and ethics involved with using robots within society.

This book, apart from giving an insight into the basic principles of robotics, accompanies it with real case studies demonstrating how theoretical concepts are actually implemented in real life. These case studies were developed from research journals and industrial implementations that actually bridge the gap between theoretical knowledge and its implementation, thus bringing concrete examples of robotics in action to the reader. From autonomous vehicles to robotic surgery, the case studies in this manuscript shed light on the successes and challenges faced by engineers and researchers in their discipline.

Throughout this book, we discuss the cross-cutting nature of the discipline of robotics, which shows the importance of collaboration and bringing together different domains-from mechanical engineering to computer science, electronics, and even cognitive science. This capability of robots in imitating human actions and decision-making processes raises the questions of their potential not only but also the ethics and society of their wider application. How will they change the workforce? What does it mean to design machines that can make decisions independently? All these are questions that continue to form the current debate about robotics.

This book aims to equip readers with a solid understanding of the current state of robotics and to instill curiosity about its future possibilities. As robotics evolves and expands into new domains, the need for educated and informed individuals in this field becomes increasingly critical. Whether you are a student, researcher, or professional working in robotics or a related field, this manuscript will equip you with the information required to interact with the challenges and opportunities that arise with this exciting and rapidly developing discipline.

1

Robotics and Automation in Food Processing, Developments Prompted by the Present Era Industrial Revolution

Abir Chakravorty

The first three industrial revolutions have indeed drastically changed the landscape of agriculture—from purely traditional farming to mechanized farming, and now, with the latest technological advances, to precision agriculture. Moving toward an industrial farming model has greatly increased productivity, but also brought on a wide range of problems that have gained strength over time. Industry 4.0 is on the threshold of revolutionizing agriculture once again, thus heralding the fourth agricultural revolution. It starts by discussing the current state of industrial agriculture and dwelling on the lessons learned from the patterns, processes, and supply chain of industrialized agricultural production. Next, we touch on five highly relevant emerging technologies that can help propel Agriculture 4.0 and beyond: IoT, robotics, AI, and big data analytics. We will focus on how these technologies are being used in the agricultural domain, and on various challenges in the research community. With this article, we hope to discover new research opportunities for the reader, especially from the academia as well as the industry side.

1.1 Abridgement of the Existing Obstacles and Opportunities Presented by the Present Era Industrial Revolution in the Food Processing Sector

The present-era of industrial technologies challenges and offers much in food processing. Below is a brief overview based on the insights from the section.

Food Engineering Automation with Robotics and AI, First Edition. Abir Chakravorty.

1.1.1 Current Challenges in Food Processing

1. Lack of Digitization: There is a lack of digitization in the food processing sector, which effectively hinders the collection and analysis of data. The lack of digital infrastructure constrains automation and highly intelligent decision-making capabilities.
2. Inefficient Agri-food Supply Chains: Agri-food supply chains are not even smartly managing their chains in terms of efficiency, which delays the delivery charges and all such waste-generating situations. This is one serious concern in food processing.
3. Ecological and Health Issues: Industrial farming, based on monoculture and intensive livestock production, can give rise to ecological health issues, public health risks, and animal welfare problems. Unsustainable food production systems are behind such practice.
4. Integration Problems: Data, element tracking, and poor problems of agricultural information systems. Such shallow integration limits the exploration of biological knowledge and makes multiscale analysis impossible.

1.1.2 Technologies in Agricultural Automation

1. IoT: IoT technologies can help offer real-time tracking and management for food processing operations, as well as efficiency and productivity gains in the supply chain. It can track resources and processes inside the chain much better.
2. Big Data Analytics: It simply involves examining large amounts of data collected from different sources that are helpful in making decisions. It enhances forecasting, quality control, and even operant performance in food processing.
3. AI: This will be used in process improvement to make predictive outputs with auto-decision-making, thereby increasing the percentage value in food processing, meaning it will yield products of higher quality and less waste generation.
4. Robotics: As long as robotics can help improve food automation on the processing fronts, it indeed makes speed and efficiency shorter for production lines. This also allows one to manage tasks that would otherwise be risky, or perhaps hazardous to people.
5. Blockchain Technology: Blockchain can create transparency and traceability of the food supply chain through a guarantee that all foods sold are safe and quality products. It provides controlled data sharing between stakeholders, very vital for any food processing standards in proper implementation (Liu et al., 2020).

Figure 1.1 The application of robotics and autonomous systems in agriculture.

Figure 1.1 depicts the timeline for the agricultural and industrial revolutions. While this might be difficult, the food-processing industry may seem severely hindered by digitization and perceptions of inefficiencies in supply chains and ecologies, but Industry 4.0 would create a commonplace where these bottlenecks are removed, and the processes are productive and sustainable.

1.2 Adoption of Technology in Food Processing and Key Industry Technologies

1.2.1 Technologies in the Present Era Food Processing

Technologies of Industry 4.0 and 5.0 transform the food industry with better efficiency, sustainability, and product quality. Some key technologies and implications are listed below:

The role of AI and machine learning

AI and ML are used in predictive analytics, optimizing the production process, and enhancing decision-making.

These technologies aid in predicting demand, thereby reducing waste and improving product quality based on data-driven insights.

The devices in IoT enable real-time monitoring and controlling of equipment and processes for better control in the production process.

This technology enhances food safety because conditions are monitored across the supply chain. Immediate responses to any issues are made possible. Big data analytics processes large volumes of data created in the food industry through different origins.

It helps identify trends and enhances operational efficiency, thereby ensuring that products are tailored according to consumers' wishes and therefore increase customer satisfaction. Blockchain technology ensures safety and transparency in tracking food products from the farm to the table.

This technology increases traceability, hence allowing safe food and confidence for consumers with this choice of credible information regarding origins. These smart sensors measure a number of parameters at the food processing stage, such as temperature, humidity, and quality. They ensure optimum conditions prevail, spoilage decreases, and, finally, product quality is assured at every step of the production cycle.

1.3 Definition of Robot and Its Various Classification

A robot is a programmable, electromechanical machine designed to perform tasks autonomously or semi-autonomously, often mimicking human or animal actions. Robots are equipped with sensors, actuators, and control systems, enabling them to interact with and adapt to their environment. Figure 1.2a represents the classification of robots; Figure 1.2b represents various types of joints between robotic links; Figure 1.2c represents various types of industrial robotic manipulator and the work volume; Figure 1.2d represents robot's reach and degree of freedom for each category of joint.

The various types of industrial robot classification are presented in Figures 1.2 and 1.3.

1.3.1 Automation and Its Various Types

Automation is the use of technology, machines, or systems to perform tasks or processes without direct human intervention. It involves a process in which functions are predetermined by guidelines or algorithms to follow, allowing for higher efficiency, accuracy, and consistency. Automation can be applied in any field—in manufacturing, data processing, transportation, and even everyday activities such as smart home management. The general objective is to minimize the need

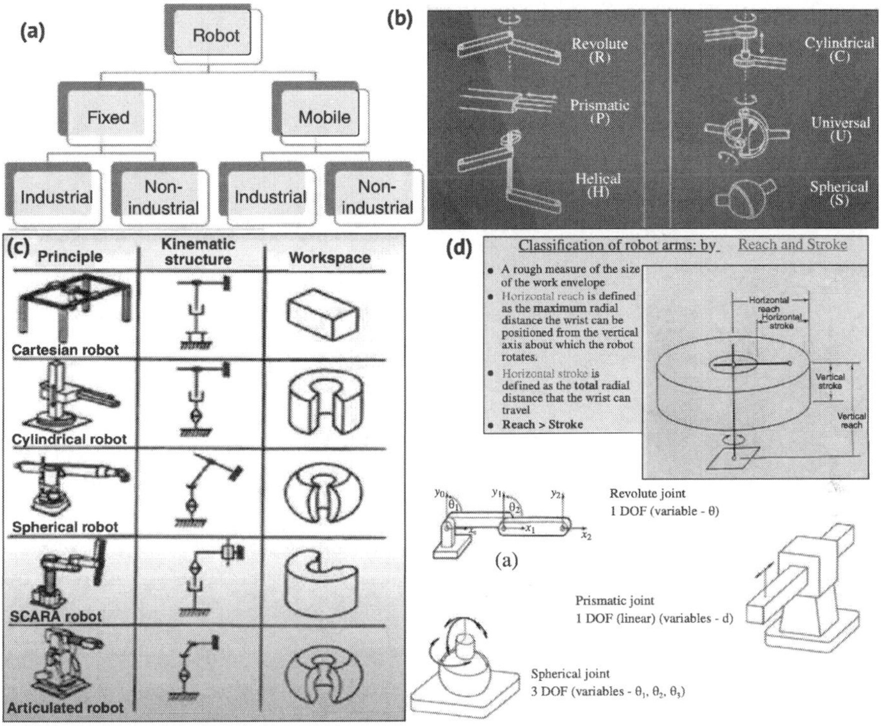

Figure 1.2 Robots classification.

for manual labor, decrease human error, and maximize productivity. Figure 1.4 describes the types of automation.

Robotics and Automation Role: There is an application of robotics to facilitate the automation of repetitive food processing, packaging, and logistics tasks. This, in turn, enhances productivity, decreases labor costs, and increases safety as human interference is limited in hazardous environments.

The industry robot classification is presented in Figure 1.3. Cyber-Physical Systems (CPS) and Digital Twins Role: CPS refers to the integration of the physical process with digital systems. This is further elaborated as digital twins that mimic real-world processes for analysis.

Impact: These technologies have upgraded the capability of food production systems for planning, monitoring, and optimization, which, in turn, leads to improved efficiency and reduced losses.

Challenge in Adopting TechnologySilo Mentality: Information is not shared among the members of this industry, which leads to a lack of collaboration and innovation.

Higher Installation Costs: The cost of installing these technologies is a limitation for small businesses.

Lack of Skills: The general lack of technical skills needed to implement and maintain these advanced technologies.

The challenges in the food domain and the solutions provided using data and AI are as shown in Figure 1.5. Industry 4.0 technologies in food processing, therefore,

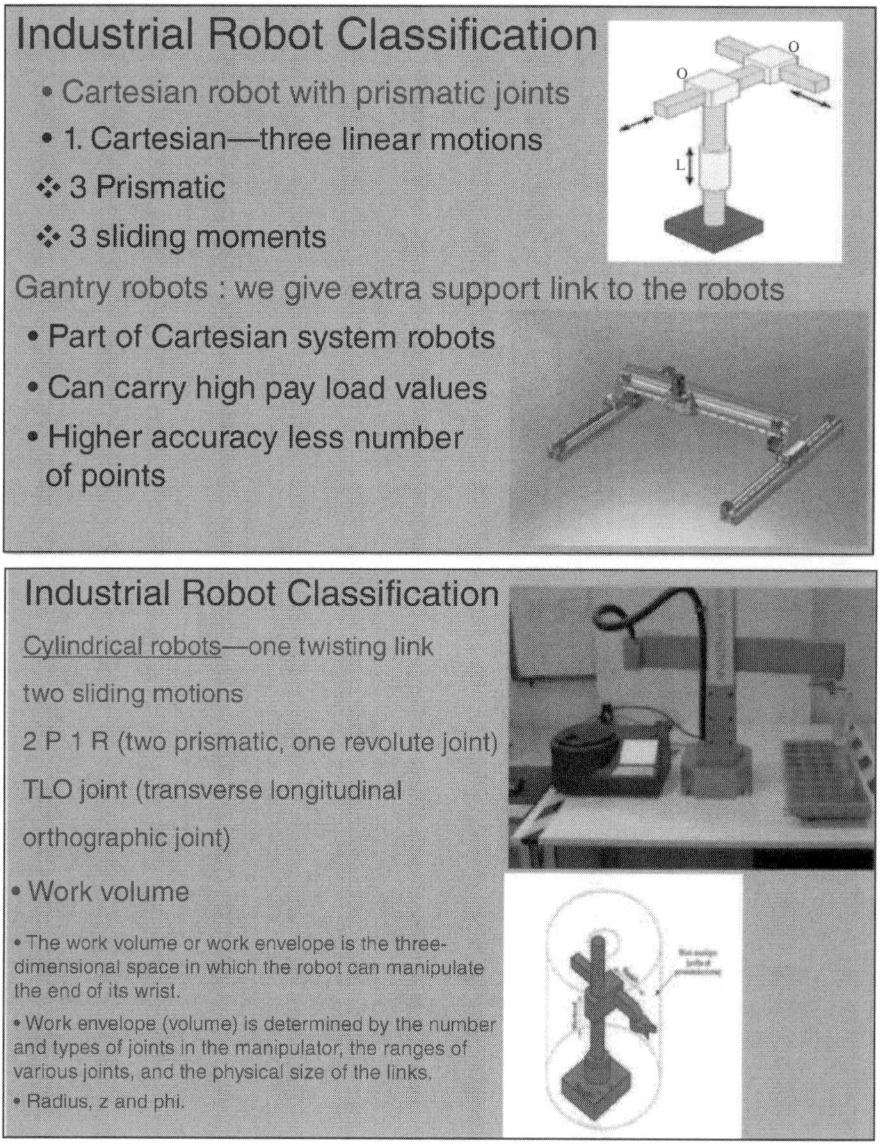

Figure 1.3 Industrial robot classification based on type of axes of motion and joints.

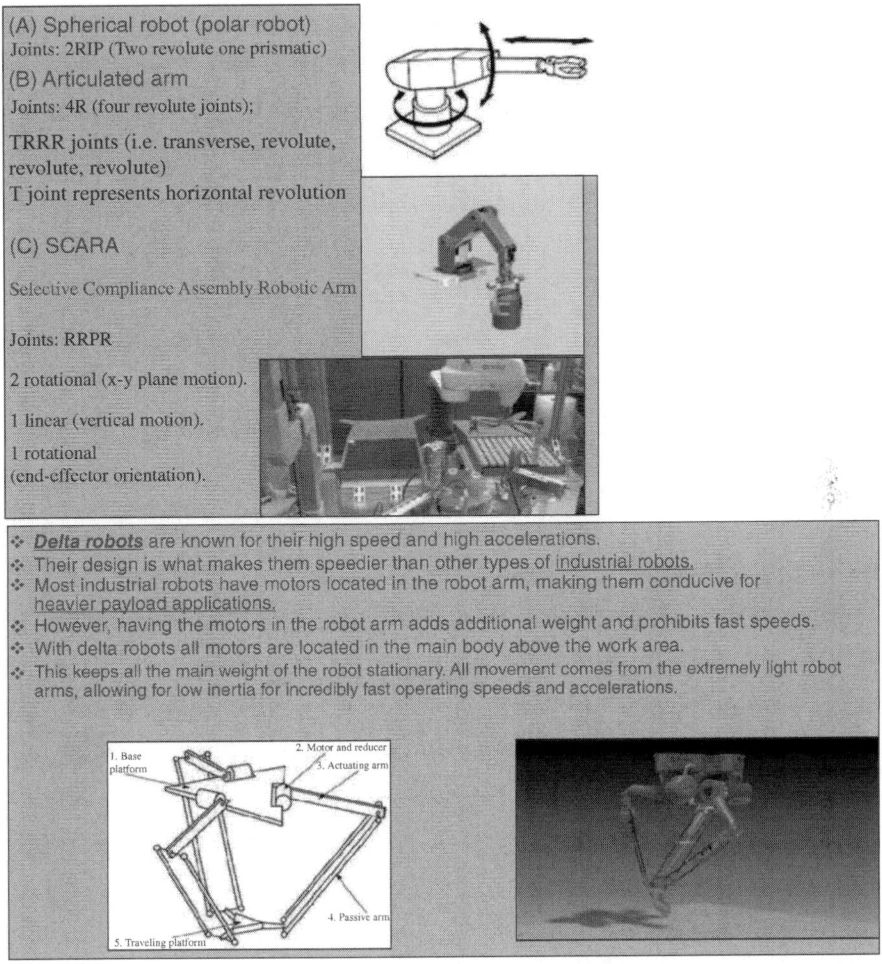

Figure 1.3 (Cont'd)

have the potential to significantly enhance sustainability and efficiency in the food industry. However, an effective response to the challenges associated with adoption is needed to bring about the intended change in the food industry. Continuous efforts at collaboration and investment in developing skills will be important in the implementation of successful Industry 4.0 solutions (Hassoun et al., 2023).

1.3.2 Overview of Industry 4.0 in Food Processing

Industry 4.0 is the integration of digital technologies in manufacturing operations, targeting higher productivity and competitiveness in food processing.

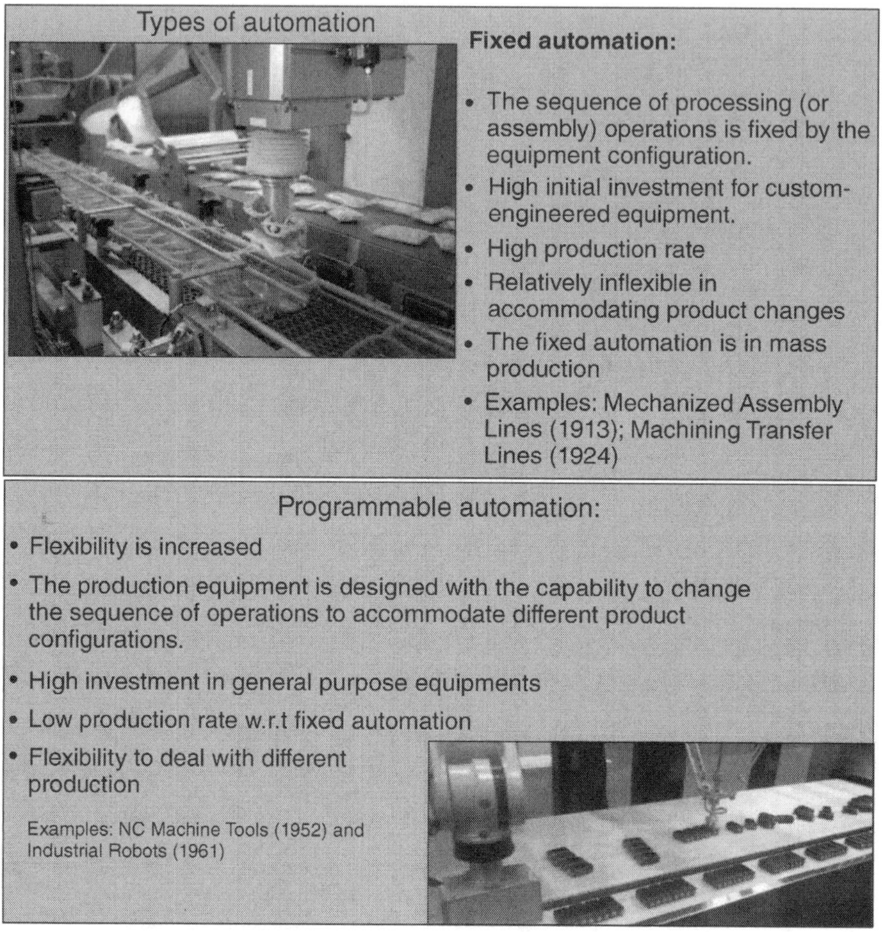

Figure 1.4 Types of automation.

Focus: This section focuses on the adaptation and implementation process in a medium-sized Italian fruit processing company of Industry 4.0 technologies on driving factors and challenges, and results thereof.

1.3.3 Driver for Adoption

1. Technological Development: Efficiency and quality can be produced through upgrades, which induce the use of Industry 4.0 technologies.
2. Competitiveness: Digital transformation is encouraged by companies to compete in the global market.
3. Sustainability: Adoption of Industry 4.0 technologies results in energy savings and sustainability improvements in the food processing field.

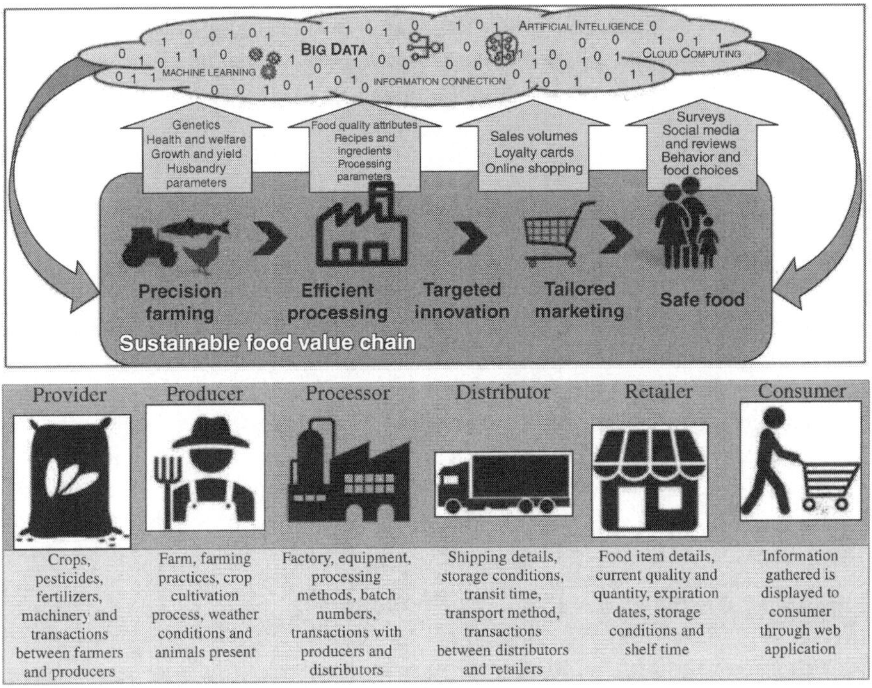

Figure 1.5 The challenges in food domain and the solution offered using data and AI.

1.3.4 Challenges

1. Implementation process: The complexity in the adoption process is caused by several facilitating and hindering factors involved in the process.
2. Human Resources Skill Gaps: Training and retraining of human resources in the new skills of data analysis and digital applications are needed to meet the new skills demanded by these application packages.
3. Management Commitment: Digital transformation will only be used to its fullest potential to overcome challenges if there is total commitment from top management.

1.3.5 Adoption Outcomes

Performance Improvement: Industry 4.0 technologies enhance firm performance, especially in the management of operations and strategy.

Organizational Benefits: Impacts on human resource management as well as on the level of economic/environmental are positive.

Value in the Long Run: The incremental adoption process has a value that lies with some technology outcomes being relevant especially in terms of traced customer information through IoT (Romanello and Veglio, 2022).

1.3.6 Future Directions

Broader Research: Future directions of study should explore the different outcomes of Industry 4.0 across various positions in the agri-food supply chain since mechanisms and outcomes are expected to vary considerably.
Policy Implications: The incentives of policymakers will focus on encouraging education programs that will have to teach emerging skills in the agri-food sector.

This case study exemplifies the potential for transformation in food processing under Industry 4.0; therefore, it requires a strategic need for managing, skilling, and continuous research with the aim of fully realizing its benefits in the agri-food sector.

1.4 Use of Robotics in Food Processing

Food processing is undergoing a revolution with robotics, increasing the industry's efficiency, quality, and safety. Below is a detailed look into how robotics is becoming a means of utilization in this area.

1.4.1 Key Benefits of Robotics in Food Processing

Cost Cut: The installation of robots results in significant reductions in labor costs, which can be a decisive factor in ensuring profitability in food processing operations.

With good efficiency, in the maneuvering of food products, robots minimize the risk of contamination and ensure superior quality standards.

Automation increases efficiency as robotics also means faster processing times, where the amount of productivity goes higher, minimizing delays due to errors from a human.

1.4.2 Application of Robotics

Sorting and Picking: They are programmed to identify fruits and vegetables based on their physicochemical properties, such as shape, size, and color. This is used very effectively for sorting and picking when fruits are fully ripe.

Packaging and Transportation: Robots are employed in packaging lines for laying products inside containers. Packaging operations have become uniform and faster with this.

Harvesting: Robots have been successfully applied to harvesting tree fruits, with leading examples in California where robots harvest berries.

1.4.3 Types of Robots Used

1. Articulated Robots: These are the ones that can be used generally for performing various tasks in multiple applications of food processing.
2. SCARA Robots: These are used for fast and precise assembly and packaging applications.
3. Delta Robots: These are ideal and suitable for high-speed picking and packing operations, and their application is mostly seen in the food industry.

1.4.4 Challenges and Considerations

1. Economic Viability: While the benefits of robotics are substantial, such a system is not economically possible for all food processing industries. Companies have to determine their demand and capacity individually before investing in robotic systems.
2. Collaboration with Human Workforce: The food processing industry generally needs a collaborative style that includes robots and human workers together to collaborate on their areas of strength. Figure 1.6 shows various applications of robotics in the food processing industry.

The potential use of robotics in the food industry is vast, with applications ranging from sorting and harvesting to packaging and transportation. As technology advances, the integration of robotics is expected to grow, leading to more efficient and reliable food processing systems (Iqbal et al., 2017).

1.5 Use of Smart Sensors in Food Processing

Smart sensors are of paramount importance in order to enhance the efficiency and the safety of food processing, especially in the beverage manufacturing industry. How these sensors have found application in the present research will be discussed in the case study below.

Direct applications of smart sensors in food processing include monitoring the quality of beverages. Beverage product quality is monitored by using smart biosensors in the form of wireless sensor networks. These sensors measure the concentrations of the key substances in real time and ensure that the products meet safety and quality standards.

Another important feature of this is wireless communication. Seamless data transfer from sensors to the monitoring system with the use of protocols such as

Figure 1.6 The various application of robotics in food processing. Generated with AI using ChatGPT (version GPT-4).

ZigBee and Bluetooth will be established. A two-way approach with such protocols is a way to achieve user convenience and flexibility in managing a production process.

1.5.1 Data Processing and Control

Smart data processing technologies, such as the ADuCM350, enable high-rate acquisition and analysis of data, enabling a production line change in accordance with real-time data, which creates an even more efficient procedure in total.

1.5.2 Environmental Monitoring

In food production plants, smart sensors can also monitor environmental parameters for the best conditions for production. These include temperature, humidity, among other factors that might affect quality.

1.5.3 Benefits of Smart Sensors

Smart sensors help automate the monitoring process through the reduced frequency of performing manual checks, thereby shortening the production cycles while also ensuring less downtime. Some of the transducers, such as phage as the bio-recognition element in various assays, are represented in Table 2.1.

Even the process of non-stop monitoring helps in better and earlier detection of any quality problems, thus ensuring low risks from the angle of foodborne diseases as well as the trust that health regulations are being followed.

Through this implementation of smart sensors, waste and resources might be reduced using various processes during the production process. Therefore, saving money will be a great outcome.

The best example of the huge potential of smart sensors and transducers (see Table 1.1) in the industry is the use of smart sensors in food processing, especially in beverage manufacturing. They have an impact through the provision of real-time data, enhanced communication, and quality control mechanisms that allow for safer and efficient practices in food production. These systems are, therefore, expected to further advance through development and application in the food industry.

1.6 Applications of AI in Food Processing

AI incorporates various machine learning algorithms and techniques to identify food contamination, which enhances food safety and quality control.

The approaches for developing these models focus on the detection of pathogens, chemicals, and foreign materials in foods. Supervised learning models are trained using large datasets with both contaminated and clean samples of any food product, which they can then recognize as indicative of contamination.

The algorithms inspect features from the training data for the identification of a clean product versus one contaminated with chemicals. This feature enables the model to learn detectability of contaminants not present in the model beforehand.

Machine learning algorithms have resulted in improved detection accuracy at levels of over 95%. This is a great improvement over the detection values that traditional methods provide. Traditional methods usually yield detection rates lower than this value.

Real-time Data Analysis: AI technologies will allow real-time monitoring and analysis of food products, ensuring quicker response times for contamination issues.

Case Study Example

Dairy Processing Plant: In a practical application, the introduction of AI systems reduced contaminating episodes by 30% at a dairy processing plant. The achievement was made by monitoring parameters that are critical, as well as temperature and humidity, which are essential in identifying sources of physical contaminants during the time of production.

The main methods and AI models that are used in contamination detection in the food industry apply primarily through machine learning algorithms, especially

Table 1.1 Various transducers in food processing industries.

Transduction	Assay type and mechanism	Target
Amperometric electrode	Components such as β-galactosidase, α-glucosidase, and β-glucosidase are released when phage-induced cell lysis occurs.	*E. coli* (K-12), *B. anthracis*, *M. Smegmatis*
Impedimetric biosensors	Target analyte-specific display peptides are produced using phage display technology.	PSMA, Antibody for P8
LAPS (light-addressable potentiometric sensor)	Engineering display peptides unique to the target analyte via phage display technology.	hPRL-3, MDAMB231
Bio-luminescence	Luciferase reporter phage.	*M. tuberculosis*, *L. monocytogenes*
Fluorescence	Phage with a fluorescent marker combined with immunomagnetic beads.	*E. coli* 0157 : H7
Quantum dots	Quantum dots coupled with biotinylated phage and streptavidin.	*E. coli* BL-21
Au-phage network	Target analyte-specific display peptides can be engineered using phage display technology.	Melanoma cells
SPR (surface plasmon resonance)	Physical adsorption and SAMs are used to immobilize affinity-selected phages.	*S. aureus*, β-galactosidase, *L. monocytogenes*
Opto-fluidic ring resonator	Engineering display peptides unique to the target analyte via phage display technology.	Streptavidin
QCM (quartz crystal microbalance)	To create display peptides unique to the target analyte, phage display technology is utilized. Physical adsorption was used to immobilize affinity-selected phages.	*S. typhimurium*
Magnetoelastic cantilever	Using phage display technology, display peptides unique to the target analyte can be created. Phage immobilization via affinity selection through physical adsorption.	*B. anthracis*, *S. typhimurium*
Magnetoelastic particle resonators	Using phage display technology, display peptides unique to the target analyte can be created. Phage immobilization via affinity selection through physical adsorption.	*B. anthracis*, *S. typhimurium*, *E. coli*

supervised learning. In addition to this, these technologies enhance not only the accuracy of detection but also the overall safety and quality of the food product. Hence, the implementation of AI in food safety practices is a gigantic stride toward addressing today's challenges that the industry would be facing in terms of contamination.

Study: Overview on AI in Food Safety and Quality Control

The application of AI has brought an evolutionary change in how the food industry manages its safety and quality control. This paper presents two cases-cum-case studies on a dairy processing plant and a meat processing facility.

1.6.1 Case Study 1: Dairy Processing Plant

Contamination Reduction: At a dairy processing plant, it was observed that contamination was reduced by 30% through the implementation of AI-based systems.

Quality Improvement: Critical parameters, such as temperature and humidity, monitored by the AI system revealed sources of physical contaminants during production.

1.6.2 Case Study 2: Meat Processing Facility

Real-time Monitoring: AI was applied to monitor hygiene practices in real-time at a meat processing facility.

Improved Compliance: AI, together with IoT technology, has permitted the continuous collection and management of hygiene parameters, and these improved significantly in terms of compliance with food safety standards.

1.6.3 Key Technologies

Computer Vision Systems: With advanced image capturing and processing techniques, they evaluate food products. These systems classify the products as either rejects or passes based on defects and contamination that have been detected in the product.

Machine Learning Algorithms: They are used in classifying, monitoring, and managing food safety and quality-related challenges. This provides more accuracy and timeliness in data provision, analysis, and risk assessment.

1.6.4 Implementation Action Steps

Step 1: Identify Needs and Challenges. Identify the specific requirements and problems of the food processing facility to determine where AI solutions would yield the most value in ROI.

Step 2: Design and Deploy AI Systems. Develop AI and IoT systems that meet the identified needs and incorporate appropriate data acquisition strategies.

1.6.5 Compliance and Reporting

1. Regulation Analysis: AI systems enable a filtering and sorting mechanism of regulatory texts and data, which also facilitates making compliance easier to read.

2. Audit Trail: This provides complete paperwork of all the rules analyzed, as well as the compliance reports, and ensures accountability in terms of meeting the food safety standards.

In addition to upgrading the quality and safety of food through its implementation, AI also facilitates compliance. The case studies suggested the importance of AI technologies in contributing to solving challenges the food industry is currently facing.

1.7 Homogeneous Transformation Matrices

We can give a description of the configuration of a body frame B in the fixed space frame A by giving both the location p of the frame B, in A coordinates, and the rotation matrix R indicating the orientation of B in coordinates for s. We collect these together in a single 4 by 4 matrix T, called a homogeneous transformation matrix, or just a transformation matrix for short. The bottom row, composed of three zeros and a one, is placed for convenience. A schematic is presented in Figure 1.7, which represents the applicability of homogeneous transformation matrices.

${}^{A}t_{B}$: Origin of frame B w.r.t. frame A $\therefore$ Point ' p ' of frame $\{B\}$ in frame $\{A\}$:

$$ {}^{A}p = {}^{A}t_{B} + {}^{A}R_{B}{}^{B}p $$

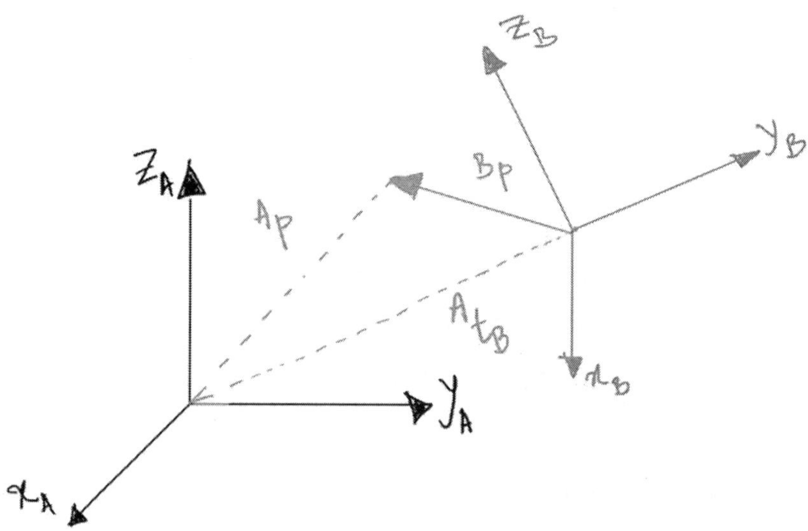

Figure 1.7 Schematic to represent the applicability of homogeneous transformation matrices.

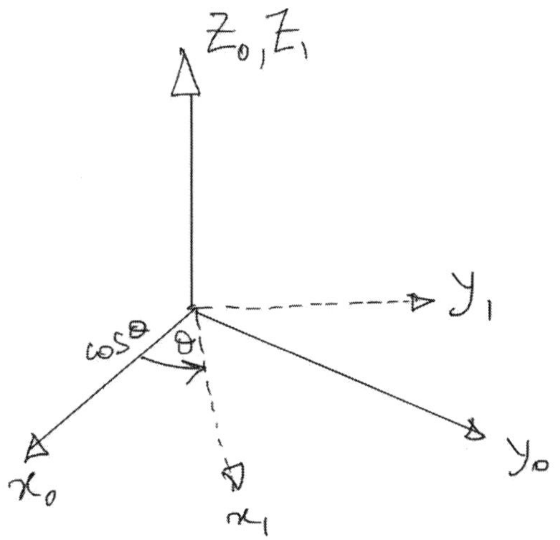

Figure 1.8 Rotation matrices.

In a more compact way:

$$\begin{bmatrix} {}^{A}p \\ 1 \end{bmatrix} = \begin{bmatrix} {}^{A}R_{B} & {}^{A}t_{B} \\ 0 & 1 \end{bmatrix} \begin{bmatrix} {}^{B}p \\ 1 \end{bmatrix}$$

where $\begin{bmatrix} {}^{A}p \\ 1 \end{bmatrix} = A_{\tilde{p}}$; and $\begin{bmatrix} {}^{B}p \\ 1 \end{bmatrix} = A_{\tilde{p}}$ are homogeneous coordinates, and $\begin{bmatrix} {}^{A}R_{B} & {}^{A}t_{B} \\ 0 & 1 \end{bmatrix} = {}^{A}T_{B}$ is the homogeneous transformation matrix.

They represent the position and orientation of a frame w.r.t. another frame rotation. Mathematically, they belong to $SE(3)\ :\ R^3 \times SO(3)$ (SE = Special Euclidean). They can also be viewed from the perspective of projective geometry. Figures 1.8 and 1.9 represent the rotation matrices, and Figure 1.10 represents the matrix for the rotation matrices.

Pure transformations:

$$T = \begin{bmatrix} R & 0 \\ 0 & 1 \end{bmatrix} \qquad T = \begin{bmatrix} I & t \\ 0 & 1 \end{bmatrix}$$

Pure rotation — Pure translation

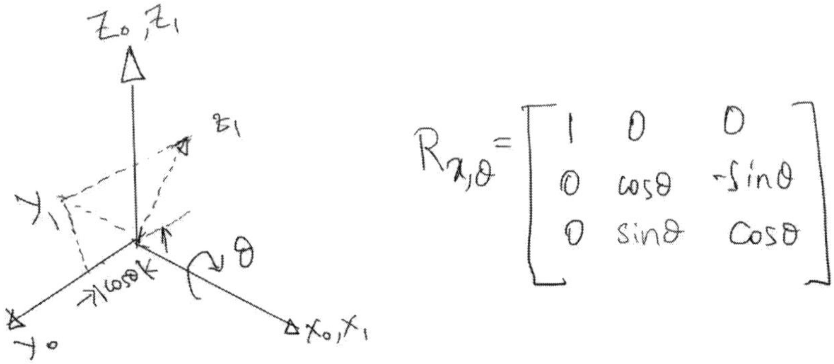

Figure 1.9 Rotation matrices.

$$P^0 \begin{bmatrix} -3S\theta_2C\theta_1 + (a_4 + d_3)\,(C\theta_2C\theta_1 + a_3C\theta_2C\theta_1 + a_2C\theta_1 \\ -3S\theta_2S\theta_1 + (a_4 + d_3)\,(C\theta_2S\theta_1) + a_3C\theta_2S\theta_1 + a_2S\theta_1 \\ (a_4 + d_3)\,S\theta_2 + a_3S\theta_2 + a_1 \end{bmatrix} \quad (13)$$

where $C\theta_1 = \text{Cos}\theta_1$; $S\theta_1 = \text{Sin}\theta_1$

Figure 1.10 Rotation matrices.

Pure rotations:

$$\text{Rot}_x(\theta) = \begin{bmatrix} 1 & 0 & 0 & 0 \\ 0 & \cos\theta & -\sin\theta & 0 \\ 0 & \sin\theta & \cos\theta & 0 \\ 0 & 0 & 0 & 1 \end{bmatrix}; \text{Rot}_y(\theta) = \begin{bmatrix} \cos\theta & 0 & \sin\theta & 0 \\ 0 & 1 & 0 & 0 \\ -\sin\theta & 0 & \cos\theta & 0 \\ 0 & 0 & 0 & 1 \end{bmatrix}$$

$$\text{Rot}_z(\theta) = \begin{bmatrix} \cos\theta & -\sin\theta & 0 & 0 \\ \sin\theta & \cos\theta & 0 & 0 \\ 0 & 0 & 1 & 0 \\ 0 & 0 & 0 & 1 \end{bmatrix}$$

Pure translations:

$$\text{Trans}_x(d) = \begin{bmatrix} 1 & 0 & 0 & d \\ 0 & 1 & 0 & 0 \\ 0 & 0 & 1 & 0 \\ 0 & 0 & 0 & 1 \end{bmatrix} \qquad \text{Trans}\, y(d) = \begin{bmatrix} 1 & 0 & 0 & 0 \\ 0 & 1 & 0 & d \\ 0 & 0 & 1 & 0 \\ 0 & 0 & 0 & 1 \end{bmatrix}$$

$$\text{Trans}_z(d) = \begin{bmatrix} 1 & 0 & 0 & 0 \\ 0 & 1 & 0 & 0 \\ 0 & 0 & 1 & d \\ 0 & 0 & 0 & 1 \end{bmatrix}$$

Example 1: Let us study the movement in a 2-revolute and one-prismatic joint manipulator.

$$\begin{array}{cc} & \begin{array}{ccc} \text{new}\,x & \text{new}\,y & \text{new}\,z \end{array} \\ R_1^0 \approx \begin{array}{r} \text{old}\,x \\ \text{old}\,y \\ \text{old}\,z \end{array} & \begin{bmatrix} 1 & 0 & 0 \\ 0 & 0 & -1 \\ 0 & 1 & 0 \end{bmatrix} \end{array}$$

$$R_1^0 = \begin{bmatrix} C\theta_1 & -S\theta_1 & 0 \\ S\theta_1 & C\theta_1 & 0 \\ 0 & 0 & 1 \end{bmatrix} \begin{bmatrix} 1 & 0 & 0 \\ 0 & 0 & -1 \\ 0 & 1 & 0 \end{bmatrix}$$

As we know, the homogeneous transformation matrices are given by

$$\left[\frac{{}^{A}R_{B}(\text{Rotation Matrix})\ |\text{Translation Matrix}|}{0\,0\,0\ (\text{Perspective Matrix})\ |\text{Scaling Matrix}|} \right]$$

$$H_1^0 = \begin{bmatrix} C\theta_1 & 0 & S\theta_1 & a_2C\theta_1 \\ S\theta_1 & 0 & -C\theta_1 & a_2S\theta_1 \\ 0 & 1 & 0 & 0 \\ 0 & 0 & 0 & 1 \end{bmatrix}$$

$$R_2^1 \approx \begin{array}{c|ccc} & \text{new } x & \text{new } y & \text{new } z \\ \hline \text{old } x & 0 & 0 & 1 \\ \text{old } y & 0 & 1 & 0 \\ \text{old } z & -1 & 0 & 0 \end{array}$$

$$R_2^1 = \begin{bmatrix} C\theta_2 & -S\theta_2 & 0 \\ S\theta_2 & C\theta_2 & 0 \\ 0 & 0 & 1 \end{bmatrix} \begin{bmatrix} 0 & 0 & 1 \\ 0 & 1 & 0 \\ -1 & 0 & 0 \end{bmatrix}$$

$$\therefore \mathrm{H}_2^1 = \begin{bmatrix} 0 & -S\theta_2 & C\theta_2 & a_3C\theta_2 \\ 0 & C\theta_2 & S\theta_2 & a_3S\theta_2 \\ -1 & 0 & 0 & 0 \\ 0 & 0 & 0 & 1 \end{bmatrix}$$

$$H_3^2 = \begin{bmatrix} 1 & 0 & 0 & 0 \\ 0 & 1 & 0 & 0 \\ 0 & 0 & 1 & a_4 + d_3 \\ 0 & 0 & 0 & 1 \end{bmatrix}; \ R_3^2 = \begin{bmatrix} 1 & 0 & 0 \\ 0 & 1 & 0 \\ 0 & 0 & 1 \end{bmatrix}$$

$\because$ There is no rotation from 2 to 3 and only translation as can be seen in Figure 1.11.

$$H_3^0 = H_1^0 H_2^1 H_3^2$$

$$H_3^0 = \begin{bmatrix} c\theta_1 & 0 & s\theta_1 & a_2c\theta_1 \\ s\theta_1 & 1 & -c\theta_1 & a_2s\theta_1 \\ 0 & 0 & 0 & a_1 \\ 0 & 0 & 0 & 1 \end{bmatrix} \begin{bmatrix} 0 & -s\theta_2 & c\theta_2 & a_3c\theta_2 \\ 0 & c\theta_2 & s\theta_2 & a_3s\theta_2 \\ -1 & 0 & 0 & 0 \\ 0 & 0 & 0 & 1 \end{bmatrix} \begin{bmatrix} 1 & 0 & 0 & 0 \\ 0 & 1 & 0 & 0 \\ 0 & 0 & 1 & a_4 + d_3 \\ 0 & 0 & 0 & 1 \end{bmatrix}$$

$$H_3^{\circ} = \begin{bmatrix} -S\theta_1 & -S\theta_2c\theta_1 & c\theta_2c\theta_1 & a_3c\theta_2c\theta_1 + a_2c\theta_1 \\ C\theta_1 & -S\theta_2S\theta_1 & c\theta_2s\theta_1 & a_3c\theta_2s\theta_1 + a_2s\theta_1 \\ 0 & 0 & S\theta_2 & a_3S\theta_2 + a_1 \\ 0 & 0 & 0 & 1 \end{bmatrix} \begin{bmatrix} 1 & 0 & 0 & 0 \\ 0 & 1 & 0 & 0 \\ 0 & 0 & 1 & a_4 + d_3 \\ 0 & 0 & 0 & 1 \end{bmatrix}$$

$$H_3^0 = \begin{bmatrix} -s\theta_1 & -s\theta_2 c\theta_1 & c\theta_2 c\theta_1 & (a_4 + d_3)(c\theta_2 c\theta_1) + a_3 c\theta_2 c\theta_1 + a_2 c\theta_1 \\ c\theta_1 & -s\theta_2 s\theta_1 & c\theta_2 s\theta_1 & (a_4 + d_3)(c\theta_2 s\theta_1) + a_3 c\theta_2 s\theta_1 + a_2 s\theta_1 \\ 0 & 0 & s\theta_2 & (a_4 + d_3)(s\theta_2) + a_3 s\theta_2 + a_1 \\ 0 & 0 & 0 & 1 \end{bmatrix}$$

The H_3^0 is the homogeneous transformation matrix of the end effector with the base frame.

Suppose a man is sitting at the end effector, his coordinates are $P^3 = \begin{bmatrix} 0 \\ 3 \\ 0 \end{bmatrix}$

The man's location with respect to the base frame can be found by

$$P^0 = \begin{bmatrix} -s\theta_1 & -s\theta_2 c\theta_1 & c\theta_2 c\theta_1 & (a_4 + d_3)(c\theta_2 c\theta_1) + a_3 c\theta_2 c\theta_1 + a_2 c\theta_1 \\ c\theta_1 & -s\theta_2 s\theta_1 & c\theta_2 s\theta_1 & (a_4 + d_3)(c\theta_2 s\theta_1) + a_3 c\theta_2 s\theta_1 + a_2 s\theta_1 \\ 0 & 0 & s\theta_2 & (a_4 + d_3)(s\theta_2) + a_3 s\theta_2 + a_1 \\ 0 & 0 & 0 & 1 \end{bmatrix} \begin{bmatrix} 0 \\ 3 \\ 0 \\ 1 \end{bmatrix}$$

Figure 1.11 Application of homogeneous transformation matrice in a 2-revolute and one-prismatic joint manipulator.

$$P^{\circ} = \begin{bmatrix} -3s\theta_2 c\theta_1 + (a_4 + d_3)(c\theta_2 c\theta_1) + a_3 c\theta_2 c\theta_1 + a_2 c\theta_1 \\ -3s\theta_2 s\theta_1 + (a_4 + d_3)(c\theta_2 s\theta_1) + a_3 c\theta_2 s\theta_1 + a_2 s\theta_1 \\ (a_4 + d_3)\, s\theta_2 + a_2 s\theta_2 + a_1 \\ 1 \end{bmatrix}$$

This is the location of the person supposed to be sitting in the end effector with respect to the base frame.

Example 2: Determine coordinates of points $(5, 3, 1)^T$ *attached to a moving frame u, v, w, undergoing (i) −90° rotation about z, (ii) translation of [5, −3, 4] about xyz, and (iii) −90° rotation about Y. Assuming the same origin of x, y, and z and u, v, w, find the final coordinates.*

$$\text{Ans:} \begin{bmatrix} C(-90) & 0 & S(-90) & 0 \\ 0 & 1 & 0 & 0 \\ -S(-90) & 0 & C(-90) & 0 \\ 0 & 0 & 0 & 1 \end{bmatrix} \begin{bmatrix} 1 & 0 & 0 & 5 \\ 0 & 1 & 0 & -3 \\ 0 & 0 & 1 & 4 \\ 0 & 0 & 0 & 1 \end{bmatrix} \begin{bmatrix} C(-90) & -S(-90) & 0 & 0 \\ S(-90) & C(-90) & 0 & 0 \\ 0 & 0 & 1 & 0 \\ 0 & 0 & 0 & 1 \end{bmatrix}$$

The first term is a rotation of −90° about the y-axis, followed by translation, and followed by a rotation of −90° about the z-axis.

$$= \begin{bmatrix} 0 & 0 & -1 & -4 \\ -1 & 0 & 0 & -3 \\ 0 & 1 & 0 & 5 \\ 0 & 0 & 0 & 1 \end{bmatrix} \begin{bmatrix} 5 \\ 3 \\ 1 \\ 1 \end{bmatrix} = \begin{bmatrix} -5 \\ -8 \\ 8 \\ 1 \end{bmatrix}$$

Example 3: A camera is attached to a gripper in a robot arm. The camera can focus on an object to determine its frame relative to the camera. Find the transformation matrix for the object relative to the robot arm using the matrices below.

$$^{\text{Cam}}_{\text{obj}}T = \begin{bmatrix} 0 & 0 & 1 & 2 \\ 1 & 0 & 0 & 2 \\ 0 & 1 & 0 & 4 \\ 0 & 0 & 0 & 1 \end{bmatrix}; \quad ^{\text{Grip}}_{\text{cam}}T = \begin{bmatrix} 0 & 0 & -1 & 3 \\ 0 & -1 & 0 & 0 \\ -1 & 0 & 0 & 5 \\ 0 & 0 & 0 & 1 \end{bmatrix}$$

$$ {}^{\text{Grip}}_{\text{arm}}T = \begin{bmatrix} 0 & -1 & 0 & 0 \\ 1 & 0 & 0 & 0 \\ 0 & 0 & 1 & 4 \\ 0 & 0 & 0 & 0 \end{bmatrix} $$

$$ {}^{\text{Arm}}_{\text{obj}}T = {}^{\text{Arm}}_{\text{Grip}}T . {}^{\text{Grip}}_{\text{cam}}T . {}^{\text{Cam}}_{\text{obj}}T $$

$$ {}^{\text{Grip}}_{\text{arm}}T = \begin{bmatrix} 0 & -1 & 0 & 0 \\ 1 & 0 & 0 & 0 \\ 0 & 0 & 1 & 4 \\ 0 & 0 & 0 & 1 \end{bmatrix}; \text{ where } {}^{\text{Grip}}_{\text{arm}}T^{-1} = {}^{\text{Arm}}_{\text{Grip}}T $$

$$ {}^{\text{Arm}}_{\text{Grip}}\text{T} = \begin{bmatrix} 0 & 1 & 0 & -(0+0+0) \\ -1 & 0 & 0 & -(0+0+0) \\ 0 & 0 & 1 & -(0+0+4) \\ 0 & 0 & 0 & 1 \end{bmatrix} = \begin{bmatrix} 0 & 1 & 0 & 0 \\ -1 & 0 & 0 & 0 \\ 0 & 0 & 1 & -4 \\ 0 & 0 & 0 & 1 \end{bmatrix} $$

$$ {}^{\text{Arm}}_{\text{obj}}\text{T} = \begin{bmatrix} 0 & 1 & 0 & 0 \\ -1 & 0 & 0 & 0 \\ 0 & 0 & 1 & -4 \\ 0 & 0 & 0 & 1 \end{bmatrix} \begin{bmatrix} 0 & 0 & -1 & 3 \\ 0 & -1 & 0 & 0 \\ -1 & 0 & 0 & 5 \\ 0 & 0 & 0 & 1 \end{bmatrix} \begin{bmatrix} 0 & 0 & 1 & 2 \\ 1 & 0 & 0 & 2 \\ 0 & 1 & 0 & 4 \\ 0 & 0 & 0 & 1 \end{bmatrix} = \begin{bmatrix} -1 & 0 & 0 & -2 \\ 0 & 1 & 0 & 1 \\ 0 & 0 & -1 & -1 \\ 0 & 0 & 0 & 1 \end{bmatrix} $$

P.S. Gaussian elimination for doing the inverse of a matrix

$$
{}_{B}^{A}T = \begin{bmatrix} a_1 & b_1 & c_1 & p_x \\ a_2 & b_2 & c_2 & p_y \\ a_3 & b_3 & c_3 & p_z \\ 0 & 0 & 0 & 1 \end{bmatrix}
$$

$$
{}_{B}^{A}T^{-1} = \begin{bmatrix} a_1 & a_2 & a_3 & -(p_x a_1 + p_y a_2 + p_z a_3) \\ b_1 & b_2 & b_3 & -(p_x b_1 + p_y b_2 + p_z b_3) \\ c_1 & c_2 & c_3 & -(p_x c_1 + p_y c_2 + p_z c_3) \\ 0 & 0 & 0 & 1 \end{bmatrix}
$$

1.8 Denavit–Hartenberg Parameters

A common way to describe the kinematics of a robotic manipulator in terms of four parameters is by using the Denavit–Hartenberg (DH) parameters. These parameters describe how a robot's kinematic chain changes from one joint to the next. Transformation matrices between the links are generated using the DH parameters, which are defined for each joint in the robot.

The four DH parameters are:

1. θ (theta): The joint angle—it represents the rotation about the Z-axis of the previous link to align with the next link.
2. d: The link offset—it represents the distance along the Z-axis between the previous and next joint, typically the translation along the Z-axis.
3. a: The link length—the distance between the Z-axes of two successive joints, measured along the X-axis.
4. α (alpha): The twist angle—the angle between the Z-axes of two successive joints, measured about the X-axis.

Given these four parameters, the transformation matrix between two frames (i.e. from one joint to the next) can be written as:

$$
T_i^{i-1} = \begin{bmatrix} \cos(\theta_i) & -\sin(\theta_i)\cos(\alpha_i) & \sin(\theta_i)\sin(\alpha_i) & a_i\cos(\theta_i) \\ \sin(\theta_i) & \cos(\theta_i)\cos(\alpha_i) & -\cos(\theta_i)\sin(\alpha_i) & a_i\sin(\theta_i) \\ 0 & \sin(\alpha_i) & \cos(\alpha_i) & d_i \\ 0 & 0 & 0 & 1 \end{bmatrix}
$$

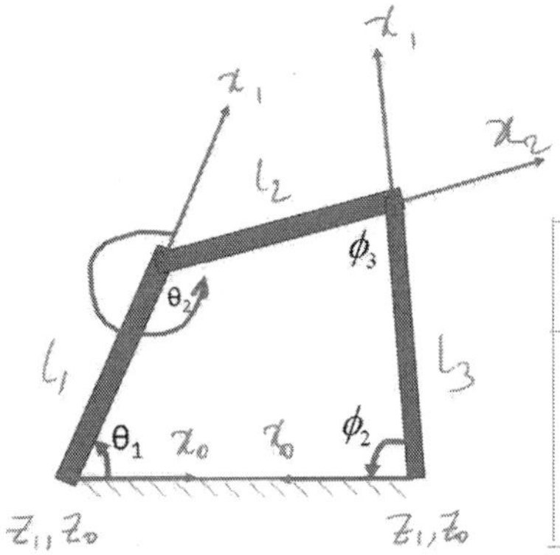

i	α_{i-1}	a_{i-1}	d_i	θ_i
1	0	0	0	θ_1
2	0	l_1	0	θ_2

Figure 1.12 Denavit–Hartenberg parameters for some example cases.

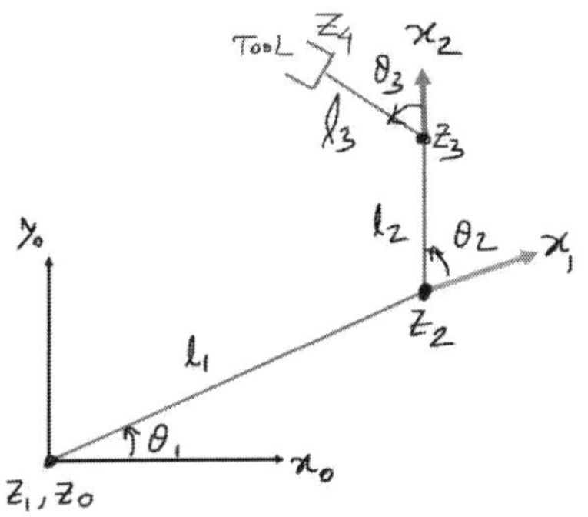

i	α_{i-1}	a_{i-1}	d_i	θ_i
1	0	0	0	θ_1
2	0	l_1	0	θ_2
3	0	l_2	0	θ_3

Figure 1.13 Denavit–Hartenberg parameters for some example cases.

Figures 1.12 and 1.13 represent the test cases of DH parameters and are as follows:

1.9 Degrees of Freedom

According to *Grübler's* formula $Dof, f = m\,(N - 1 - J) + \sum_{i=1}^{J} f_i$ where N = number of bodies, including ground J = number of joints, $m = 6$ for spatial, 3 for planar.

Joint type	Degrees of freedom, f	Constraint 'c' between two planar rigid bodies	Constraint 'c' between two spatial rigid bodies
Revolute (R)	1	2	5
Prismatic (P)	1	2	5
Helical (H)	1	N/A	5
Cylindrical (C)	2	N/A	4
Universal (U)	1	N/A	4
Spherical (S)	3	N/A	3

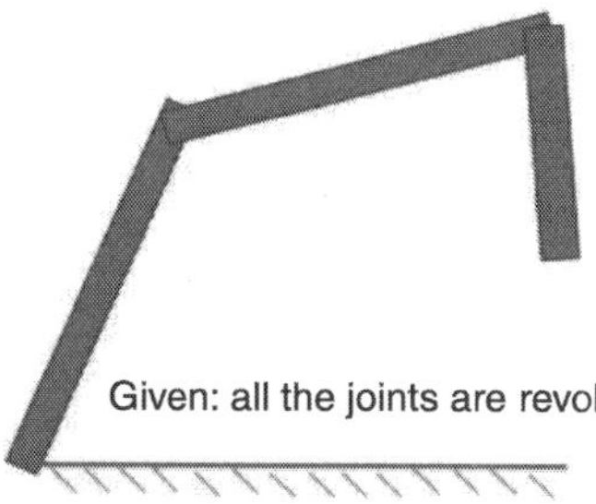

Figure 1.14 Example case degrees of freedom.

An example of degrees of freedom is represented in Figure 1.14. Let us calculate the degrees of freedom. For example in Figure 1.7, where $m = 3$, $N = 4$, $J = 3$; $\sum_{i=1}^{J} f_i = 3$, i.e. one degree of freedom for each joint.

Degree of freedom $= 3(4{-}1{-}3) + 3 = 3$. Therefore, the number of free movements or coordinates that describe how a robot and its parts may move. To each degree of freedom, corresponds a particular freedom of movement of the robot, either translating along an axis or rotating about an axis.

1.10 Applications of IoT in Food Processing

Study: Revolutionizing the Food Supply Chain with IoT Technologies

1.10.1 Reviewing IoT Implementation in FSCs

IoT Definition: The IoT refers to the worldwide framework that links physical and digital objects across interoperable data communication technologies. This has played a huge role in changing the functionality of several industries, especially in FSCs.

Present Scenario: IoT technology has advanced dramatically over the years, with several benefits offered, and far outweighing its disadvantages in the food chain supply.

1.10.2 Benefits of Integration of IoT in FSCs

Increase Sustainability: The use of IoT-based solutions within the supply chain increases its sustainability. It leads to better resource utilization and reduced waste.

Real-time Monitoring: IoT allows detailed tracking of the utilization of energy and resources in real time, significantly opening up avenues for cost savings and augmentation of efficiency in operations. For further details, refer to the IoT framework depicted in Figure 1.15.

Improved Translucency: This technology promotes increased transparency and control over food operational activities. It allows better decision-making and quick response to arising problems.

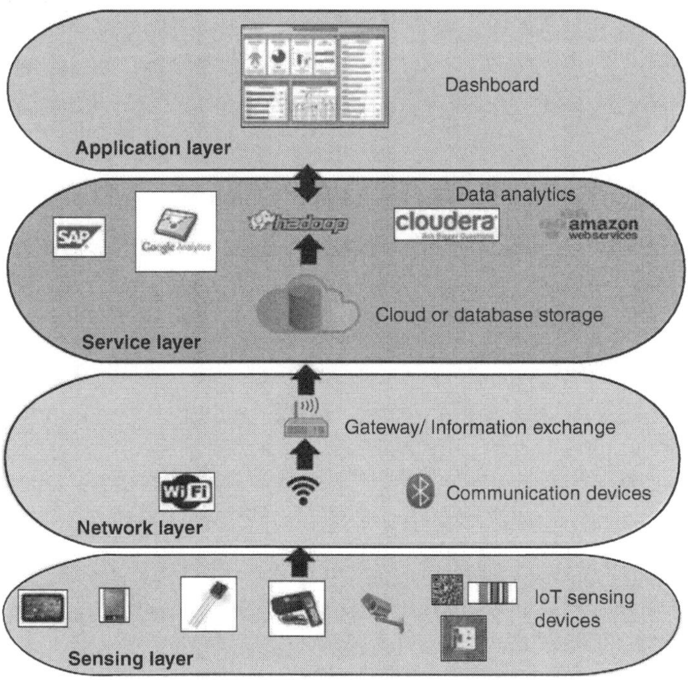

Figure 1.15 IoT architecture (Jagtap et al. [2021]/Elsevier).

1.10.3 Case Study: Simulation of a Brewery

Simulation Software: Various simulation tools such as ARENA and FlexSim were used for simulating the supply chain of the brewery under question. Diverse stakeholders were going to test parameters and optimize the same before actual production.Based on simulation, the cost plan enhancement and downtimes reduction during pre-production will also appear prominently with IoT.

1.10.4 Challenges and Considerations

Cybersecurity Risks: A food supply chain experiences a variety of cyber risks, such as data theft and production shutdown. To secure sensitive information, safeguarding against loss and interruption through monetary loss and operational losses becomes a prime concern.

Cost to Invest: While the cost of realizing digital supply chains has declined, initial investment in IoT technologies may be a challenge for many food manufacturers.

Future of IoT in FSCs: The IoT technology will thrive with long-run outcomes and its competitive advantage in the market. The companies that adopt these technologies first will secure most of the market share.

1.11 Prediction Models in Measuring Color and Food Quality

1.11.1 Introduction to Color Measurement of Food Quality

Color is one of the key sensory traits that decisively determines the buying decision by a consumer, as it reflects the food's physical, chemical, or microbiological quality. The application of hyperspectral imaging (HSI) and analysis of images is represented in Figure 1.16.

Determinations of color by classical methods, such as colorimeters and chromameter, are not the easiest tasks and cause very big difficulties when examining a nonhomogeneous sample.

HSI emerged recently as one of the fastest and least invasive technologies to investigate the sensory properties, one of which is color, in many food products.

Systems applied in HSI operate in the Vis-NIR spectral range, spanning from 400 to 1000 nm, which is often used in sensory evaluation.

1.11.2 Color Measurement with Predictive Models

Many multivariate methods of analysis have been developed to predict color characteristics such as

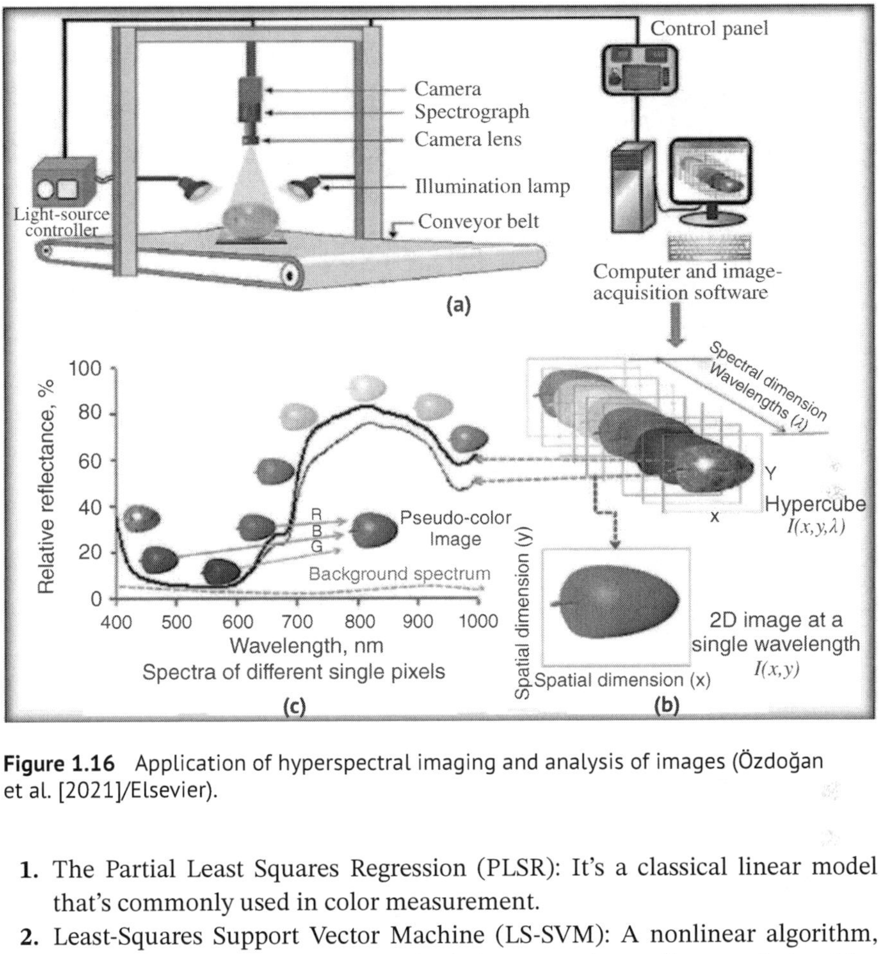

Figure 1.16 Application of hyperspectral imaging and analysis of images (Özdoğan et al. [2021]/Elsevier).

1. The Partial Least Squares Regression (PLSR): It's a classical linear model that's commonly used in color measurement.
2. Least-Squares Support Vector Machine (LS-SVM): A nonlinear algorithm, performed better in capturing both linear and non-linear relationships between color and spectral data.

Another linear method used in color prediction is multiple linear regression (MLR).

1.11.3 Performance of Prediction Models

Experiments revealed that LS-SVM models were generally better for the purpose of prediction than PLSR and MLR, specifically when the dataset was complex. For instance, the color features of potatoes were predicted using CARS-LS-SVM. The R^2 was estimated as 0.851 for L*, 0.957 for a*, and 0.924 for b*.

1.11.4 Challenges and Concerns

Even though HSI supplies a high quantity of spectral information, proper wavelength selection is required since incorrect selection may lead to loss of valuable information. High hyperspectral precision can be achieved by combining spectral and spatial information obtained from hypercubes. The integration of advanced predictive models such as LS-SVM with HSI technology provides one of the promising ways to improve accuracy in the color measurement for quality assessment in food.

The food industry continues to seek non-invasive and efficient methods for quality control, and the potential application of these models will expand with greater consumer satisfaction and product quality.

The case study results show the importance of predictive models regarding quantifying color and food quality, exhibiting developments in the technology that allow for more precise and efficient measurement.

1.12 AI/ ML Prediction Modeling in Food Processing and Food Safety

1.12.1 Introduction

This research proposed a developing image analysis flow chart to predict banana quality indices using AI and machine learning techniques. The process involves several significant steps based on the methodologies described in the paper.

Image Acquisition: Start by capturing RGB images of bananas using a suitable imaging system for quality assessment.

Image Enhancement: After capturing the images, median filtering should be performed on the RGB images to reduce noise and improve quality, thereby increasing the accuracy of further steps.

Image Segmentation: Image segmentation will be observed to detect the ROI, which is basically the banana. That is where the segmentation and recognition of the banana from the background takes place.

Color Feature Extraction: Next is the extraction of the color features from the images. Among the most essential colors for distinguishing bananas from the background is green. The Optimal Threshold for Green Intensities using Otsu's Algorithm is therefore determined.

Binary Image Creation: Create a binary image by superimposing the binary mask of the banana surface over the original color image. This helps in isolating the banana's color information from the background.

Data Preparation: Prepare the extracted color features for analysis. This includes organizing the data into a format suitable for modeling.

Model Development: Utilize Support Vector Regression (SVR) to develop predictive models. The radial basis function (RBF) kernel is employed for this purpose, and various parameters are optimized using leave-one-out cross-validation to ensure model robustness.

Quality Indices Prediction: Use the trained SVR model to predict quality indices such as total soluble solids, pH, titratable acidity, and firmness based on the extracted color features (Sanaeifar et al., 2016).

Model Evaluation: Evaluate the performance of the SVR model against other models, such as artificial neural networks (ANN), to determine the best predictive accuracy.

1.13 AI/ ML Prediction Modeling in Food Processing and Food Safety

Achievement of Predictive Maintenance through Machine Learning

Predictive maintenance was achieved in this study through a systematic approach which involved several key steps and methodologies.

Data collection through large data arrays gathered from sensor-enabled machinery comes forth; this collects real-time parameters such as oil temperature, pressure flow, motor speed, and motor vibration, which are critical.

For the study, it employed various machine learning algorithms, including SVM, RF, and LSTM. These algorithms were used to predict equipment failures correctly by training historical datasets.

Using RMSE, R-squared (R^2), and MAE metrics, the developed models were compared to determine the most effective method for the prediction.

Data Preprocessing and Feature Engineering: Before training the models, data was heavily preprocessed, including cleaning, conversion, and standardization. During this step, the most relevant feature was selected to maximize the information in the dimension to improve prediction accuracy related to machine failure.

Integration with IoT and AI: The predictive maintenance software was developed to integrate any IoT sensors and analytics powered by AI. This has facilitated the processing of data in real time, thus triggering alarms to notify the maintenance crew of the breakdown.

Implementation in Real-world Environments: The study highlighted that this was the first implementation of an AI-driven predictive maintenance system in an actual factory within the beverage manufacturing industry. This practical application demonstrated the system's efficiency in handling complex data in real-time, which is crucial for effective predictive maintenance.

Results and Performance Evaluation: The results indicated that the SVM model outperformed the other models in terms of prediction accuracy, as evidenced by the alignment of forecasted and actual data on graphical trends.

The overall goal was to enhance operational efficiency and reduce maintenance costs by 5–10% through the successful deployment of the best-performing model.

1.14 Case Studies

Schematic Diagram for a Random Forest Model on Food Product Categorization

A random forest model is indeed a very robust machine learning technique that can classify food products with several attributes. Here is a step-by-step description of how the model works, especially regarding its application in food product categorization:

The model starts by taking root nodes portraying the core characteristics of the food items, as noted by Khan et al. (2022). To illustrate, based on the categorization of food items, the root nodes can consist of variables describing the level of sweetness, crunchiness, etc.

Inner nodes can be formed under a root node through many other measurable parameters. Here, the inner nodes have been utilized to enhance the process of categorization. Taste parameters may be taken as inner nodes, and then that inner node can be further subdivided into specific taste characteristics, such as sourness and bitterness, etc.

If there are leaf nodes, these are the internal nodes representing terminating categories of the food items. This means that for every leaf node, there is one specific categorization based on the quality parameters defined earlier. For example,

Category 1: Products meeting all criteria for quality parameters.
Category 2: Products that only partially meet quality parameters, for instance 80%.
Category 3: Products that do not meet the quality criteria.

Model Complexity: Even though the model can tolerate many variables, optimal selection of the number of trees in the forest is necessary. Too many trees lead to very complex models, thus requiring much more computational power to run, and are often not improved much upon accuracy. Care must therefore be taken in selecting an optimized number of trees for better performance.

Practical Application Example: This model can really be utilized to classify fresh apples. Parameters, such as color, taste, size, and shape may be used as root categories, which can lead to more details concerning classification in terms of apple quality.

The model for food product categorization with the use of random forest utilizes a systematic approach of root nodes, internal nodes, and leaf nodes for the purpose of categorizing products based on measurable attributes. Given the complexity involved with the model and improving upon its computational efficiency, this method increases the accuracy of the categorization and is eventually used in decision-making processes within the food industry. The figure Algorithm employed for the color measurement of the fruit sample is presented in

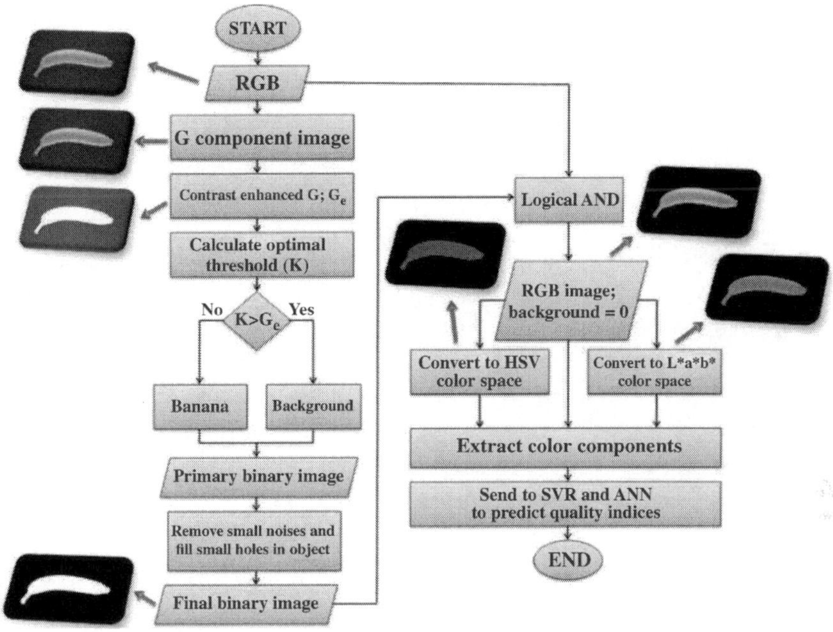

Figure 1.17 Algorithm employed for the color measurement of the fruit sample (Sanaeifar et al. [2016]/Elsevier).

Figure 1.17, and Figure 1.18 represents ML-based strategies for optimizing the processed food quality.

1.15 Review Questions

Q1 What is Food Processing 4.0?

Q2 How does IoT enhance food safety in processing?

Q3 What role does AI play in food processing?

Q4 How can robotics improve efficiency in food processing?

Q5 What are the sustainability benefits of Food Processing 4.0?

Q6 What challenges does the food processing industry face in adopting Industry 4.0 technologies?

Q7 How is big data utilized in food processing?

Q8 What future trends are anticipated in Food Processing 4.0?

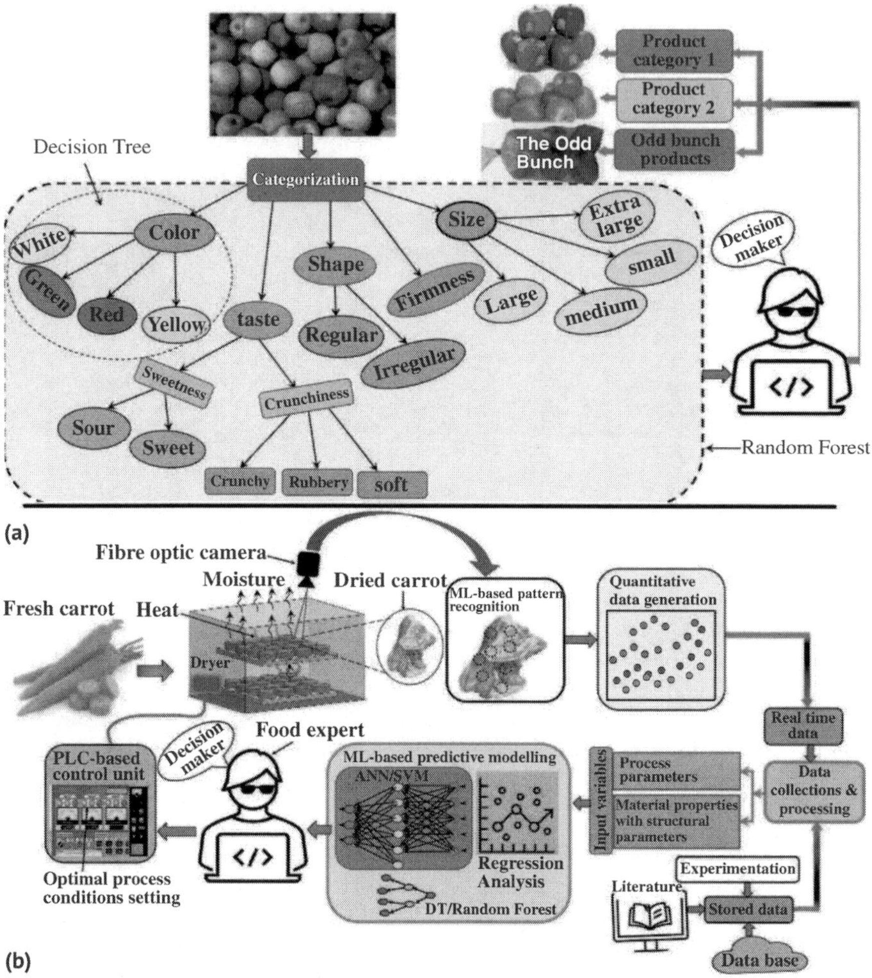

Figure 1.18 ML-based strategies for optimizing the processed food quality (Khan et al. [2022]/John Wiley & Sons).

Bibliography

A Hassoun, A Aït-Kaddour, AM Abu-Mahfouz, NB Rathod, F Bader, FJ Barba, A Biancolillo, J Cropotova, CM Galanakis, and AR Jambrak. The fourth industrial revolution in the food industry—part I: industry 4.0 technologies. *Critical Reviews in Food Science and Nutrition*, 63(23):6547–6563, 2023.

J Iqbal, ZH Khan, and A Khalid. Prospects of robotics in food industry. *Food Science and Technology (Campinas)*, 37(2):159–165, 2017.

S Jagtap, L Duong, H Trollman, F Bader, G Garcia-Garcia, G Skouteris, J Li, P Pathare, W Martindale, M Swainson et al. IoT technologies in the food supply chain. In: *Food Technology Disruptions*. Elsevier, 175–211, 2021.

MIH Khan, SS Sablani, RNayak, and Y Gu. Machine learning-based modeling in food processing applications: state of the art. *Comprehensive Reviews in Food Science and Food Safety*, 21(2):1409–1438, 2022.

S Li, A Simonian, and BA Chin. Sensors for agriculture and the food industry. *The Electrochemical Society Interface*, 19(4):41, 2010.

Y Liu, X Ma, L Shu, GP Hancke, and AM Adnan Mahfouz. From industry 4.0 to agriculture 4.0: current status, enabling technologies, and research challenges. *IEEE Transactions on Industrial Informatics*, 17(6):4322–4334, 2020.

G Özdoğan, X Lin, and DW Sun. Rapid and noninvasive sensory analyses of food products by hyperspectral imaging: recent application developments. *Trends in Food Science & Technology*, 111:151–165, 2021.

R Romanello and V Veglio. Industry 4.0 in food processing: drivers, challenges and outcomes. *British Food Journal*, 124(13):375–390, 2022.

A Sanaeifar, A Bakhshipour, and M De La Guardia. Prediction of banana quality indices from color features using support vector regression. *Talanta*, 148:54–61, 2016.

2

Automation in Food Processing Applications

After reading this chapter, you should be able to understand:

- Full range of automation over foodstuffs: all steps including retail display, catering service, cold store, as well as all transport, from retail to shop floor, right through transport. Each poses different specific challenges, and each needs optimized control strategies—whether just optimizing refrigeration in retail food displays, or optimizing efficiency within refrigerated transport systems.
- The first drivers for automation, which are energy efficiency, food quality preservation, and regulatory compliance, are always changing. Advances in research are responding to these needs, and the trends lean toward smarter, energy-efficient, and adaptive control systems that dynamically respond to food safety requirements and operational efficiencies.
- Practical applications of examples, especially case studies, will demonstrate an understanding of how automation drives the effective operation of food safety levels in a supply chain. Such examples will be successful implementations, experiences of problems, and benefits clearly identifiable with the automated solutions in real food industries.

The consumption of perishable fruits and vegetables is vital in reducing the incidence of chronic diseases, yet a considerable quantity of FV is disposed of annually, mainly due to the lack of post-harvest storage supervision. Among all present systems, those that monitor cold storage primarily track temperature and humidity; other critical factors, such as light exposure and gas levels, to maintain the quality of stored FVs, are neglected. Activation functions can be classified into several types, including but not limited to sigmoid, linear, Gaussian, exponential, Tanh, and Softmax. Most deep learning frameworks use the Softmax activation function to generate class probabilities at the output layer.

Food Engineering Automation with Robotics and AI, First Edition. Abir Chakravorty.

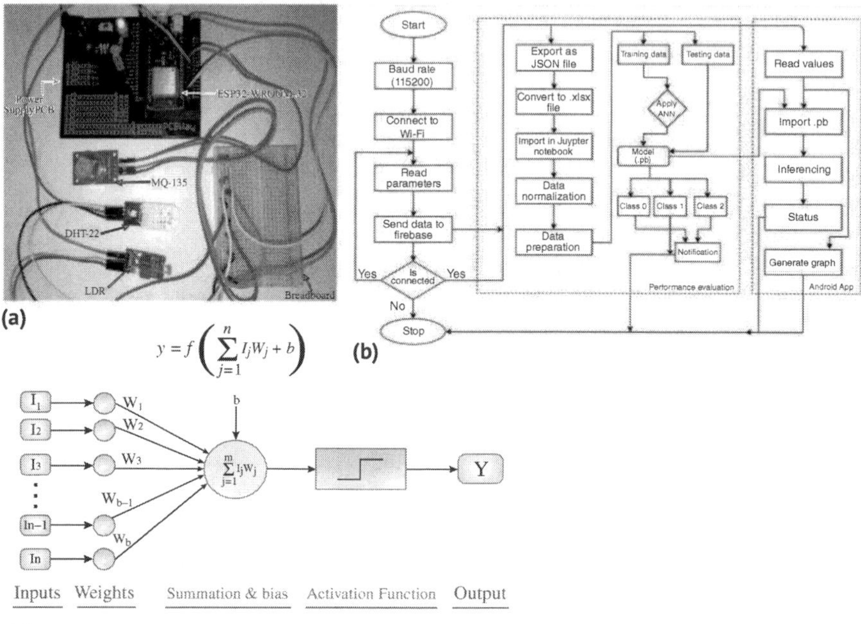

Figure 2.1 (a) Experimental setup of RT-IMNS, (b) schematic description of complete workflow of RT-IMNS, and (c) schematic of an artificial neuron (Afreen and Bajwa [2021]/IEEE/CC BY 4.0).

It uses Internet of Things (IoT) technology in a new way to establish a real-time intelligent monitoring and notification system (RT-IMNS). The experimental setup and complete workflow of the RT-IMNS system is presented in Figure 2.1. It monitors environmental parameters such as temperature, humidity, luminosity, and gas concentrations in cold storage continuously. It signals the concerned authorities if one of these environmental conditions goes beyond limits. At last, it takes advantage of an ANN working on forward propagation as a decision support system for considering the quality of the stored commodities by classifying them into three categories: good, unsatisfactory, and alarming (Onim et al., 2022).

The most accurate prediction was observed by the ANN-based prediction model in comparison with CS, ANB, XGBoost, and DM models. It achieved an accuracy level of 99%. The model established an accuracy record that surpassed the CS, ANB, XGBoost, and DM models with their accuracies at 95.60, 87.50, 93.59, and 90%, respectively. In other words, in the proposed systems, values for all classes on precision, recall, and F1-score were excellent: 100% in the "good" class; 98% for the "unsatisfactory" with precision; 99% with recall; and 100% for precision and 98% for "alarming" class with 99% for F1 (Afreen and Bajwa, 2021).

2.1 Automatic Control of Food Chilling and Freezing

Implement an IoT-based, real-time, intelligent monitoring and notification system for cold storage, with key steps directed towards minimizing the loss of perishable commodities. Here's a short description of the procedure:

2.1.1 Installation of IoT Sensors

In cold storage, sensors are mounted to measure pertinent environmental variables such as temperature, humidity, luminosity, and gas concentrations. This is because these variables are fundamental toward the attainment of timely information of factors that may influence perishable goods such as FVs. Table 2.1 represents freezing technologies and characteristics.

2.1.2 Data Acquisition and Transmission

Data from the sensors is harvested uninterruptedly and transmitted to a cloud database. This ensures that the information can be accessed effortlessly and applied during analysis and monitoring.

2.1.3 Data Preprocessing and Analysis

The collected data requires preliminary cleanup and normalization to prepare it for use in the analysis. This is arguably the most critical step in preparing the data for analysis, and for this reason, features are extracted for use in machine learning models.

2.1.4 Model Training and Prediction

The above data was exposed to an ANN for learning the weights toward classification of commodities status, which falls under one of the following three categories: good category, unsatisfactory, or alarming. The model was highly accurate and ensured clear enhancement in the prediction of possible spoilage.

2.1.5 Alerting System

It features a notification mechanism that will alert people when environmental parameters are over the safety limits. It will be an important feature to facilitate timely interventions so that spoilage can be avoided.

Table 2.1 Freezing technologies and characteristics.

Freezing technology	Characteristics
Air-blast freezer	In a blast of cold air circulation, the products are frozen at temperatures ranging from −35 to −45 °C under forced circulation. The freezing time typically ranges between 12 and 48 hours.
Tunnel freezer	Trays of goods are arranged in racks or trolleys and frozen in a tunnel with cold air circulation at a temperature of −35 °C.
Belt freezer	Inside the blast rooms, designed to continuously provide precooled air flow at about −40 °C with the use of a wire mesh conveyor.
Fluidized bed freezer	Air between −25 and −35 °C is passed at high velocity (2 – 6 m/s) in modified blast freezers, across a conveyor belt or perforated tray with a 2–13 cm bed of food.
Contact freezer	The freezing media completely encloses the frozen product, increasing the effectiveness of heat transfer.
Immersion freezer	Food is transported on a submerged mesh conveyor through a bath of chilled calcium chloride solution, brine, glycerol, or propylene glycol.
Indirect contact freezer	Usually, a steel plate, a conducting material, separates the items from the refrigerant.
Plate freezer	A series of hollow plates, either vertical or horizontal, make up these freezers. Refrigerant is pushed through them at −40 °C.
Cryogenic freezer	These freezers use liquid nitrogen or carbon dioxide, either solid or liquid, in close proximity to food. A pre-cooled substance is used for refrigeration, and food is exposed to an atmosphere that is below −60 °C.
Liquid nitrogen freezer	A perforated belt carries food through a tunnel, where it is frozen by liquid nitrogen spray and chilled by gaseous nitrogen at −196 °C.
Liquid carbon dioxide freezer	Applied prior to the product being subjected to nitrogen spray as a pre-freezing treatment.

2.1.6 Continuous Improvement

The system is designed to learn from new data with the ability to continuously improve on monitoring and prediction accuracy. Further improvements include improving the system in terms of monitored parameters, thus the interfaces become more user-friendly to increase access.

2.1.7 Future Improvements

Further, there is work that is ongoing where monitoring abilities are being expanded to include more environmental parameters as well as gases in general, for the further enhancement of the overall effectiveness of the system.

2.2 Introduction: Main Forces Behind and Obstacles to Automated Food Freezing and Chilling Control

Automatic food chilling and freezing require several elements in order to efficiently improve its safety and quality. These elements include advanced sensor technology, automation systems, as well as energy management.

2.2.1 Advanced Sensor Technology

Temperature sensors and current sensors play a crucial role in monitoring the conditions inside refrigeration systems to ensure real-time adjustments are made for optimal temperatures (Koritsoglou et al., 2020). Systems use thermocouples to monitor frosting and defrosting states, thereby regulating compressors and evaporators.

2.2.2 Automations

Automatic controllers regulate the refrigeration operation through signals from temperature action, which enable optimal operating cycles in cooling and defrosting and avoids compressor overloading.

Innovative designs, such as the application of saline water pools combined with stirring devices, help enhance the practicality and energy efficiency in refrigeration systems (Martins et al., 2017).

2.2.3 Energy Management Techniques

Liquid carbon dioxide as an immersion chiller is used with regard to exogenous climatic conditions, thereby optimizing the utilization of energy during storage and transport, posits.

Elnour et al. (2023) reported that the use of temperature compensation devices in chill chambers improved the freezing efficiency and operational cost.

These developments significantly enhance food safety and quality but pose challenges in integrating these technologies into multifaceted food chains; these might weigh against widespread implementation.

2.3 Challenges to Automation in Food Chilling and Freezing Include

1. Training: Training operators and tech support personnel is cumbersome, especially when new equipment is introduced into the business. In such instances, much time and money will have to be spent on training for the automation to be a success (Ramanathan et al., 2022).
2. Employee Turnover: High employee turnover can be adverse to business operation as a stable group has to be trained incessantly.
3. Automation Versus Day-to-Day Operations: Balancing the desire for automation against the needs of day-to-day business operations can be difficult to achieve (Miller and Knudson, 2014).
4. Equipment Manufacturer: The equipment manufacturer must provide a kit of training materials and procedures for new equipment.

Some extra issues related to the food cold chain:

Broken Freezer Doors: Doors cannot be checked at a moisture-tight seal due to this issue, hence products get spoiled because the necessary temperature cannot be maintained (Badia-Melis et al., 2018).

There are also some downsides connected with energy: the energy costs for cold storage warehouses are rather high due to preserving a low temperature, which makes the utility bills pretty costly.

Labor Costs: Working conditions for employees may be unattractive due to strict temperature control, this can make it difficult to engage and maintain people.

2.4 Automation in Refrigerated Food Retail Display

Let us take the case study of the processes at Yumchop Foods (see Figure 2.2):

Step 1: Problem Definition
Food Waste Problems: Yumchop Foods had a high amount of food waste due to bad storage conditions and human monitoring deficits.

Step 2: Implementation of Technology
Installation of IoT Sensors: Yumchop Foods fitted sensors to monitor real-time temperature and humidity.
Collecting Data: Data is transmitted to the cloud server for analysis.

Step 3: Data AnalysisData Analysis
Analyze the Collected Data: Yumchop Foods compared the collected data with optimal and non-optimal storage conditions.
Set Up Alert System: They set up an alert system that gives a warning to employees if there is a deviation from optimal conditions or not.

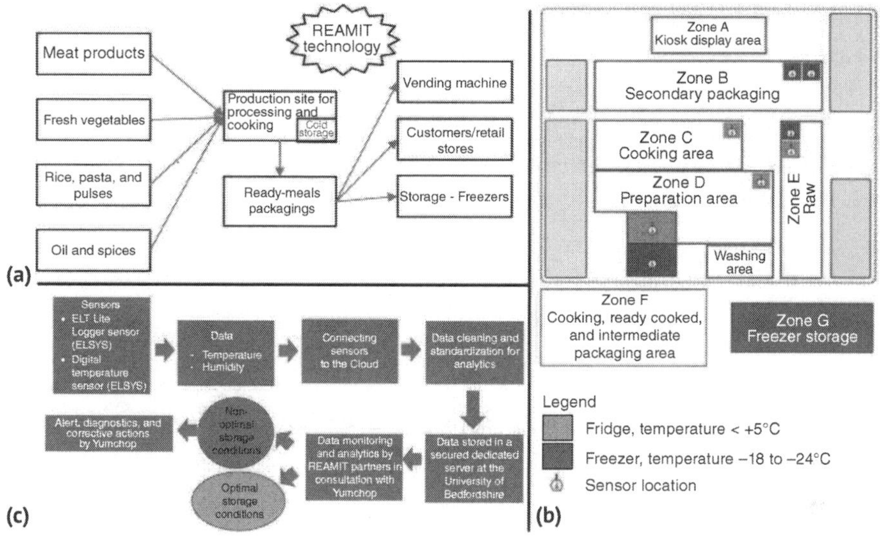

Figure 2.2 (a) Yumchop's processing diagram (Ramanathan et al., 2022), (b) an architecture for implementing digital technologies for Yumchop (Ramanathan et al., 2022), and (c) location of sensors in Yumchop's premises (Afreen and Bajwa [2021]/IEEE/CC BY 4.0).

Step 4: Monitor and Improve

Monitor Continuously in Real-time: Non-stop monitoring of temperature and humidity to ensure food quality and safety.

Corrective Action: Yumchop Foods followed corrective actions for deviations through alerts and data analysis.

Step 5: Outcome Measurement

Less Food Waste: They measured the amount of food waste produced due to better monitoring and storage.

Better Quality Products: Yumchop Foods measured the retention of food quality and also whether the legal standards of food safety are met.

Step 6: Future Ideas for Research

Seasonal Variation: They measured the effect of seasonal changes on temperature and energy usage.

Observed Patterns of Compliance: In relation to food safety regulation compliance, Yumchop Foods analyzed the temperature records.

Structured Approach: This flowchart depicts the approach adopted by Yumchop Foods to adopt digital innovation in their business activities, reduce food waste, and enhance operational efficiency.

2.5 Automation of Refrigeration and Freezing Operations in Food Catering

The research focuses on the automation of temperature control processes in catering facilities, founded on the principles of HACCP, which is the key to food safety and quality management.

2.5.1 Main Findings

Demand for Automation: There is a need to automate catering operations that monitor temperature manually—the process takes a while to identify critical control points.

Automation of temperature mode in premises, refrigerating, and freezing installations facilitates adherence to the norms of food safety standards.

Technical Selection of Equipment: The researchers tested various controllers and sensors, and they chose Arduino Mega 2560 because it was not very costly and had a sufficient number of ports.

For the temperature sensor, the researchers used the TMP36 due to its capability to measure negative temperatures and cost-effectiveness.

2.5.2 Flow of the Automation Process in Temperature Control System

Step 1: Identify Requirements

Regulatory Compliance: Getting to know what TRCU 021/2011 and HACCP principles entail concerning food safety.

Current Monitoring Practices: Knowing the existing methods of temperature monitoring being used by the catering businesses3.

Step 2: Technical Specification Development

Selecting Equipments: Suitable technical equipments such as Arduino Mega 2560, TMP36 temperature sensor.

System Architecture Design: Design a diagram of how the thermometer can work and be integrated into the software.

Step 3: Software Development

Programming: The code is written in the Arduino IDE to program the Arduino Mega for temperature monitoring.

Setting up Web Server and Application for data transmission using MQTT protocol.

Step 4: Implementation

Testing: Test at A la fourchette LLC to ensure all the testing of the automated system made at A la fourchette LLC would be compliant with ISO 10012–2003.

Training the Users: Staff should be given training to use the system. The system should be easy to operate and fully user-friendly.

Step 5: Ongoing Monitoring

Real-time Data Acquisition: Successful continuous monitoring of the temperature modes throughout the premises and the units of refrigeration.

Feedback Loop: Open a feedback control system so that the system can be monitored and altered if required.

Step 6: Assessing Outcomes

Performance Assessment: Monitor the effectiveness of the automated system toward enhancing food safety and productivity.

Report Results: Record increases in safe temperature levels and conformity with standards.

2.6 Automation in Refrigerated Food Transport Systems

Flowchart of the Study on Energy-efficient Frozen Food Transports

Creating a flowchart for the study on energy-efficient frozen food transport involves outlining the key components and processes described in the paper. Here's a structured breakdown of the study:

Introduction to the Problem

Definition of the Refrigerated Routing Problem (RRP) for multi-drop deliveries of frozen food from a central depot to clients.

Objective of the Study

Aim to minimize fuel consumption for both traction and refrigeration during transport (Chernova and Veloso, 2004).

Modeling Considerations

Traffic Congestion: Speed variations due to traffic conditions are considered, with peak hours affecting travel speed.

Load Variations: The model accounts for decreasing load on board as deliveries are made.

Temperature Effects: Outdoor temperature variations are modeled, including transmission and infiltration loads.

Data Collection

Use of real local network scenarios for frozen bread dough distribution to supermarkets.

Parameters recorded for each temperature time slot, leading to a comprehensive dataset for analysis.

Model Implementation

The model is coded in MiniZinc and solved using the Gecode solver, with computational efficiency noted.

Simulation and Results

Multiple scenarios tested, focusing on fuel minimization, total distance, and duration of routes.

Results indicate significant variations in refrigeration consumption based on localities and seasonal conditions.

Conclusions Drawn

The study highlights the importance of considering both traction and refrigeration requirements in routing problems.

Energy savings are influenced by factors such as demand distribution, departure time, and seasonality.

Understanding the Refrigerated Routing Problem (RRP) graph can be effectively represented using a graph, which is a fundamental concept in operations research and logistics. Here's a detailed explanation of the RRP graph based on the information provided in the paper:

Graph Structure: The RRP is modeled as a graph where:

Nodes: Represent the central warehouse (depot) and the clients (delivery points). Each client is identified by an index, with the depot being the initial and final node of the route.

Edges: The arcs connecting every pair of nodes represent the travel distances (d) between them. This is crucial for calculating the optimal routes for deliveries.

Characteristics of the Graph:

Each node is characterized by the quantity (Q) of unit loads that need to be delivered to that client. This means that the graph not only shows the locations but also the demands at each location.

The graph can be visualized with the central depot at the center, branching out to various clients, each connected by edges that denote the distance or travel time required to reach them.

Objective of the RRP:

The primary goal of the RRP graph is to determine the most efficient route that minimizes fuel consumption while considering the constraints of refrigeration and load variations. This involves analyzing the graph to find the optimal path that balances these factors.

Complexity and Considerations:

The RRP graph is complex due to the multiple variables involved, such as traffic conditions, load changes, and temperature effects. These factors can significantly influence the route selection and overall fuel efficiency.

The graph representation allows for the application of advanced solving techniques, such as Constraint Programming, which can handle the intricate relationships and constraints present in the RRP.

2.7 Automation in Food Chilling and Freezing Systems

Flowchart Study of Smart Freezing Technology in Frozen Foods

Creating a flowchart study based on the paper "Critical Reviews in Food Science and Nutrition" continues with the steps in outlining major components and processes discussed within the paper regarding the smart freezing technology. Below follows a structured version of the flowchart study:

2.7.1 Introduction to Smart Freezing Technology

Focus on the quality of frozen foods with fresh-like qualities.

There is great importance of emerging technologies in the freezing processes.

Keystone technologies in smart freezing:

1. Computer Vision:
   Auto-grades, and sorts food materials.
   Increases efficiency in the processing, which encompasses cleaning of fishes and weighing them.
2. Electronic Tongue:
   Assesses food freshness, especially regarding fishes;
   It is combined with an electronic nose to nondestructively identify freshness.
3. Digital Simulation Technology:
   Uses mathematical models to model an efficient freezing process. Predicts the freezing times and optimizes the efficiency of rapid freezing.
4. Nuclear Magnetic Resonance Technology:
   This measures the moisture mobility and distribution in foods.
   It provides information on the quality of frozen products.

2.7.2 Freezing Technology Applications

Monitors the freezing process in real time.

Freezing is controlled in stages to maximize product quality in an automated manner.

Using various sensing technologies for appropriate quality control.

2.7.3 Challenges and Innovations

Frozen food technology involves several problems regarding flavor and high energy consumption.

2.7.4 Quality and Nutritional Value

Infinite innovation is required to achieve consumer demand for quality and nutrition. Smart freezing technologies may become significant in ensuring improved quality and productivity in processing frozen food. The future studies should aim at fully integrating the applied technologies.

2.8 Automation in Food Cold Storage Systems

Flowchart Case Study: Large Cold Store Refrigeration Systems with Automated Fault Detection and Optimization of Energy Usage

This case study introduces the optimization of large cold store refrigeration systems for high efficiency through automated fault detection and energy usage optimization. The study addresses some common faults that affected the defrosting level and carbon emissions.

Step 1: Identify Common Faults

Defrosting Problem: It should be admitted that in the use of hot gas in the refrigeration machine, there exist defrosting problems. In time keeping checking on them would ensure efficiency with a minimal carbon emission rate.

Gas-driven Valve Issues: Determine the fact that anomalies exhibited by the gas-driven suction stop valve lead to malfunctions associated with the defrosting process.

Step 2: Data Collection and Analysis

Apply Machine Learning: Apply SOM-based clustering analysis of operational data. In this case, its diagnostic accuracy rates were 88.6% in the case of DHL, 91.2% in the case of DML, and 93.8% in the case of DLL.

Assess Carbon Savings: Measure the amount of carbon reduced by getting rid of defrosting faults that demonstrate an abatement at a pace of around 18.3% or 1484 kg per day.

Step 3: Fault Diagnosis

Theoretical and Structural Analysis: Identify the potential sources of faults through theoretical analysis and structural study. Validate the source of faults by data mining.

Automate Fault Diagnosis: Use SOM-based machine learning algorithms to automate fault diagnosis to enhance the efficiency of equipment and pipelines.

Step 4: Implement Solutions

Optimize Energy Usage: Design intelligent energy optimization strategies based on the analysis of the operational states and results of the fault diagnosis.

Monitor and Adapt: The performance of the refrigeration system must be monitored for its efficiency and carbon emissions. There must be the necessary adjustments that ensure the system is at maximum efficiency with the least carbon emission possible.

Step 5: Outcome Evaluation

Calculating Efficiency Gains: Overall, measure the systems' capability for efficiency gains in the refrigeration system and the reduction of carbon emissions from the solutions that have been executed.

Feedback Cycle: Form a loop where feedback accumulates to make improvements on the identification of faulty areas and energy optimization through continued performance data.

This flowchart case study makes the refrigeration systems of cold stores more efficient in a structured manner by considering fault identification and energy optimization, which aid in producing environmental sustainability.

2.9 Advances in Research and Future Trends

Recent Advances in the Automation of Food Chilling and Freezing

This section deals with recent advances in food chilling and freezing-related aspects of automation, emphasizing fault detection and energy optimization in refrigeration systems. The main points are as follows:

1. Automated Fault Identification: This would therefore introduce a new approach in the identification of faults by automatic means in refrigeration systems; the identification, locating, and determination of key fault-related parameters leading to improved efficiency in large cold storage systems.
2. Intake of Machine Learning: The very important feature is the intake of the SOM clustering algorithm; this is machine learning on an unsupervised basis, organizing system states into different groups, thereby allowing better fault detection and energy consumption optimization. Through state changes during a defrosting process, the SOM algorithm simplifies the fault analysis, therefore improving safety and energy efficiency in refrigeration systems.
3. Real-time Monitoring and Control: Implementation of real-time frost detection systems is considered an important aspect of achieving optimal refrigeration performance; for example, a maximum energy conservation of 10% by such systems using real-time frost detection systems. Real-time monitoring

helps in advanced intervention, avoiding energy waste and thereby enhancing overall system efficiency.

4. Thawing Strategies: Innovation strategies to thaw include hybrid systems that associate real-time detection with control methods to reduce the occurrence of defrosting. Zhao and his coworkers optically detected the defrosting control method that drastically reduced the frequency of defrosting by 75%. Such an automation strategy can increase the performance of the system (see Figure 2.2c).
5. Energy Consumption and Carbon Emissions: The chapter also presents the importance of assessing energy consumption and carbon emissions before and after implementing automated solutions. This evaluation is crucial for understanding the environmental impact of refrigeration systems and for promoting sustainable practices in the food chilling and freezing industry.

These advances reflect a growing trend toward automation in refrigeration systems, aiming to enhance efficiency, reduce energy consumption, and minimize environmental impact in the food industry.

2.10 Case Study

Flowchart of IoT in the Food Supply Chain

Even though the paper does not explicitly mention the IoT as an element of the optimization model for refrigerated automated warehouses, it is possible to identify how IoT can be included in the food supply chain based on the context of this paper. Below is a flowchart that describes a potential IoT framework that might complement the optimization model discussed. Figure 2.3 represents the intelligent cold chain logistics system.

1. IoT Device Deployment
   Temperature Sensors: Install sensors to gather real-time temperature monitoring and assure the quality and safety of food.
   Humidity Sensors: Use of humidity sensors to maintain the optimal condition in refrigerated storage.
2. Data Collection
   Real-time Data Gathering: Collect data from sensors about temperature, humidity, and energy consumption.
   Supply Chain Tracking: Monitor the location and condition of products along the supply chain.
3. Data Analysis
   Predictive Analytics: Analyze the collected data to predict possible degradation in quality and energy utilization.
   Decision Support: Use analytics in informing decisions about configurations of storage and inventory management.

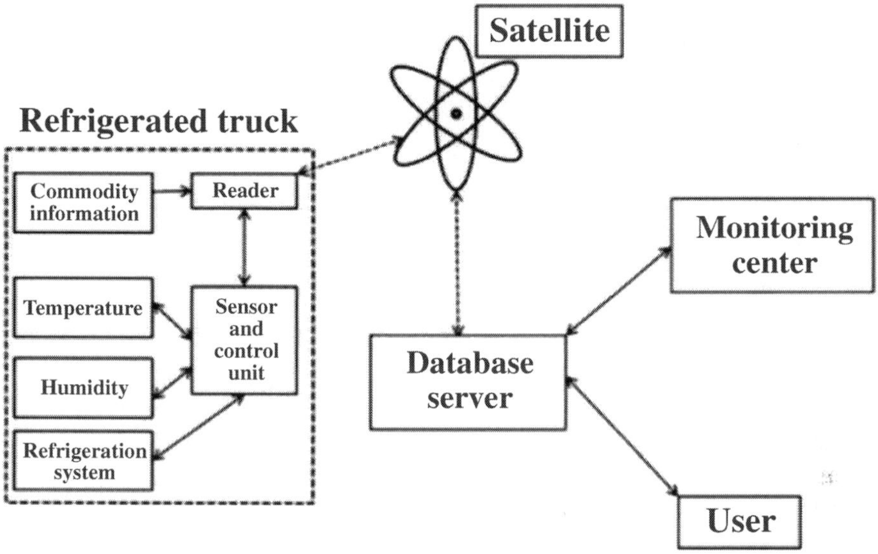

Figure 2.3 Schematic diagram of an intelligent cold chain logistics system (Xu et al., [2017]/Reproduced with permission of Informa UK Limited).

4. Integration Model Optimization
   Feedback Loop: Incorporate the inputs from IoT into the optimization model for adjusting configurations under real-time conditions.
   Adaptive Optimizations: The in-time adjustment of the storage temperatures and configurations of refrigeration systems in order to ensure that energy usage is reduced while food quality is maximized.
5. Energy Optimization
   Energy Consumption Monitoring: It monitors and tracks the energy consumption of the refrigeration systems so that wastage can be identified.
   Sustainability Indices: It measures carbon dioxide emissions and energy efficiencies of a specific food supply chain to support the sustainability concepts in this industry as well.
6. User Experience
   Monitoring Dashboard: It must be designed to facilitate an easy user interface for stakeholder monitoring and making decisions based on the same.
   Alert and Notification: Leverage deviation of temperature or humidity for taking prompt corrective action.
7. Continuous Improvement
   Feedback Mechanism: Ensure the optimization model and the practice in operation are improved continually in view of insights from IoT data.

2.11 Review Questions

Q1 A food processing plant uses manual labor for packaging, with each worker able to package 100 units of product per hour. After automating the packaging process using a robotic system, the plant can now package 500 units per hour with the same number of workers. If the plant operates for 8 hours a day, calculate the increase in production capacity per day due to automation.

Q2 A food processing company spends $10 000 per month on labor costs for quality control inspections. After implementing an automated system that costs $5000 per month to operate, the company estimates a 20% reduction in labor costs. Calculate the monthly savings in labor costs after automation and the net savings for the company.

Q3 What are the key challenges faced by the food processing industry when implementing automation technologies?

Q4 What role do robotics and AI play in enhancing food safety and quality control in food processing?

Bibliography

H Afreen and IS Bajwa. An IoT-based real-time intelligent monitoring and notification system of cold storage. *IEEE Access*, 9:38236–38253, 2021.

R Badia-Melis, U Mc Carthy, L Ruiz-Garcia, J Garcia-Hierro, and JR Villalba. New trends in cold chain monitoring applications—a review. *Food Control*, 86:170–182, 2018.

S Chernova and M Veloso. An evolutionary approach to gait learning for four-legged robots. In *2004 IEEE/RSJ International Conference on Intelligent Robots and Systems (IROS)(IEEE Cat.*, 3:2562–2567, 2004. IEEE No. *04CH37566)*.

M Elnour, H Haller, and M Martin. Life cycle assessment of a retail store aquaponic system in a cold-weather region. *Frontiers in Sustainability*, 3:1051091, 2023.

K Koritsoglou, V Christou, G Ntritsos, G Tsoumanis, MG Tsipouras, N Giannakeas, and AT Tzallas. Improving the accuracy of low-cost sensor measurements for freezer automation. *Sensors*, 20(21):6389, 2020.

LFB Martins, PB de Araujo, EV Fortes, and LH Macedo. Design of the pi–upfc–pod and pss damping controllers using an artificial bee colony algorithm. *Journal of Control, Automation and Electrical Systems*, 28(6):762–773, 2017.

SR Miller and WA Knudson. Nutrition and cost comparisons of select canned, frozen, and fresh fruits and vegetables. *American Journal of Lifestyle Medicine*, 8(6):430–437, 2014.

MSH Onim, ZM Md Sakif, A Ahnaf, A Kabir, AK Azad, AMT Oo, R Afreen, ST Hridy, M Hossain, T Jabid, et al. Solnet: a convolutional neural network for detecting dust on solar panels. *Energies*, 16(1):155, 2022.

U Ramanathan, R Ramanathan, A Adefisan, TD Costa, X Cama-Moncunill, and G Samriya. Adapting digital technologies to reduce food waste and improve operational efficiency of a frozen food company—the case of yumchop foods in the UK. *Sustainability*, 14(24):16614, 2022.

JC Xu, M Zhang, AS Mujumdar, and B Adhikari. Recent developments in smart freezing technology applied to fresh foods. *Critical Reviews in Food Science and Nutrition*, 57(13):2835–2843, 2017.

3

Robot Types and Their Classification

After reading this chapter, you should be able to:

- Robotics Kinematics for Applications in Food Processing: Fundamentals of robotic kinematics—the chapter covers forward and reverse kinematics, which are fundamentally two parts of the calculation toward computing the positions and orientations for a robot's end effector to handle food efficiently. Transformation matrices are established as the primary tools involved in spatial transformations with which to move and manipulate with precision robotic arms in their tasks, such as those done in food processing and packaging applications.
- These efficient approaches of advanced path planning and control algorithms, Dijkstra's and A* for path planning, fuzzy logic control strategies, bring intelligent capabilities to robotic navigation, mapping, and obstacle avoidance. AI/ML and artificial neural network (ANN) applications bring adaptability, smart control, and decision-making capabilities to real-world food processing automation.
- Industry-specific Robotics Applications and Case Studies: Practical insights in the chapters on robotic grippers and control strategies in food packaging and handling give real examples of robotic applications. The integration of kinematic analysis, transformation matrices, and algorithms for control is demonstrated in these applications, showing how theoretical principles convert into effective solutions in the automation of the food industry.

Food Engineering Automation with Robotics and AI, First Edition. Abir Chakravorty.

3.1 Categories of Robots

Robot classification is generally a very difficult task. Certain conditions determine a partial classification. Different kinds of robots can be produced using different criteria. A large number of references only address industrial robot categorization. Since the work aims to cover a wide range of categorization domains, the following are included in the proposed criteria: robot location depicted in Figure 3.1, robot applications, robot movement, robot architecture, generation of robots, size, type of controller, different types of sensors, and robot design. Robots can be classified according to their axis of motion and degrees of freedom; the complete classification trees are presented in Figure 3.1.

Cartesian robot

Joints: 3P (three prismatic [i.e. up, down, left, right in a straight line] joints)

Cylindrical robot

Joints: 2P1R (two Prismatic, one revolute joint)

Spherical robot

Joints: 2R1P (two revolute, one prismatic joint)

Articulated arm

Joints: 4R (four revolute joints); TRRR joints (i.e. transverse, revolute, revolute, revolute)

T joint represents horizontal revolution.

Classification based on degrees of freedom (DoF)	**Based on actuator**
A) DoF<6 Constrained Manipulator There are several types of robotic manipulators. These types vary based on the combination of joints and include **cartesian, cylindrical, polar, articulated, SCARA, and delta configurations.**	A) Hydraulically actuated robotic system
B) DoF=6 spatial manipulator	B) Pneumatic actuated robotic system
C) DOF>6 redundant manipulator	C) Electrically actuated robotic system

Figure 3.1 Robotics classification based on degrees of freedom.

SCARA (Selective Compliance Assembly Robotic Arm)
Joints: RRPR
2 rotational (x-y plane motion).
1 linear/prismatic (vertical motion).
1 rotational (end-effector orientation).

3.2 Design of IoT-based Pick-and-place Robot

3.2.1 Design of an Affordable IoT Open-Source Robot Arm for Online Teaching

This is a low-cost prototype of the robot, using IoT components as power. The system supports a novice student, researcher, and hobbyist without a robotics lab or even multiple materials and actuators to build, study, and validate the motion behavior of a robot fabricated from scratch. The links of this robot are constructed out of PLA filament to use the least possible amount of material while making the most of the flexibility in design provided by 3D printing. Additionally, this amount of minimal usage of material also allows the servo motors to carry the weight of every single component in the prototype. The robot is controlled from a mobile phone over an HMI interface, which adjusts the angle of each servo motor using Denavit–Hartenberg (DH) parameters for their position generated from a direct kinematics matrix. This interface allows for the reproduction of motion trajectories by way of inverse kinematics via geometrical relations between the links with a button.

We are developing a replication protocol for the robot to trace. To illustrate this, we have chosen a circle, a five-petal flower, and a hand-drawn spiral. All of these have been tested under open-source MATLAB code so that the robot can trace its trajectory. Equations related to the following have been tested on the robot with the help of MATLAB, so that the dimensions and trajectories in its physical prototype can be simulated. We have made the program available as an open-source project and adapted the MATLAB code to be compatible with the Arduino IDE, to fit the controller with wireless communication. Because ESP32 has integrated Wi-Fi and dual-mode Bluetooth, and the program has been loaded into this hardware.

3.3 Direct Kinematics

We first define the initial position of the robot and the available workspace with the servo motors. For this prototype, we assume that both servos are aligned at 0°. As can be seen in the image, S_0 and S_1 correspond to the two positions of the servo motors. It is also assumed that S_2 is the position at the end of the second link. Also, a_1 and a_2 depict lengths of each link.

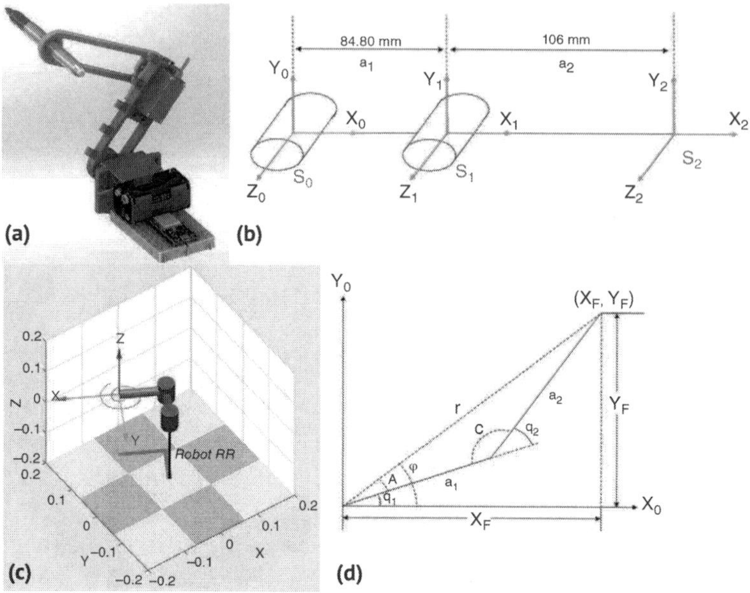

Figure 3.2 (a) The CAD design of the robot, (b) DH parameters demonstrated (home position), (c) robot replicating the handwritten spiral, and (d) the robot position for inverse kinematic analysis (Benitez et al. [2020]/Elsevier/CC BY 4.0).

This is also helpful in building a parameter table to see the prior configuration in a sectioned way and understand the connections between the actuators' actions. The following are the parameters shown in the table: (i) q_n: The input angle of the servo motor. (ii) θ_n: Preferred angle of the robot. (iii) a_n: Arm or link length. Since the robot is planar, in this case, both linkages have an offset of d_n. (iv) α_n: Twist (applied when there is a non-aligned Z-axis in the D–H configuration). The direct kinematics matrix is found once the configuration is determined. In general form, it is as follows: Figure 3.2 represents the CAD design of the robot, the DH parameters, the robot replicating the handwritten spiral, and the robot position for inverse kinematic analysis.

$$T_n = \begin{bmatrix} \cos(\theta_n) & -\sin(\theta_n)\cos(\alpha_n) & \sin(\theta_n)\sin(\alpha_n) & a_n\cos(\theta_n) \\ \sin(\theta_n) & \cos(\theta_n)\cos(\alpha_n) & -\cos(\theta_n)\sin(\alpha_n) & a_n\sin(\theta_n) \\ 0 & \sin(\alpha_n) & \cos(\alpha_n) & d_n \\ 0 & 0 & 0 & 1 \end{bmatrix} \tag{3.1}$$

Table 3.1 Parameters for D–H from the selected configuration.

Link	q_n	a_n	d_n	α_n
S1	$q_1 = \theta_1$	$a_1 = 84.8\,\text{mm}$	$d_1 = 0$	$\alpha_1 = 0$
S2	$q_2 = \theta_2$	$a_2 = 106\,\text{mm}$	$d_2 = 0$	$\alpha_2 = 0$

Thus, a matrix for each connection is produced by considering the data from Table 3.1 and Eq. (3.1).

$$T_1 = \begin{bmatrix} \cos(\theta_1) & -\sin(\theta_1) & 0 & 84.8\cos(\theta_1) \\ \sin(\theta_1) & \cos(\theta_1) & 0 & 84.8\sin(\theta_1) \\ 0 & 0 & 1 & 0 \\ 0 & 0 & 0 & 1 \end{bmatrix}$$
$$T_2 = \begin{bmatrix} \cos(\theta_2) & -\sin(\theta_2) & 0 & 106\cos(\theta_2) \\ \sin(\theta_2) & \cos(\theta_2) & 0 & 106\sin(\theta_2) \\ 0 & 0 & 1 & 0 \\ 0 & 0 & 0 & 1 \end{bmatrix} \tag{3.2}$$

Multiplying T_1 and T_2 the following general direct kinematics matrix is obtained:

$$T = \begin{bmatrix} \cos(\theta_1+\theta_2) & -\sin(\theta_1+\theta_2) & 0 & 84.8\cos(\theta_1) + 106\cos(\theta_1+\theta_2) \\ \sin(\theta_1+\theta_2) & \cos(\theta_1+\theta_2) & 0 & 84.8\sin(\theta_1) + 106\sin(\theta_1+\theta_2) \\ 0 & 0 & 1 & 0 \\ 0 & 0 & 0 & 1 \end{bmatrix} \tag{3.3}$$

Eq. (3.3) is evaluated as Eq. (3.4) as follows:

$$\text{where } \theta_1 = 0^\circ \text{ and } \theta_2 = 0^\circ \rightarrow T = \begin{bmatrix} 1 & 0 & 0 & 190.8 \\ 0 & 1 & 0 & 0 \\ 0 & 0 & 1 & 0 \\ 0 & 0 & 0 & 1 \end{bmatrix} \tag{3.4}$$

It can be seen in the previous table that the first three rows and columns represent the current axes of the final effector in relation to the global axes of the system, as defined in the DH configuration shown in Figure 3.4. For example, the number in the first row and column is 1, meaning the X-axis is aligned in the natural direction of both servo motors at 0. The last column indicates the coordinates of the final effector. In the previous configuration, the effector is at (190.8, 0), or 190.8 mm to the right from the position (0, 0).

3.4 Inverse Kinematics

Joint space $(\theta_1, \theta_2) \rightarrow$ Forward Kinematics $\rightarrow$ World space (x, y, ϕ)

World space $(x, y, \phi) \rightarrow$ Inverse Kinematics $\rightarrow$ Joint space (θ_1, θ_2)

The inverse kinematics for single- and two-degrees-of-freedom **planar manipulator** is shown in Figures 3.3 and 3.4.

3.5 Inverse Kinematics of a Three-degrees-of-freedom Spatial Manipulator with a Unique Geometry

A regular three-revolute Yaskawa GP12 robot inverse kinematics is calculated. α is the default angle, which is always present. On top of that, if we are moving by an angle θ_3 then it becomes

$$\theta_{3\alpha} = \theta_3 + \alpha$$

$$x = [a_1 + a_2 \cos\theta_2 + a_3 \cos(\theta_2 + \theta_3 + \alpha)] \cos\theta_1$$

$$y = [a_1 + a_2 \cos\theta_2 + a_3 \cos(\theta_2 + \theta_3 + \alpha)] \sin\theta_1$$

$$z = d_1 + a_2 \sin\theta_2 + a_3 \sin(\theta_2 + \theta_3 + \alpha)$$

$$\theta_1 = \tan^{-1}\frac{y}{x}$$

The inverse kinematics of a three-degrees-of-freedom spatial manipulator is shown in Figure 3.5.

For Yaskawa GP 12, robotic manipulator, $d_1 = 0$ and $\alpha = 17.354°$

$$e'_x = {}^{x}/_{\cos\theta} - a_1$$

$$e'_z = z$$

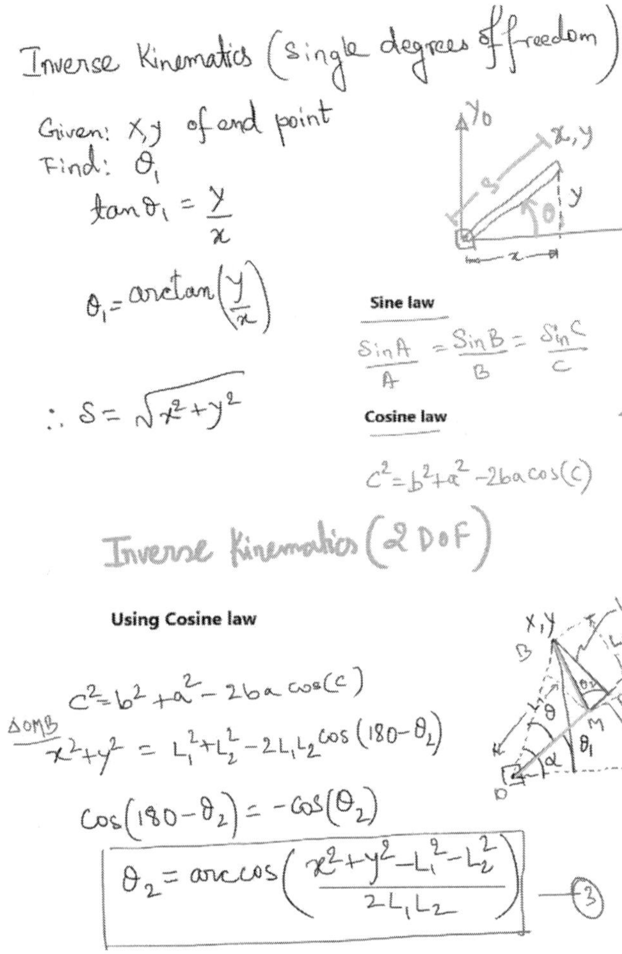

Figure 3.3 Inverse kinematics of a single- and two-degrees-of-freedom planar manipulator.

Analogous to 2R manipulator

$$(x,y)_{2R} \equiv \left(e_x', e_z'\right)_{3R}, \text{ and } (\theta_1, \theta_2)_{2R} \equiv (\theta_2, \theta_{3\alpha})_{3R}$$

This is equivalent to 2R with link 2 and link 3 in the plane $x'z'$, which is rotated along z by an angle. From Figure 3.5c, we can get

$$e_x' = a_2 \cos\theta_2 + a_3 \cos(\theta_2 + \theta_3 + \alpha)$$

$$e_z' = a_2 \sin\theta_2 + a_3 \sin(\theta_2 + \theta_3 + \alpha)$$

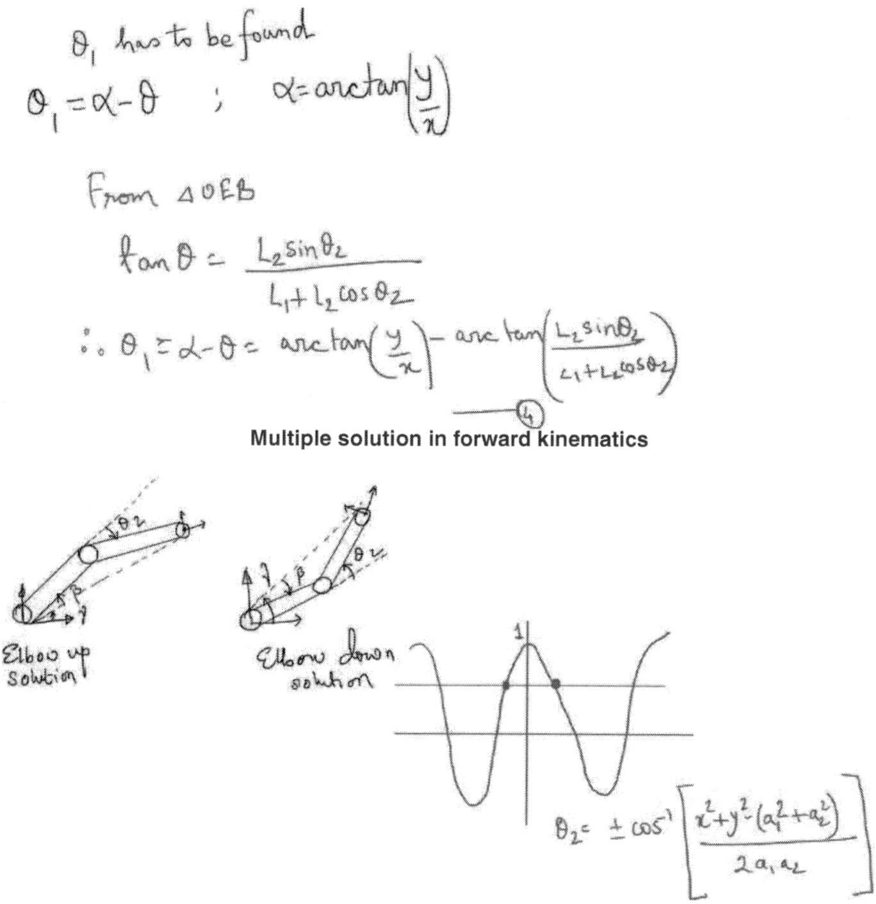

Figure 3.4 Inverse kinematics of a two-degrees-of-freedom planar manipulator.

$$\theta_{3\alpha} = \cos^{-1}\left(\frac{e_x^2 + e_z^2 - (a_2^2 + a_3^2)}{2a_2a_3}\right)$$

$$\theta_2 = \tan^{-1}\left(\frac{e_z'}{e_x'}\right) - \tan^{-1}\left(\frac{a_3 \sin\theta_{3\alpha}}{a_2 + a_3 \cos\theta_{3\alpha}}\right)$$

If we see for a 2DoF planar manipulator, two solutions exist, which are well known and are called the “elbow up” and “elbow down” solution, as shown in Figure 3.4.

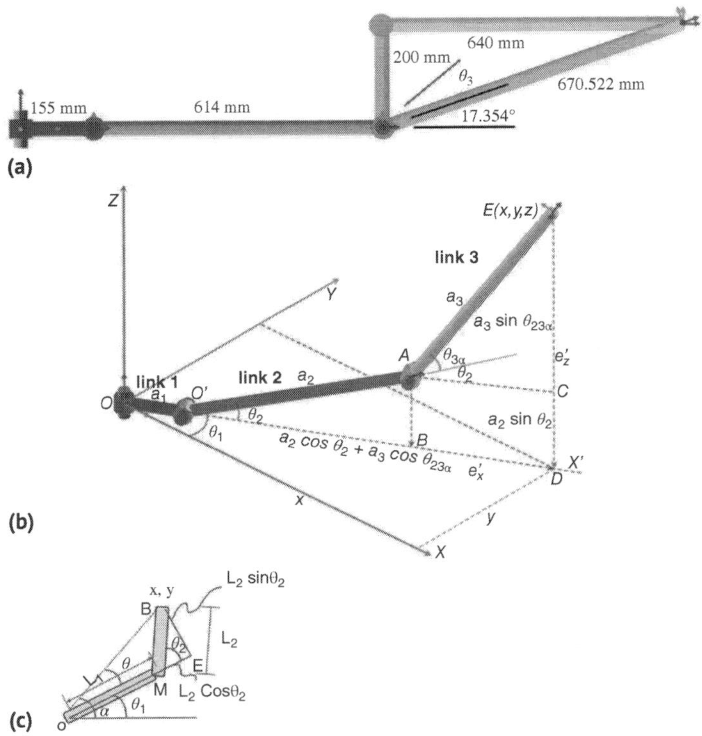

Figure 3.5 Inverse kinematics of a three-degrees-of-freedom spatial manipulator.

3.5.1 Workspace Analysis

It is defined as the set of values x, y, ϕ for which an inverse kinematics solution exists

$$-1 \leq \left(\frac{e_x^2 + e_z^2 - (a_2^2 + a_3^2)}{2a_2a_3} \right) \leq 1$$

Since the value of cos lies between –1 and 1; therefore, the workspace of the robotic manipulator is defined as

$$(a_2 - a_3)^2 \leq e_x^2 + e_z^2 \leq (a_2 + a_3)^2$$

Where the length of the links are: $a_1 > a_2 > a_3$; therefore, the maximum and minimum reach and approach of a robot is

$Maximum = a_1 + a_2 + a_3$, and $Minimum = a_1 - a_2 - a_3$

3.5.2 Inverse Kinematics

For a geometric relation-based approach to correctly use inverse kinematics, it is important that the equations of such a method contain all the elements necessary to determine the angle one is looking for his set of coordinates. More than one solution may result for one instruction due to the method of obtaining solutions from the equations. In order to make a suggestion, the robot must be posed in a position where all the angles are visible. In the HOME position, the angles cannot be viewed, since the home position is one where all angles are enclosed. Thus, to start off with the suggestion, the work on the inverse kinematics solution for this robot should be initiated first and then progressed to the second link, as shown in Figure 3.2.

(i) a_1 and a_2 are the lengths of each link. (ii) q_1 and q_2 are the angles desired for the robot to calculate. (iii) X_F, Y_F, r, and φ are auxiliary variables for finding relations between the links. (iv) A and C are auxiliary variables to find relations between the links using trigonometric relationships.

After defining the model of Figure 3.5, the equations of q_1 and q_2 can be found using trigonometric and algebraic knowledge.

$$r^2 = X_F^2 - Y_F^2 \tag{3.5}$$

q_1 can be found as:

$$\begin{aligned} q_1 &= \varphi - A \\ \varphi &= \tan^{-1}\left(\frac{Y_F}{X_F}\right) \end{aligned} \tag{3.6}$$

Applying the law of Cosines for Auxiliary Angle A:

$$\begin{aligned} a_2^2 &= r^2 + a_1^2 - 2ra_1\cos(A) \\ \cos(A) &= \frac{r^2 + a_1^2 - a_2^2}{2ra_1} \end{aligned} \tag{3.7}$$

Using trigonometric identity to find the sine of the auxiliary angle A:

$$\sin(A) = \sqrt{1 - \left(\frac{r^2 + a_1^2 - a_2^2}{2ra_1}\right)^2} \tag{3.8}$$

Equations (3.7) and (3.8) are used to find the tangent of auxiliary angle A:

$$A = \tan^{-1}\left(\frac{\sqrt{1 - \left(\frac{r^2+a_1^2-a_2^2}{2ra_1}\right)^2}}{\frac{r^2+a_1^2-a_2^2}{2ra_1}}\right) \tag{3.9}$$

By substituting Eqs. (3.5) and (3.9) into Eq. (3.6), it is possible to find the generalized coordinate q_1 as:

$$q_1 = \tan^{-1}\left(\frac{Y_F}{X_F}\right) - \tan^{-1}\left(\frac{\sqrt{1-\left(\frac{r^2+a_1^2-a_2^2}{2ra_1}\right)^2}}{\frac{r^2+a_1^2-a_2^2}{2ra_1}}\right) \tag{3.10}$$

From Figure 3.2, it is possible to find q_2:

$$q_2 = \pi - C \tag{3.11}$$

The law of cosines for auxiliary angle C:

$$r^2 = a_1^2 + a_2^2 - 2a_1a_2\cos(C) \tag{3.12}$$

$$\cos(C) = \frac{a_1^2 + a_2^2 - r^2}{2a_1a_2} \tag{3.13}$$

Using trigonometric identity to find the sine of auxiliary angle C:

$$\sin(C) = \sqrt{1-\left(\frac{a_1^2 + a_2^2 - r^2}{2a_1a_2}\right)^2} \tag{3.14}$$

Equations (3.13) and (3.14) are used to find the tangent of auxiliary angle C:

$$C = \tan^{-1}\left(\frac{\sqrt{1-\left(\frac{a_1^2+a_2^2-r^2}{2a_1a_2}\right)^2}}{\frac{a_1^2+a_2^2-r^2}{2a_1a_2}}\right) \tag{3.15}$$

By substituting Eq. (3.15) into Eq. (3.11),

$$q_2 = \pi - \tan^{-1}\left(\frac{\sqrt{1-\left(\frac{a_1^2+a_2^2-r^2}{2a_1a_2}\right)^2}}{\frac{a_1^2+a_2^2-r^2}{2a_1a_2}}\right) \tag{3.16}$$

3.6 Splines in Robotics

Spline in Robotics: It is a curve that forms the mathematical expression used for path planning, motion control, or trajectory generation. It enables the tracing of desired trajectories by a robot in smooth motion, along with speed, acceleration, and jerk. This is important in applications such as failure cases, where unwanted turns could lead to instability or collision.

Trajectory generation would be done using spline techniques so as to have a smooth and controlled movement of the robot. It interpolates its key frames or positions at a specific time. This will be particularly crucial for designs like robotic arms that should have smooth movements, so that the design itself goes on to operate properly.

Motion Control: Splines are useful for expressing the position of a robot with respect to time. This allows the robot to vary velocity and acceleration efficiently and safely during operations.

Animation: Splines are useful for animation in smooth movements of robots, making them possible to be comparable to natural movements, such as in a simulation or an animation.

Types of Splines Used with Robots:

Some applications of splines are as follows:

1. Linear Splines: Such applications where data does not require complex movement and consist of severe direction changes.
2. Cubic Splines: Used more in robotics due to the smooth continuity of position, velocity, and acceleration.
3. B-splines and NURBS: Flexibility to shape according to the requirement and can be followed for complex trajectories.

Advantages of Splines:

1. Smoothness is one of the implications of splines, wherein wear and tear are reduced and performance is improved.
2. Flexibility: Spline curves are smooth and easy to follow, as per any constraint or requirement, which aids in motion profiles.
3. Efficiency: Algorithms for the evaluation and synthesis of splines are computationally efficient, making spline real-time applications feasible.

3.6.1 Linear Splines

A linear spline connects each pair of adjacent data points with a straight line. Given a set of points $(x_0, y_0), (x_1, y_1), \ldots, (x_n, y_n)$, the linear spline $S(x)$ can be defined in segments:

$$S(x) = \begin{cases} y_0 & \text{if } x = x_0 \\ y_i + \dfrac{(y_{i+1} - y_i)}{(x_{i+1} - x_i)}(x - x_i) & \text{if } x_i < x < x_{i+1} \end{cases} \tag{3.17}$$

This is simple and computationally inexpensive, but it can lead to sharp corners or discontinuities in the first derivative at the data points.

3.6.2 Cubic Splines

Let us consider the task as moving the tool from its initial position to the goal position within some time period. Inverse kinematics allows to calculate the set of joint angles corresponding to the goal position and orientation (Aribowo and Terashima, 2014). The initial position of the manipulator is also known in the form of a set of joint angles. What is needed is a function for each joint whose value at t_0 is the initial position of the joint, and whose value at t_f is the desired goal position of that joint. As illustrated in Figure 3.6, there are many smooth functions, $\theta(t)$, that might be used to interpolate the joint value.

In making a single smooth motion, at least four constraints on $\theta(t)$ are evident. Two constraints on the function's value come from the selection of initial and final values:

$$\begin{aligned} \theta(0) &= \theta_0 \\ \theta(t_f) &= \theta_f \end{aligned} \tag{3.18}$$

An additional two constraints are that the function be continuous in velocity, which, in this case, means that the initial and final velocities are zero:

$$\begin{aligned} \dot{\theta}(0) &= 0 \\ \dot{\theta}(t_f) &= 0 \end{aligned} \tag{3.19}$$

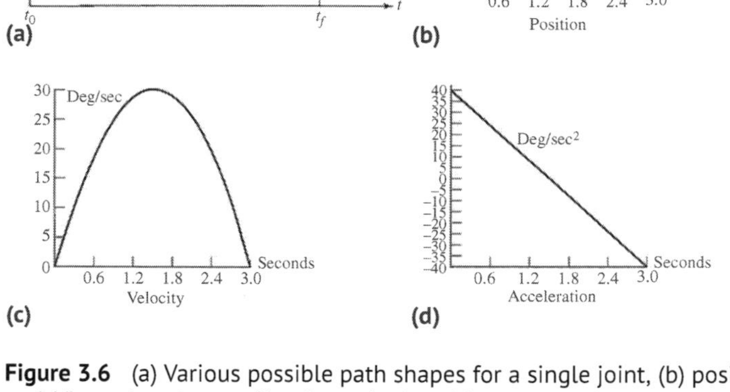

Figure 3.6 (a) Various possible path shapes for a single joint, (b) position, (c) velocity, and (d) acceleration profiles for a single cubic segment that starts and ends at rest.

A polynomial of at least the third degree can satisfy these four. Since a cubic polynomial has four coefficients, one of the cubic polynomials that satisfies the specified requirements can be taken in the form of Eqs. (3.18) and (3.19).

$$\theta(t) = a_0 + a_1 t + a_2 t^2 + a_3 t^3 \tag{3.20}$$

so the joint velocity and acceleration along this path are clear

$$\begin{aligned} \dot{\theta}(t) &= a_1 + 2a_2 t + 3a_3 t^2 \\ \ddot{\theta}(t) &= 2a_2 + 6a_3 t. \end{aligned} \tag{3.21}$$

Combining Eqs. (3.20) and (3.21) with the four desired constraints yields four equations in four unknowns:

$$\begin{aligned} \theta_0 &= a_0 \\ \theta_f &= a_0 + a_1 t_f + a_2 t_f^2 + a_3 t_f^3 \\ 0 &= a_1 \\ 0 &= a_1 + 2a_2 t_f + 3a_3 t_f^2 \end{aligned} \tag{3.22}$$

Solving these equations for the a_i, we obtain

$$\begin{aligned} a_0 &= \theta_0 \\ a_1 &= 0 \end{aligned} \tag{3.23}$$

$$\begin{aligned} a_2 &= \frac{3}{t_f^2}(\theta_f - \theta_0) \\ a_3 &= -\frac{2}{t_f^3}(\theta_f - \theta_0) \end{aligned} \tag{3.24}$$

Using Eq. (3.24), we can calculate a cubic polynomial relating any position of the joint to any desired final position. This is the solution to the problem where the joint starts and finishes at zero velocity.

Example 1: A one-link robot with a rotary joint is stationary at $\theta = 15$ degrees. It is to be driven smoothly from rest to $\theta = 75$ degrees in 3 s. Determine the coefficients of a cubic that does this and returns the manipulator to rest at the destination. Plot position, velocity, and acceleration of the joint as a function of time.

Substituting into Eq. (3.24), we obtain

$$\begin{aligned} a_0 &= 15.0 \\ a_1 &= 0.0 \\ a_2 &= 20.0 \\ a_3 &= -4.44 \end{aligned} \tag{3.25}$$

Using Eqs. (3.20) and (3.21), we obtain

$$\begin{aligned} \theta(t) &= 15.0 + 20.0t^2 - 4.44t^3 \\ \dot{\theta}(t) &= 40.0t - 13.33t^2 \\ \ddot{\theta}(t) &= 40.0 - 26.66t \end{aligned} \tag{3.26}$$

Figure 3.6 plots the position, velocity, and acceleration functions for this motion, sampled at 40 Hz. Note that the velocity profile for any cubic function is a parabola, and that the acceleration profile is linear.

Example 2: Determine the coefficients of two cubics connected at the intermediate via point of a two-segment spline with continuous acceleration. θ_0 is the starting angle, θ_v is the via point, and θ_g is the target point.

The first cubic is

$$\theta(t) = a_{10} + a_{11}t + a_{12}t^2 + a_{13}t^3 \tag{3.27}$$

and the second is

$$\theta(t) = a_{20} + a_{21}t + a_{22}t^2 + a_{23}t^3 \tag{3.28}$$

Each cubic will be evaluated over an interval starting at $t = 0$ and ending at $t = t_{fi}$, where $i = 1$ or $i = 2$.

The constraints we wish to enforce are

$$\begin{aligned} \theta_0 &= a_{10} \\ \theta_v &= a_{10} + a_{11}t_{f1} + a_{12}t_{f1}^2 + a_{13}t_{f1}^3 \\ \theta_v &= a_{20} \\ \theta_g &= a_{20} + a_{21}t_{f2} + a_{22}t_{f2}^2 + a_{23}t_{f2}^3 \\ 0 &= a_{11} \end{aligned} \tag{3.29}$$

$$
\begin{aligned}
0 &= a_{21} + 2a_{22}t_{f2} + 3a_{23}t_{f2}^2 \\
a_{11} + 2a_{12}t_{f1} + 3a_{12}t_{f1}^2 &= a_{21} \\
2a_{12} + 6a_{13}t_{f1} &= 2a_{22}
\end{aligned}
\tag{3.30}
$$

These constraints specify a linear-equation problem having eight equations and eight unknowns. Solving the case $t_f = t_{f1} = t_{f2}$, we obtain

$$
\begin{aligned}
a_{10} &= \theta_0 \\
a_{11} &= 0 \\
a_{12} &= \frac{12\theta_v - 3\theta_g - 9\theta_0}{4t_f^2} \\
a_{13} &= \frac{-8\theta_v + 3\theta_g + 5\theta_0}{4t_f^3} \\
a_{20} &= \theta_v \\
a_{21} &= \frac{3\theta_g - 3\theta_0}{4t_f} \\
a_{22} &= \frac{-12\theta_v + 6\theta_g + 6\theta_0}{4t_f^2} \\
a_{23} &= \frac{8\theta_v - 5\theta_g - 3\theta_0}{4t_f^3}
\end{aligned}
\tag{3.31}
$$

3.7 Automation and Its Types

Automation is a technological function that uses technology to execute given functions with minimal human intervention. It incorporates a large number of control systems for operating equipment in processes, manufacturing, transportation, among other applications. Automation encourages efficiency, eliminates errors, and enhances safety. Automation is of different types and levels. Basically, automation alters the nature of processes in industries. The category of automation can broadly be placed under manual, semi-automatic, and fully automatic systems, which are defined by the level of human input and control. This can further be categorized into types of automation; based on specific applications and user requirements, the types are fixed, programmable, flexible, integrated, and cognitive.

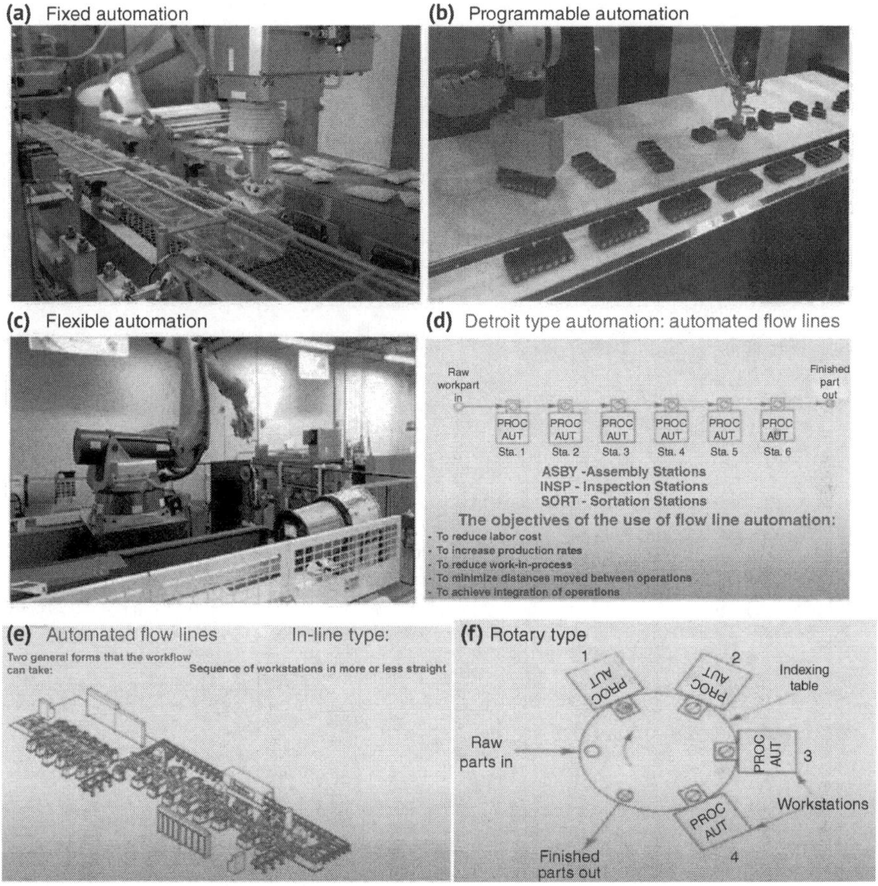

Figure 3.7 Various automation processes.

3.7.1 Levels of Automation

- Fixed Automation: The sequence of processing (or assembly) operations is defined by the design of the equipment. Significant initial investment for equipment that is custom-built. Very high production rate. Rigid with respect to changes in product. Fixed Automation Mass production Application. Figure 3.7 depicts a few different types of automation.
- Programmable Automation: It is designed in such a manner that it can be easily changed with different product configurations by changing the sequence of operations of production equipment. High investment in general-purpose equipment. Low production level compared to the fixed required automation. Flexibility to handle various product configurations.

- Flexible (Soft) Automation: Extension of Programmable Automation: It is a system capable of producing a variety of products with virtually no time loss for changeovers from one product to the next. High investment for a custom-engineered system. Continuous production of variable mixtures of products.
- Industrial Automation Definition: These are control systems that monitor equipment in various types of industries; this may be from small machines and processes up to bigger systems.

  Examples: Manufacturing automation, process control systems.

  Benefits: High productivity, quality, and safety.

  Disadvantages: These require massive capital investment and costly maintenance.

  Description: This is the use of hardware and software to create, store, manipulate, and relay office information in a digital format.

  Examples include document management systems, email automation, and workflow software.

  Benefits: It increases efficiency and communication while reducing paperwork.

  Disadvantages: Dependent on technology; security threats.
- Home Automation Description: This is the control of home appliances and systems using smart devices and networks.

  Examples include smart thermostats, security systems, and lighting control.

  Advantages: Greater convenience, energy efficiency, and security.

  Disadvantages: Expensive to set up initially; potential glitches with the functions of technology.
- Test Automation
  Definition: It is the use of software to command and execute tests and compare real results with expected results.

  Example: Automated tools used in software testing.

  Advantages: Test automation thus saves time while test running; with test automation, more extensive testing becomes possible.

  Disadvantages: Test automation requires initial investment in test tools; it does not detect all defects.
- RPA or Robotic Process Automation
  Definition: The application of the software robot to automate simple, repetitive human tasks that are usually performed by business processes.

Examples: Data entry, invoicing, etc.

Benefits: Manpower saving; good accuracy; and rapid working.

Drawbacks: The end is applicable only for rule-based jobs; it has to be updated periodically.

Degrees of Automation

- Manual: Human control with no automation.
- Semi-automatic: Human and machine combined control, which requires human intervention for certain tasks.
- Fully Automatic: Complete machine control without human intervention.

Types of Automation

- Static Automation: Inflexible processes with zero flexibility.
- Adaptive Automation: Systems that learn from their immediate feedback and reliability based on Avril et al. in the year 2021.
- Test Automation Frameworks: Keyword-driven, data-driven, and behavior-driven frameworks, each with unique functionalities and trade-offs. Automation is efficient and reliable, though it has brought some of controversies related to issues of trust and dependence, especially in high-risk environment where human oversight is still significant.

3.8 Basic Image Processing Techniques

Digital image processing refers to all those techniques that permit the processing of digital images using computer algorithms. Considered a very vital preprocessing technique, it has been generally applied to many applications such as face recognition, object detection, and compressing images. Image processing may be to enhance the quality of an image or to retrieve significant information from it. This process is crucial for many deep learning and computer vision applications since proper preprocessing can significantly boost model performance. In the application of image manipulation, adding or removing items from images becomes very significant, especially in the entertainment industry. This paper discusses one of the challenges involving medical image segmentation and discusses image inpainting in the preprocessing stage to remove artifacts from dermoscopy images, giving examples of how that is done.

3.8.1 Binary Image

Images that include only two unique values of pixel intensity, 0 (representing black) and 1 (representing white), are termed as binary images. Such images are usually applied for highlighting a discriminating portion of a colored image.

3.8.2 Gray Scale Image

Grayscale or 8-bit images represent 256 different colors, of which the pixel intensity of 0 represents black color and the pixel intensity of 255 defines white color. The remaining 254 values between them serve as different intensities of gray shades.

3.8.3 RGB Color Image

The images we are used to in this modern world are RGB or colored images that represent, as far as computers are concerned, 16-bit matrices. In other words, for each pixel, 65 536 different colors may appear. "RGB" is short for Red, Green, and Blue "channels" of an image. Till this point, we have images with single-channel color. That is, two coordinates can determine the position of any value of a matrix. Now, three identically sized matrices (also known as channels), whose values are in the range from 0 to 255, are stacked over each other, and thus we need three different coordinates to identify the value of a matrix element.
Therefore, a pixel in an RGB image will be black when the pixel value is (0, 0, 0), and white when it is (255, 255, 255). Any mix of numbers between these gives all the various colors that exist in nature. For instance, (255, 0, 0) is red since only the red channel is activated for this pixel. Similarly, (0, 255, 0) is green, and (0, 0, 255) is blue.

3.8.4 RGBA Image

RGBA images are colored RGB images with an extra channel known as "alpha" that depicts the opacity of the RGB image. Opacity ranges from a value of 0–100% and is essentially a "see-through" property. It is the percentage of light passing through a substance in physics. For example, cellophane paper is translucent, frosted glass is semi-transparent while wood is opaque. In RGBA images, the alpha channel is trying to mimic this. Here's an example below.

Steps in image processing

Phases of Image Processing: The basic steps in any conventional digital image processing pipeline are as follows.

3.8.5 Image Acquisition

The image is captured by a camera and digitized (if the camera output isn't digitized automatically) using an analogue-to-digital converter for further processing in a computer.

3.8.6 Image Enhancement

In this process, the image captured is altered in such a manner that it satisfies the needs of the specific application for which the image is to be utilized. Such methods are primarily aimed at the definition of latent or crucial information of an image, such as contrast and brightness adjustment, etc. Image enhancement is highly subjective in nature.

3.8.7 Image Restoration

This step is about the enhancement of an image's aesthetics and is an objective operation, as the degradation of an image can be related to any mathematical or probabilistic model. Noise removal or elimination of blur in images would be an example.

3.8.8 Color Image Processing

This step is about dealing with the processing of colored images, such as 16-bit RGB, RGBA images, and so on, and includes color correction or color modeling of images.

3.8.9 Wavelets and Multi-resolution Processing

Wavelets are the atoms for representing images at different levels of resolution. Images are subdivided successively into smaller regions for compression of data as well as for pyramidal representation.

3.8.10 Image Compression

The pictures need to be compressed, so can't be maintained in their actual size, as they have to be transferred to other gadgets or because of some computational memory constraints. This also comes in handy while portraying images on the internet, for instance, on Google, a tiny thumbnail of the picture is an extremely compressed version of the original. Only when you click on the picture, it shows the actual resolution. This saves bandwidth on the servers.

3.8.11 Morphological Processing

The elements of the image, those that are semantic in the image representation and description of form, have to be extracted for further processing or in downstream tasks. Morphological Processing offers the tool kit, which in turn are mathematical operations to serve just that purpose. For example, there is an operation erosion and dilation that blur and sharpen the edges of objects in the image.

3.8.12 Image Segmentation

This step is about the segmentation of an image into key parts so as to simplify and/or modify the representation of an image into a form meaningful and easier to interpret. Image segmentation enables computers to focus attention on the more meaningful parts of an image while discarding other parts, thus improving the performance of automated systems.

3.8.13 Representation and Description

Next, this step includes image segmentation processes where the task for representation is determining whether the segmented region should be represented as a boundary or an entire region. Description concerns extracting attributes that lead to some quantitative information of interest or are fundamental to distinguish one class of objects from another.

3.8.14 Object Detection and Recognition

Once the images are segmented into objects, and the representation and description stages are complete, the automatic system is required to attach a label, to indicate to the human users what object was detected in the scene, for example, "vehicle" or "person."

3.8.15 Knowledge Base

Knowledge may be as simple as the bounding box coordinates for an object of interest that has been located in the image, together with the object label assigned to it. All that would be useful in the solution of the problem for that particular task can be encoded into the knowledge base.

The challenges concerning the Food Processing are illustrated in Table 3.2.

Table 3.2 AI/ML applications in the food sector.

Challenge	AI/ML applications affected	Elaboration of application-specific challenges	References
High computational requirements	Computer vision	Applications encounter challenges due to substantial computational power requirements for tasks like automated inspection and rapid contaminant detection.	Brosnan and Sun (2004)

Challenge	AI/ML applications affected	Elaboration of application-specific challenges	References
Overfitting	Computer vision	Overfitting occurs when models become too tailored to training data, compromising generalizability. Balancing model complexity is crucial.	Rokhva et al. (2024)
Data security concerns	IoT-enabled sensors	Face challenges related to data security, especially in real-time monitoring, necessitating robust encryption and authentication measures.	McAllister et al. (2018)
Integration challenges	IoT-enabled sensors	Integrating sensors into existing systems poses challenges. Compatibility and seamless integration require meticulous planning.	Zhang et al. (2014)
Privacy concerns	Blockchain integration	Introduces privacy concerns, as technology stores data in an immutable and decentralized manner, requiring a balance between transparency and privacy.	Zhang et al. (2014)
Regulatory acceptance	Blockchain integration	Achieving regulatory acceptance in the food industry is a multifaceted challenge that requires overcoming legal and regulatory barriers for widespread adoption.	Brosnan and Sun (2004)
Interpretability challenges	Predictive modeling	Encounters challenges related to interpretability. Understanding and explaining model decisions is vital for gaining trust.	Elnour et al. (2023)
Dependence on data	Predictive modeling	Faces challenges regarding dependence on data. Ensuring availability of high-quality and diverse data is crucial for reliability and effectiveness.	Kondakci and Zhou (2017)

3.9 Robot Kinematics

Figure 3.8 displays some simple joints in robots.
Forward kinematics (angles to positions) What you have: The length of each link and the angle of each joint.

What you can solve for: The position of any point (i.e. it's (x, y, z) coordinates.

Inverse kinematics (positions to angles) What you have: The length of each link and the position of some point on the robot. What's here: The angle of each joint required to get that configuration.
Dot Product: Geometric Representation:

$$\overline{\mathbf{A}} \bullet \overline{\mathbf{B}} = \|\mathbf{A}\|\|\mathbf{B}\| \cos\theta$$

Matrix Representation:

$$\bar{A} \cdot \bar{B} = \begin{bmatrix} a_x \\ a_y \end{bmatrix} \bullet \begin{bmatrix} b_x \\ b_y \end{bmatrix} = a_x b_x + a_y b_y$$

Unit vector as depicted in Figure 3.9, vector in the direction of a chosen vector but whose magnitude is 1.

$$\overline{\mathbf{u}}_{\mathrm{B}} = \frac{\overline{\mathbf{B}}}{\|\mathbf{B}\|}$$

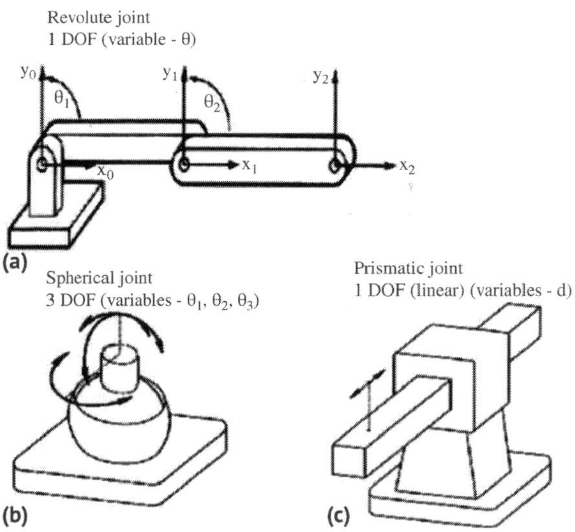

Figure 3.8 Basic joints in robots.

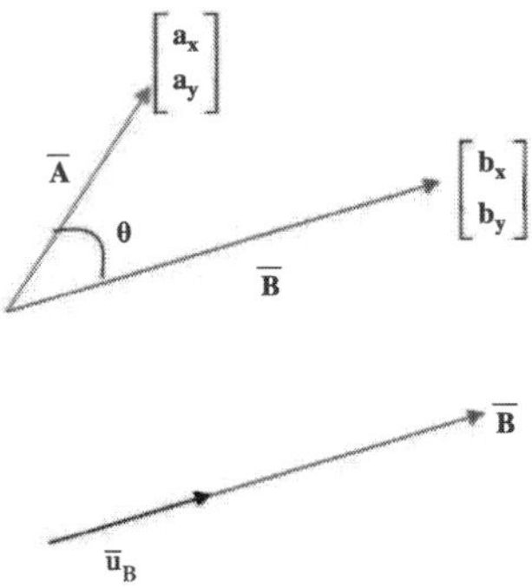

Figure 3.9 Dot product of two vectors and unit vector.

Matrix Multiplication: It is possible to multiply *a* (*n x*) matrix B by *a* ($m \times n$) matrix A since the number of rows in *B* equals the number of columns in *A*. When multiplying non-commutatively, AB is not equal to BA.

$$\begin{bmatrix} a & b \\ c & d \end{bmatrix} \times \begin{bmatrix} e & f \\ g & h \end{bmatrix} = \begin{bmatrix} (ae+bg) & (af+bh) \\ (ce+dg) & (cf+dh) \end{bmatrix}$$

Matrix Addition:

$$\begin{bmatrix} a & b \\ c & d \end{bmatrix} + \begin{bmatrix} e & f \\ g & h \end{bmatrix} = \begin{bmatrix} (a+e) & (b+f) \\ (c+g) & (d+h) \end{bmatrix}$$

Translation along x-axis: Figure 3.10 represents the translation along x-axis P_x = distance between the XY and NO coordinate planes

Notation:

$$\overline{\mathbf{V}}^{XY} = \begin{bmatrix} \mathbf{V}^X \\ \mathbf{V}^Y \end{bmatrix} \quad \overline{\mathbf{V}}^{NO} = \begin{bmatrix} \mathbf{V}^N \\ \mathbf{V}^O \end{bmatrix} \quad \overline{\mathbf{P}} = \begin{bmatrix} \mathbf{P_X} \\ \mathbf{0} \end{bmatrix}$$

Further the translation transformation can be represented as Figure 3.11, as follows:

Writing $\overline{\mathbf{V}}^{XY}$ in terms of $\overline{\mathbf{V}}^{NO}$

$$\overline{\mathbf{V}}^{XY} = \begin{bmatrix} \mathbf{P_X} + \mathbf{V}^N \\ \mathbf{V}^O \end{bmatrix} = \overline{\mathbf{P}} + \overline{\mathbf{V}}^{NO} \tag{3.32}$$

The translation along X and Y-axis represents as depicted in Figure 3.12.

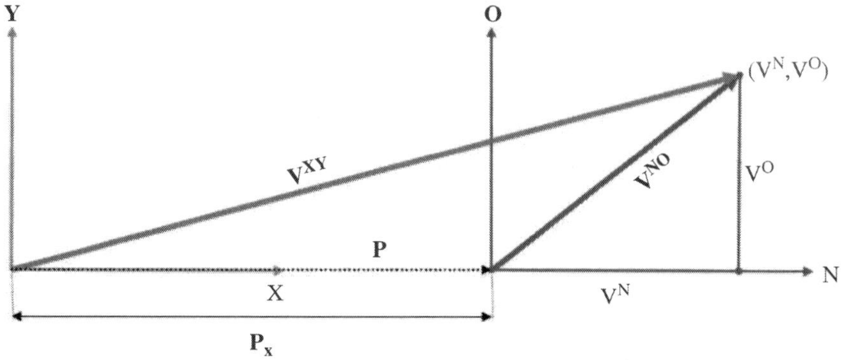

Figure 3.10 Basic transformations: Translation.

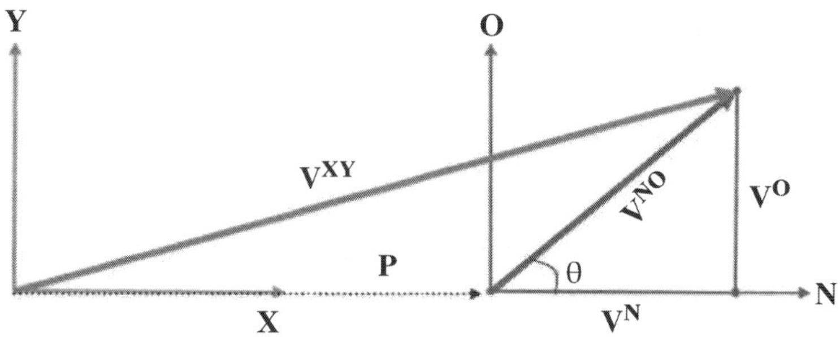

Figure 3.11 Basic transformations: Translation.

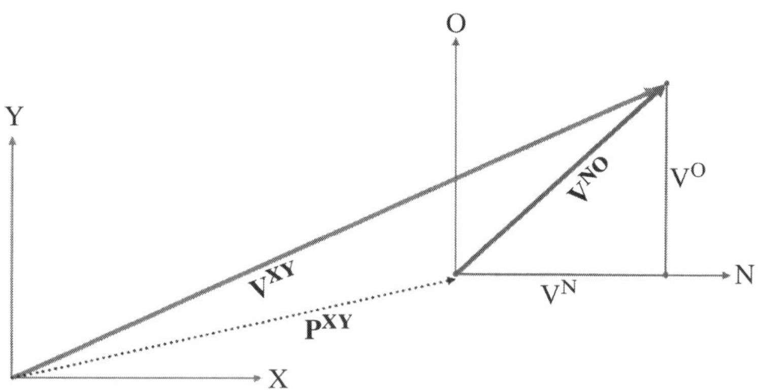

Figure 3.12 Translation along X and Y-axis.

$$\overline{\mathbf{P}}^{XY} = \begin{bmatrix} \mathbf{P_X} \\ \mathbf{P_Y} \end{bmatrix}$$

$$\overline{\mathbf{V}}^{XY} = \overline{\mathbf{P}} + \overline{\mathbf{V}}^{NO} = \begin{bmatrix} \mathbf{P_X} + \mathbf{V^N} \\ \mathbf{P_Y} + \mathbf{V^O} \end{bmatrix}$$

The unit basis vector is depicted in Figure 3.13.

Using Basis Vectors (unit vectors that point along a coordinate axis): $\overline{\mathbf{n}}$ Unit vector along the N-axis; $\overline{\mathbf{0}}$ Unit vector along the N-axis; and $\left\|\mathbf{V}^{NO}\right\|$ Magnitude of the $\mathbf{V}^{NO}$ vector.

$$\overline{\mathbf{V}}^{NO} = \begin{bmatrix} \mathbf{V^N} \\ \mathbf{V^O} \end{bmatrix} = \begin{bmatrix} \left\|\mathbf{V}^{NO}\right\| \cos\theta \\ \left\|\mathbf{V}^{NO}\right\| \sin\theta \end{bmatrix} = \begin{bmatrix} \left\|\mathbf{V}^{NO}\right\| \cos\theta \\ \left\|\mathbf{V}^{NO}\right\| \cos(\mathbf{90} - \theta) \end{bmatrix} = \begin{bmatrix} \overline{\mathbf{V}}^{NO} \bullet \overline{\mathbf{n}} \\ \overline{\mathbf{V}}^{NO} \bullet \overline{\mathbf{0}} \end{bmatrix}$$

Figure 3.14 represents the rotation along the Z-axis.

$$\Theta = \text{ Angle of rotation between the XY and NO coordinate axis}$$

$$\overline{\mathbf{V}}^{XY} = \begin{bmatrix} \mathbf{V^X} \\ \mathbf{V^Y} \end{bmatrix} \quad \overline{\mathbf{V}}^{NO} = \begin{bmatrix} \mathbf{V^N} \\ \mathbf{V^O} \end{bmatrix}$$

The rotation of the vectors is represented by Figure 3.15.

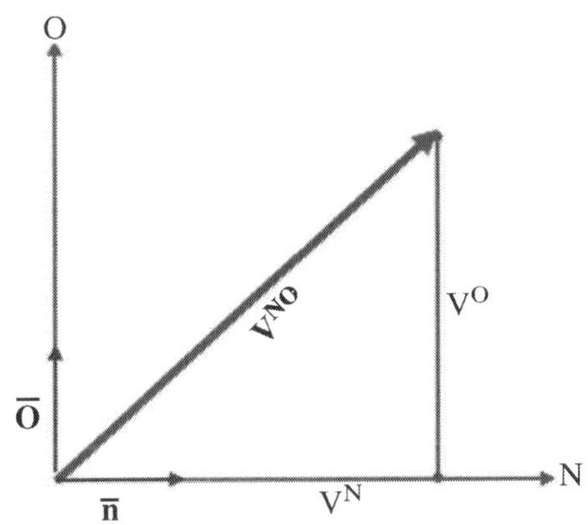

Figure 3.13 Unit basis vector.

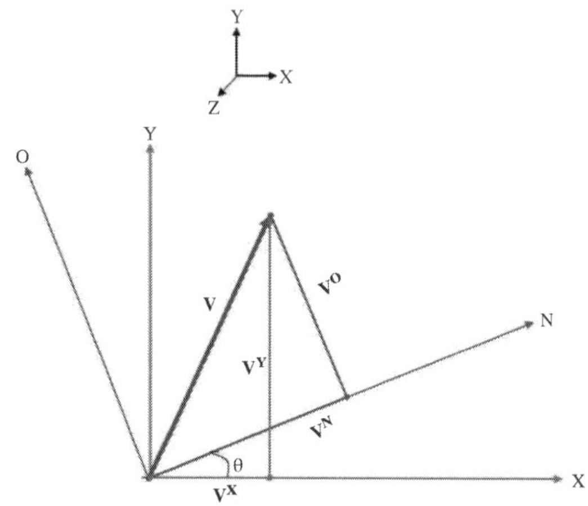

Figure 3.14 Rotation vector.

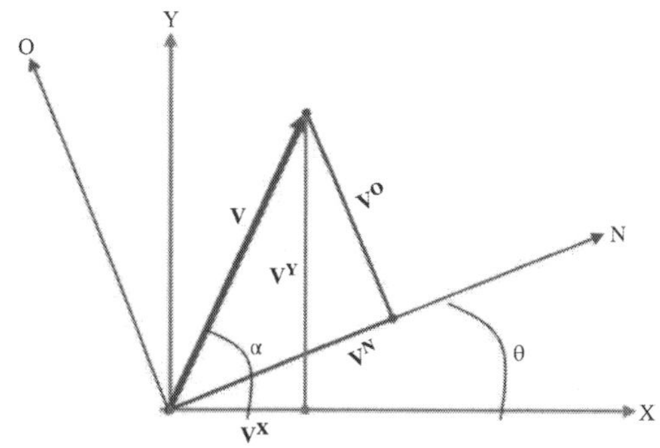

Figure 3.15 Rotation vector.

$\widehat{X}$ unit vector, along X-axis Λ, can be considered with respect to the XY coordinates or NO coordinates.

$$\left\|\overline{\mathbf{V}}^{xy}\right\| = \left\|\overline{\mathbf{V}}^{no}\right\|$$

$$V^x = \left\|\overline{V}^{xy}\right\| \cos\alpha = \left\|\overline{\mathbf{V}}^{no}\right\| \cos\alpha = \overline{V}^{no} \cdot \overline{\mathbf{x}}$$

(components of the vector)

$$\mathbf{V^x} = \mathbf{V^N}(\overline{\mathbf{x}} \bullet \overline{\mathbf{n}}) + \mathbf{V^O}(\overline{\mathbf{x}} \bullet \overline{\mathbf{0}})$$

$$= V^N(\cos\theta) + V^O(\cos(\theta + 90))$$

$$= V^N(\cos\theta) - V^O(\sin\theta)$$

Similarly,

$$\mathbf{V}^{\mathbf{Y}} = \left\| \overline{\mathbf{V}}^{NO} \right\| \sin\alpha = \left\| \overline{\mathbf{V}}^{NO} \right\| \cos(90 - \alpha) = \overline{\mathbf{V}}^{NO} \bullet \overline{\mathbf{y}}$$

$$\mathbf{V}^{\mathbf{Y}} = \left(\mathbf{V}^{N} \times \overline{\mathbf{n}} + \mathbf{V}^{O} \times \overline{\mathbf{0}}\right) \bullet \overline{\mathbf{y}}$$

$$\mathbf{V}^{\mathbf{Y}} = \mathbf{V}^{N}(\overline{\mathbf{y}} \bullet \overline{\mathbf{n}}) + \mathbf{V}^{O}(\overline{\mathbf{y}} \bullet \overline{\mathbf{o}})$$

$$= \mathbf{V}^{N}(\cos(90 - \theta)) + \mathbf{V}^{O}(\cos\theta)$$

$$= \mathbf{V}^{N}(\sin\theta) + \mathbf{V}^{O}(\cos\theta)$$

So....

$$\mathbf{V}^{x} = \mathrm{V}^{N}(\cos\theta) - \mathbf{V}^{O}(\sin\theta)$$

$$\mathbf{V}^{Y} = \mathbf{V}^{N}(\sin\theta) + \mathbf{V}^{O}(\cos\theta)$$

$$\overline{\mathbf{V}}^{XY} = \begin{bmatrix} \mathbf{V}^{X} \\ \mathbf{V}^{Y} \end{bmatrix}$$

Written in matrix form:

$$\overline{\mathbf{V}}^{XY} = \begin{bmatrix} \mathbf{V}^{X} \\ \mathbf{V}^{Y} \end{bmatrix} = \begin{bmatrix} \cos\theta & -\sin\theta \\ \sin\theta & \cos\theta \end{bmatrix} \begin{bmatrix} \mathbf{V}^{N} \\ \mathbf{V}^{O} \end{bmatrix}$$ Rotation Matrix about the z-axis

Figure 3.16 represents translation along P followed by rotation θ

$$\mathbf{V}^{XY} = \begin{bmatrix} \mathbf{V}^{X} \\ \mathbf{V}^{Y} \end{bmatrix} = \begin{bmatrix} \mathbf{P}_{x} \\ \mathbf{P}_{y} \end{bmatrix} + \begin{bmatrix} \cos\theta & -\sin\theta \\ \sin\theta & \cos\theta \end{bmatrix} \begin{bmatrix} \mathbf{V}^{N} \\ \mathbf{V}^{O} \end{bmatrix}$$

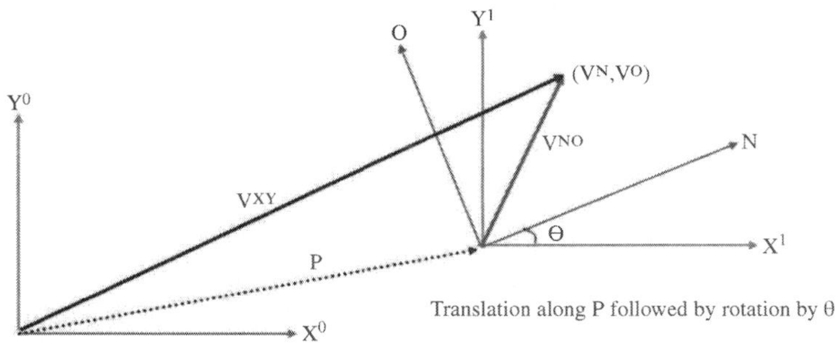

Figure 3.16 Translation followed by rotation.

Note: P_x, P_y are relative to the original coordinate frame. Translation followed by rotation is different from rotation followed by translation.

In other words, knowing the coordinates of a point (V^N, V^O) in some coordinate frame (NO) you can find the position of that point relative to your original coordinate frame (X^0Y^0).

Homogeneous Representation

Putting it all into a Matrix

$$V^{XY} = \begin{bmatrix} V^X \\ V^Y \end{bmatrix} = \begin{bmatrix} P_x \\ P_y \end{bmatrix} + \begin{bmatrix} \cos\theta & -\sin\theta \\ \sin\theta & \cos\theta \end{bmatrix} \begin{bmatrix} V^N \\ V^O \end{bmatrix}$$

What we found by doing a translation and a rotation

$$= \begin{bmatrix} V^x \\ V^y \\ 1 \end{bmatrix} = \begin{bmatrix} P_x \\ P_y \\ 1 \end{bmatrix} + \begin{bmatrix} \cos\theta & -\sin\theta & 0 \\ \sin\theta & \cos\theta & 0 \\ 0 & 0 & 1 \end{bmatrix} \begin{bmatrix} V^N \\ V^o \\ 1 \end{bmatrix}$$

Padding with 0 's and 1 's

$$= \begin{bmatrix} V^X \\ V^Y \\ 1 \end{bmatrix} = \begin{bmatrix} \cos\theta & -\sin\theta & P_x \\ \sin\theta & \cos\theta & P_y \\ 0 & 0 & 1 \end{bmatrix} \begin{bmatrix} V^N \\ V^O \\ 1 \end{bmatrix}$$

Simplifying into a matrix form

$$H = \begin{bmatrix} \cos\theta & -\sin\theta & P_x \\ \sin\theta & \cos\theta & P_y \\ 0 & 0 & 1 \end{bmatrix}$$

Homogenous matrix for a translation in

XY plane, followed by a rotation around

the z-axis

3.9.1 Forward and Inverse Kinematics for a 6-DoF Manipulator

A forward kinematics problem is to determine the end-effector position and orientation for a given set of joint angles and the D–H parameters of the robot. The following chapter gives an analytical solution to this forward kinematics problem of the KUKA KR60. The forward kinematics problem of a robotic manipulator can be solved by attaching a single frame to all joints except the base of the robot. The coordinate so obtained in relation to the position and the alignment of the robot's joint with respect to the base or any further worldwide coordinate provides information about each frame. Placing such frames on the joints reduces the computation of robot placement and direction concerning an end effector in relation to a coordinate translation issue that transformation matrices solve. As such, every joint thus possesses both a position and an orientation relative to its preceding joint. These relations are described using matrices of transformations. The DH parameter table is given in the table below,

i	a	α	d	θ
0	350	0	–	–
1	0	$\frac{\pi}{2}$	−815	$0+\theta_1$
2	850	0	0	$-\frac{\pi}{2}+\theta_2$
3	145	$\frac{\pi}{2}$	0	$0+\theta_3$
4	0	$-\frac{\pi}{2}$	−820	$\pi+\theta_4$
5	0	$\frac{\pi}{2}$	0	$\pi+\theta_5$
6	–	–	170	$0+\theta_6$

$$ {}_{i}^{i-1}T = R_X(\alpha_{i-1})\,D_X(a_{i-1})\,R_Z(\theta_i)\,D_Z(d_i) $$

in which $R_X(\alpha_{i-1})$ is a rotation matrix about the X axis by α_{i-1}, $D_X(a_{i-1})$ is translation matrix along the X axis by a_{i-1}, $R_Z(\theta_i)$ is a rotation matrix about the Z axis by θ_i, $D_Z(d_i)$ is translation matrix along the Z axis by d_i, and α, a, θ, and d are D–H parameters of the robot. So we have:

$$ R_X(\alpha_{i-1}) = \begin{bmatrix} 1 & 0 & 0 & 0 \\ 0 & \cos\alpha_{i-1} & -\sin\alpha_{i-1} & 0 \\ 0 & \sin\alpha_{i-1} & \cos\alpha_{i-1} & 0 \\ 0 & 0 & 0 & 1 \end{bmatrix} $$

$$D_X(a_{i-1}) = \begin{bmatrix} 1 & 0 & 0 & a_{i-1} \\ 0 & 1 & 0 & 0 \\ 0 & 0 & 1 & 0 \\ 0 & 0 & 0 & 1 \end{bmatrix}$$

$$R_Z(\theta_i) = \begin{bmatrix} \cos\theta_i & -\sin\theta_i & 0 & 0 \\ \sin\theta_i & \cos\theta_i & 0 & 0 \\ 0 & 0 & 1 & 0 \\ 0 & 0 & 0 & 1 \end{bmatrix}$$

$$D_Z(d_i) = \begin{bmatrix} 1 & 0 & 0 & 0 \\ 0 & 1 & 1 & 0 \\ 0 & 1 & 1 & d_i \\ 0 & 0 & 0 & 1 \end{bmatrix}$$

and ${}^{i-1}T$ is,

$$\begin{bmatrix} \cos\theta_i & -\sin\theta_i & 0 & a_{i-1} \\ \cos\alpha_{i-1}\sin\theta_i & \cos\alpha_{i-1}\cos\theta_i & -\sin\alpha_{i-1} & -d_i\sin\alpha_{i-1} \\ \sin\alpha_{i-1}\sin\theta_i & \sin\alpha_{i-1}\cos\theta_i & \cos\alpha_{i-1} & d_i\cos\alpha_{i-1} \\ 0 & 0 & 0 & 1 \end{bmatrix}$$

Hence, the following matrix multiplication computes the final transformation matrix that gives the position and orientation of the robot's end effector relative to its base.

$${}^0_6T = {}^0_1T\,{}^1_2T\,{}^2_3T\,{}^3_4T\,{}^4_5T\,{}^5_6T$$

Let

$${}^0_6T = \begin{bmatrix} r_{11} & r_{12} & r_{13} & r_{14} \\ r_{21} & r_{22} & r_{23} & r_{24} \\ r_{31} & r_{32} & r_{33} & r_{34} \\ 0 & 0 & 0 & 1 \end{bmatrix}$$

The position and the orientation of the end effector in roll-pitch-yaw representation are as follows:

$$^{0}\mathrm{P}_6 = \begin{bmatrix} r_{14} \\ r_{24} \\ r_{34} \end{bmatrix}$$

$$\text{pitch} = \mathrm{Atan2}\left(r_{13}, \sqrt{r_{23}^2 + r_{33}^2}\right)$$

$$\text{roll} = \begin{cases} 0 & \text{pitch} = \dfrac{\pi}{2}, -\dfrac{\pi}{2} \\ \mathrm{Atan2}\left(-\dfrac{r_{23}}{\cos(\text{pitch})}, \dfrac{r_{33}}{\cos(\text{pitch})}\right) & \text{o.w} \end{cases}$$

$$\text{yaw} = \begin{cases} \mathrm{Atan2}\ (r_{32}, r_{22}) & \text{pitch} = \dfrac{\pi}{2} \\ -\mathrm{Atan2}\,(r_{32}, r_{22}) & \\ \mathrm{Atan2}\left(-\dfrac{r_{12}}{\cos(\text{pitch})}, \dfrac{r_{11}}{\cos(\text{pitch})}\right) & \text{o.w} \end{cases}$$

The various robots and links are given in Figure 3.17.

3.9.1.1 Inverse Kinematics

A. Joint *I*

As illustrated in Figure 3.17, the first joint angle is calculated by projecting a vector to the X–Y plane of frame K_0 that starts at the origin of frame K_0 and ends at the origin of frame K_4 $\left({}^{4}_{4}\mathrm{P_s}\right)$. Suppose that ${}^{0}_{\mathrm{G}}$. The target position and orientation are defined by the target transformation matrix, *T*, with respect to the base.

$${}^{0}_{G}T = \begin{bmatrix} {}^{0}_{G}T_{11} & {}^{0}_{G}T_{12} & {}^{0}_{G}T_{13} & {}^{0}_{G}T_{14} \\ {}^{0}_{G}T_{21} & {}^{0}_{G}T_{22} & {}^{0}_{G}T_{23} & {}^{0}T_{24} \\ {}^{0}_{G}T_{31} & {}^{0}_{G}T_{32} & {}^{0}_{G}T_{33} & {}^{0}_{G}T_{34} \\ 0 & 0 & 0 & 1 \end{bmatrix} \Rightarrow \overrightarrow{{}^{0}_{6}N_{K_0}} = \begin{bmatrix} {}^{0}_{G}T_{13} \\ {}^{0}_{G}T_{23} \\ {}^{0}_{G}T_{33} \end{bmatrix}$$

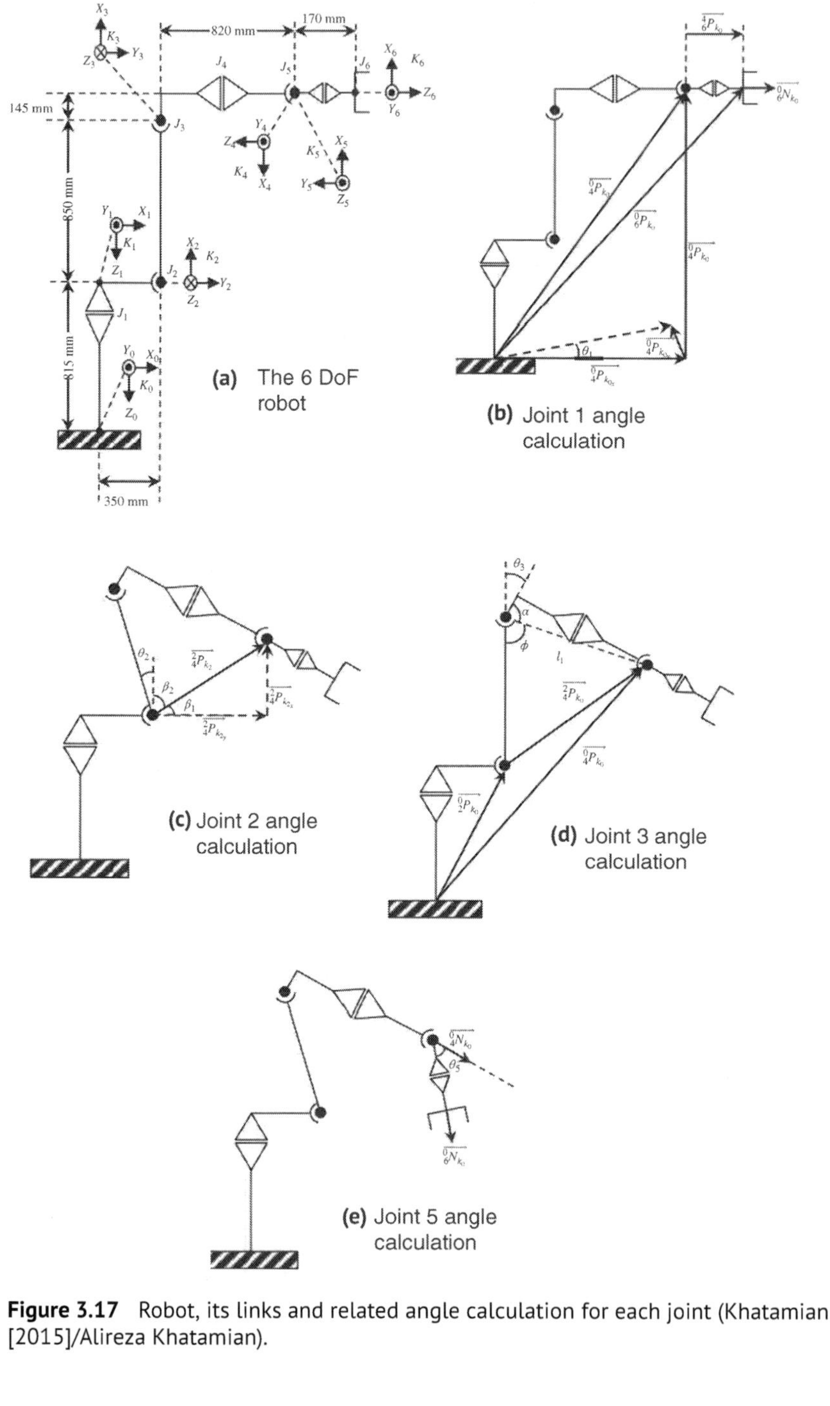

Figure 3.17 Robot, its links and related angle calculation for each joint (Khatamian [2015]/Alireza Khatamian).

Then we have

$$\begin{cases} \overrightarrow{{}^{4}_{6}P_{K_0}} = d_6 \times \overrightarrow{{}^{0}_{6}N_{K_0}} \\ \overrightarrow{{}^{0}_{6}P_{K_0}} = \begin{bmatrix} {}^{0}_{6}T^{0}_{14} \\ {}^{0}_{G}T_{24} \\ {}^{0}_{G}T_{34} \end{bmatrix} \end{cases}$$

$$\Rightarrow \overrightarrow{{}^{0}_{4}P_{K_0}} = \overrightarrow{{}^{0}_{6}P_{K_0}} - \overrightarrow{{}^{4}_{6}P_{K_0}} = \begin{bmatrix} {}^{0}T_{14} - d_6{}^{0}_{G}T_{13} \\ {}^{0}_{G}T_{24} - d_6{}^{0}_{G}T_{23} \\ {}^{0}_{G}T_{34} - d_6{}^{0}_{G}T_{33} \end{bmatrix}$$

So,

$$\theta_1 = \begin{cases} \text{Atan2}\left({}^{0}_{G}T_{24} - d_6{}^{0}_{G}T_{23}{}^{0}_{6}T_{14} - d_6{}^{0}_{G}T_{13}\right) \\ \text{Atan2}\left({}^{0}_{G}T_{24} - d_6{}^{0}_{G}T_{23}\cdot{}^{0}_{G}T_{14} - d_6{}^{0}_{G}T_{13}\right) + \pi \end{cases}$$

B. Joint 3

As depicted in Figure 3.17, to find θ_3, first ${}^{2}_{4}\mathrm{P}_{\mathrm{K}_0}$ must be found. To determine $\overline{{}^{2}_{4}P_{K_0}}$, $\overrightarrow{{}_{2}P_{K_0}}$ must be known beforehand. Having ${}^{2}_{4}P_{K_0}$ and I_1, ϕ can be determined and then using a simple geometric rule, which is used to calculate the angle between two edges of a triangle, α will be measured.

Let $\theta_2 = 0$ and

$${}^{0}_{2}T = \begin{bmatrix} {}^{0}_{2}T_{11} & {}^{0}_{2}T_{12} & {}^{0}_{2}T_{13} & {}^{0}_{2}T_{14} \\ {}^{0}_{2}T_{21} & {}^{0}_{2}T_{22} & {}^{0}_{2}T_{23} & {}^{0}_{2}T_{24} \\ {}^{0}_{2}T_{31} & {}^{0}_{2}T_{32} & {}^{0}_{2}T_{33} & {}^{0}_{2}T_{34} \end{bmatrix}$$

$$\Rightarrow \overline{{}^{0}_{2}P_{K_0}} = \begin{bmatrix} {}^{0}_{2}T_{14} \\ {}^{0}_{2}T_{24} \\ {}^{0}_{2}T_{34} \end{bmatrix}$$

Thus, $\phi =$

$$\text{Asin}\left(\frac{\left(l_1^2 - a_2^2 + \left|\overrightarrow{{}_4^2P_{K_0}}\right|^2\right)}{2\left|\overrightarrow{{}_4^2P_{K_0}}\right| l_1}\right) + \text{Asin}\left(\frac{\left|\overrightarrow{{}_4^2P_{K_0}}\right| - \frac{l_1^2 - a_2^2 + \left|\overrightarrow{{}_4^2P_{K_0}}\right|^2}{2\left|{}_4^2P_{K_0}\right|}}{a_2}\right)$$

$$\alpha = \text{Atan2}\,(-d_4, a_3)$$

So,

$$\theta_3 = \begin{cases} \pi - \phi - \alpha \\ \pi + \phi - \alpha \end{cases}$$

C. Joint 2
θ_2 is computed by $\overrightarrow{{}_4^2P_{K_2}}$, β_1 and β_2.

$$\overrightarrow{{}_4^2P_{K_2}} = {}_0^2R\overrightarrow{{}_4^2P_{K_0}} = {}_2^0R^{-1}\overrightarrow{{}_4^2P_{K_0}}$$

$${}_2^0T = \begin{bmatrix} {}_2^0R & {}_2^0P_{\text{ORG}} \\ 0 & 1 \end{bmatrix} \Rightarrow$$

$${}_2^0R = \begin{bmatrix} {}_2^0T_{11} & {}_2^0T_{12} & {}_2^0T_{13} \\ {}_2^0T_{21} & {}_2^0T_{22} & {}_2^0T_{23} \\ {}_2^0T_{31} & {}_2^0T_{32} & {}_2^0T_{33} \end{bmatrix} = {}_0^2R^{-1}$$

$$\overrightarrow{{}_4^2P_{K_2}} = \begin{bmatrix} {}_2^0T_{11} & {}_2^0T_{12} & {}_2^0T_{13} \\ {}_2^0T_{21} & {}_2^0T_{22} & {}_2^0T_{23} \\ {}_2^0T_{31} & {}_2^0T_{32} & {}_2^0T_{33} \end{bmatrix} \begin{bmatrix} {}_4^2P_{K_0x} \\ {}_4^2P_{K_0y} \\ {}_4^2P_{K_0y} \\ {}_4^2P_{K_0z} \end{bmatrix} \quad (24)$$

Thus,

$$\beta_1 = \text{Atan2}\left(\overrightarrow{{}_4^2P_{K_2x}}, \overrightarrow{{}_4^2P_{K_2y}}\right) \quad (25)$$

$$\beta_2 = \text{Asin}\left(\frac{a_2^2 - \left|\overrightarrow{{}_4^2P_{K0}}\right|^2 + l_1^2}{2l_1a_2}\right) + \text{Asin}\left(\frac{l_1 - \frac{a_2^2 - \left|\overrightarrow{{}_4^2P_{K0}}\right|^2 + l_1^2}{2l_1}}{\left|\overrightarrow{{}_4^2P_{K0}}\right|}\right) \quad (26) \tag{3.33}$$

and then,

$$\theta_2 = \begin{cases} \frac{\pi}{2} - (|\beta_1| + \beta_2) \\ \frac{\pi}{2} + (|\beta_1| - \beta_2) \end{cases}$$

D. Joint 5

In order to calculate Θ_5, 4^0T is computed by assuming that θ_4 is equal to 0. Then, by using the definition of the dot product of two normal vectors, θ_5 is obtained.

$$ {}_4^0T = \begin{bmatrix} {}_4^0T_{11} & {}_4^0T_{12} & {}_4^0T_{13} & {}_4^0T_{14} \\ {}_4^0T_{21} & {}_4^0T_{22} & {}_4^0T_{23} & {}_4^0T_{24} \\ {}_4^0T_{31} & {}_4^0T_{32} & {}_4^0T_{33} & {}_4^0T_{34} \\ 0 & 0 & 0 & 1 \end{bmatrix} \Rightarrow \overrightarrow{{}_4^0N_{K_0}} = \begin{bmatrix} {}_4^0T_{13} \\ {}_4^0T_{23} \\ {}_4^0T_{33} \end{bmatrix}$$

So, we have

$$\theta_5 = \pi - \text{A}\cos\left(\overrightarrow{{}_4^0N_{K_0}} \cdot \overrightarrow{{}_6^0N_{K_0}}\right)$$

E. Joints 4 and 6

To obtain θ_4 and θ_6, rotation matrix ${}_6^4R$ is used. On the one hand, ${}_6^4R$ is:

$${}_6^4R = {}_4^0R \cdot {}_6^{-10}R = {}_0^4R{}_6^0R$$

and on the other hand

$${}_6^4R = \text{Rot}_z(\theta_4)\,\text{Rot}_y(\theta_5 + \pi)\,\text{Rot}_z(\theta_6)$$

in which

$$\mathrm{Rot}_y(\theta_5+\pi) = \begin{bmatrix} \cos(\pi+\theta_5) & 0 & \sin(\pi+\theta_5) \\ 0 & 1 & 0 \\ -\sin(\pi+\theta_5) & 0 & \cos(\pi+\theta_5) \end{bmatrix}$$

$$\mathrm{Rot}_z(\theta_6) = \begin{bmatrix} \cos(\theta_6) & -\sin(\theta_6) & 0 \\ \sin(\theta_6) & \cos(\theta_6) & 0 \\ 0 & 0 & 1 \end{bmatrix}$$

thus,

$${}^4_6R = \begin{bmatrix} -c_4c_5c_6 - s_4s_6 & c_4c_5s_6 - s_4c_6 & -c_4s_5 \\ -s_4c_5c_6 + c_4s_6 & s_4c_5s_6 + c_4c_6 & -s_4s_5 \\ s_5c_6 & -s_5s_6 & -c_5 \end{bmatrix}$$

in which c_4 is corresponding to $\cos(\theta_4)$, s_4 is $\sin(\theta_4)$ and so forth.

For the sake of simplicity (Khatamian, 2015), let

$${}^4_6R = \begin{bmatrix} {}^4_6R_{11} & {}^4_6R_{12} & {}^4_6R_{13} \\ {}^4_6R_{21} & {}^4_6R_{22} & {}^4_6R_{23} \\ {}^4_6R_{31} & {}^4_6R_{32} & {}^4_6R_{33} \end{bmatrix}$$

So, we have,

$$\theta_4 = \mathrm{Atan2}\left(-{}^4_6R_{23}, -{}^4_6R_{13}\right)$$

$$\theta_6 = \mathrm{Atan2}\left(-{}^4_6R_{32}, {}^4_6R_{31}\right)$$

3.10 AI/ML Basics for Food Processing

3.10.1 Artificial Neural Network

The artificial neural network shown in Figure 3.18 is a machine-learning method that attempts to mimic the working of the human brain with millions of artificial neurons. Neurons of ANNs are mostly less connected compared with the biological neurons. All neurons in ANN take as inputs a number of values.

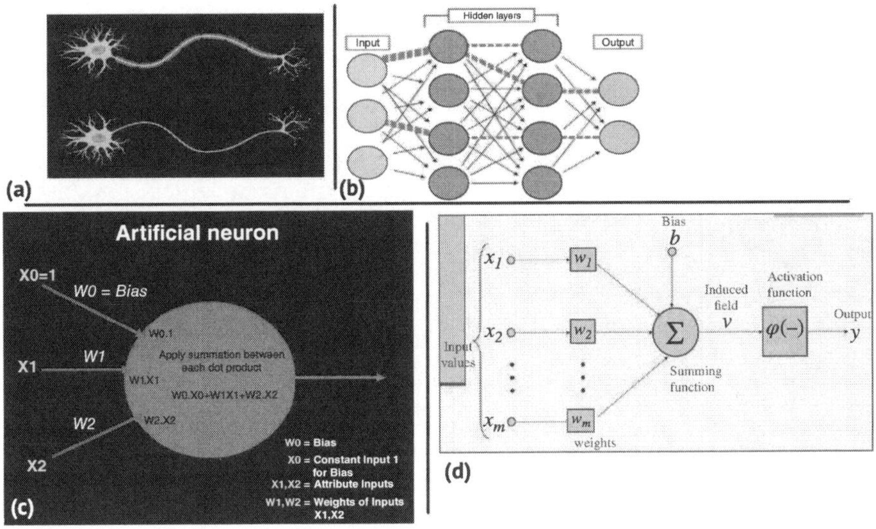

Figure 3.18 (a) Figure of a biological neuron, (b) the figure of an ANN, and panels (c) and (d) represent the neuron diagram in details with weights and biases.

These inputs have an application of an activation function that provides them an activation level of the neuron that is the output value of the neuron. Examples of how the learning task would be fulfilled constitute knowledge about the learning task, which is referred to as training examples.

What is an activation function, and why use it?

The activation function determines whether a neuron shall be activated or not based on computing the weighted sum, and further bias addition to it. The role of the activation function is to introduce non-linearity in the output of a neuron.

Explanation: We know that the neural network (NN) has neurons that work together with weight, bias and their associated activation function. In an NN, we update the weights and biases of the neurons based on error coming at the output. We call this procedure back-propagation. What makes back-propagation possible is the activation functions since their gradients are provided along with the error for weight and bias updation.

An ANN is defined by: an architecture, which consists of a collection of neurons and links between them; and a neuron model, which is the information processing unit of the NN. Every link includes a weight and a learning algorithm that is used to train the NN by adjusting the weights to accurately represent a particular learning task on the training instances. It is intended to train an NN and improve its generalization skills. It ought to respond appropriately to fresh iterations of the learning task.

Neuron

The fundamental information processing unit of an NN is the neuron. It includes:

1. A collection of links with weights $W_1, W_2, \ldots, W_m$ that describe the neuron inputs newline.
2. A linear combiner adder function to get the inputs' weighted sum: (actual figures).

$$u = \sum_{j=1}^{m} wj_j x$$

3. Activation function φ for limiting the amplitude of the neuron's output. Here "b" denotes bias. Figure 3.19 represents various activation function.

$$y = \varphi(u + b)$$

Bias of Neuron

The bias b has the effect of applying a transformation to the weighted sum u

$$v = u + b$$

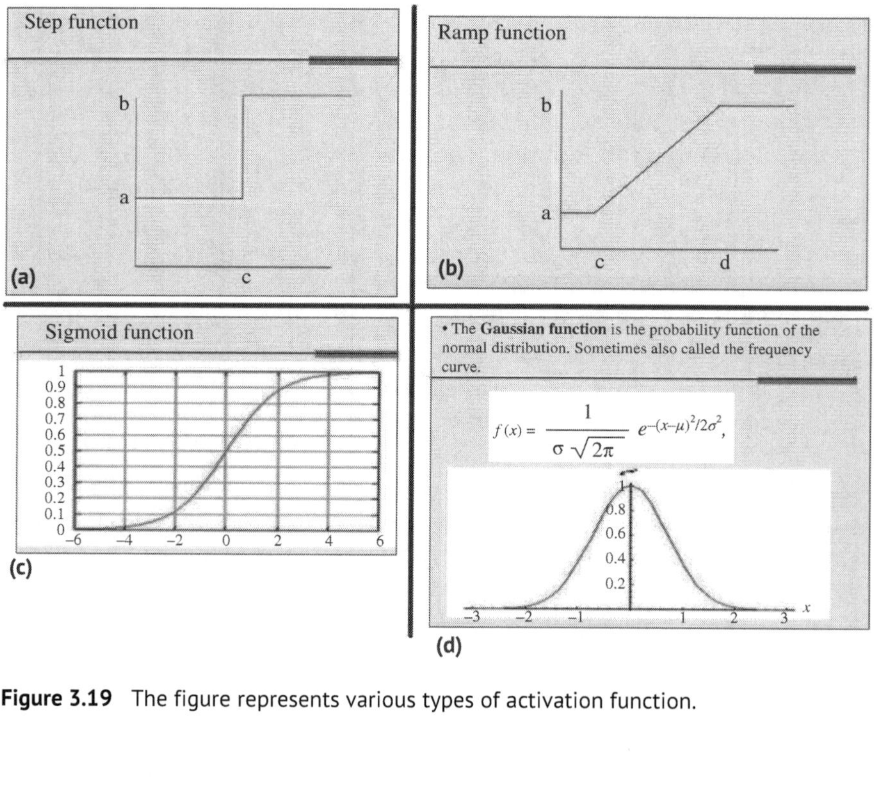

Figure 3.19 The figure represents various types of activation function.

The bias is an external parameter of the neuron. It can be modeled by adding an extra input. v is called the induced field of the neuron.

$$v = \sum_{j=0}^{m} w_j x_j$$

$$w_0 = b$$

Bias is one of the important terms in machine learning. In any model built in an ANN, bias is added very often.

Let's look into the nature of this mysterious bias.

We are given the input (x) in the NN, and we have to predict the output (y) for that input. In this case, we develop a model that predicts the outcome, $mx + c$.

The model itself determines the proper value of the constants m and c during training.

Suppose that the model is $y = mx$ instead of $y = mx + c$.

Neuron Models: The neuron model is determined by the activation function *varphi*. Examples:

Step function:

$$\varphi(v) = \begin{cases} a \text{ if } v < c \\ b \text{ if } v > c \end{cases}$$

Ramp function:

$$\varphi(v) = \begin{cases} a \text{ if } v < c \\ b \text{ if } v > d \\ a + ((v - c)(b - a)(d - c)) \text{ otherwise} \end{cases}$$

Sigmoid function with z,x,y parameters:

$$\varphi(v) = z + \frac{1}{1 + \exp(-xv + y)}$$

Gaussian function:

$$\varphi(v) = \frac{1}{\sqrt{2\pi\sigma}} \exp\left(-\frac{1}{2}\left(\frac{v - \mu}{\sigma}\right)^2\right)$$

This is the feature that introduces the nonlinearity of the model, thereby enabling the network to learn more complex patterns. Some of the activation functions include the following:

(i) Sigmoid, (ii) Tanh, (iii) ReLU (rectified linear unit), (iv) SoftMAX, etc.: Outputs values between 0 and 1. Useful for binary classification but may suffer from vanishing gradient problems.

The right choice of the activation function is of very great importance in achieving desired performance of an NN because it affects the capacity of the model to learn from and generalize for the training data.

What is a Sigmoid Function?

A sigmoid function is any mathematical function in the form of S that has a characteristic S-shaped curve. There are several known sigmoid functions, including the logistic function, hyperbolic tangent, and the arctangent. The sigmoid function is by far most often used in the context of machine learning to specifically denote the logistic function or the logistic sigmoid function.

One application of a sigmoid function is to transform a real value into one that can be understood as a probability because all sigmoid functions have the ability to map the entire number line onto a limited range, such as between 0 and 1, or −1 and 1. The most commonly used sigmoid function is actually the logistic function, which maps any real number to the interval (0, 1). Notice also the characteristic S-shape, taken from the Greek letter sigma, whence sigmoid functions get their name.

In deep learning, sigmoid functions are highly popular; they are actually the most used choices for the activation function in an ANN. Biological brain networks have been their source of inspiration regarding activation potential. Many machine learning applications that would convert a real number to a probability also benefit from using sigmoid functions. Applying a sigmoid function to the output of a final layer of the model transforms that output into the score of some probability. Therefore, it simplifies the interpretation and usage of the model. Logistic and sigmoid functions are essential parts of logistic regression. It is a two-class classification-specific version of the linear regression technique that transforms one or more real-valued inputs into a quantity that resembles a probability, such as the likelihood that a client will purchase a product. The logistic function, which enables the model to provide a probability, is often the last step in logistic regression model implementations. The various types of functions are presented in Figure 3.20.

Formula for the Sigmoid Function: All sigmoid functions are monotonic and have a bell-shaped first derivative. There are many sigmoid functions, and some of the most common are listed below. The three most commonly occurring sigmoid functions: the logistic function, the hyperbolic tangent, and the arctangent. They have very similar overall S forms.

$$\begin{aligned} S(x) &= \frac{1}{1+e^{-x}} \\ &= \frac{e^x}{e^x+1} \end{aligned}$$

Logistic Sigmoid Function Formula: The most frequently utilized sigmoid function is the logistic sigmoid function. It is known as sigmoid function in

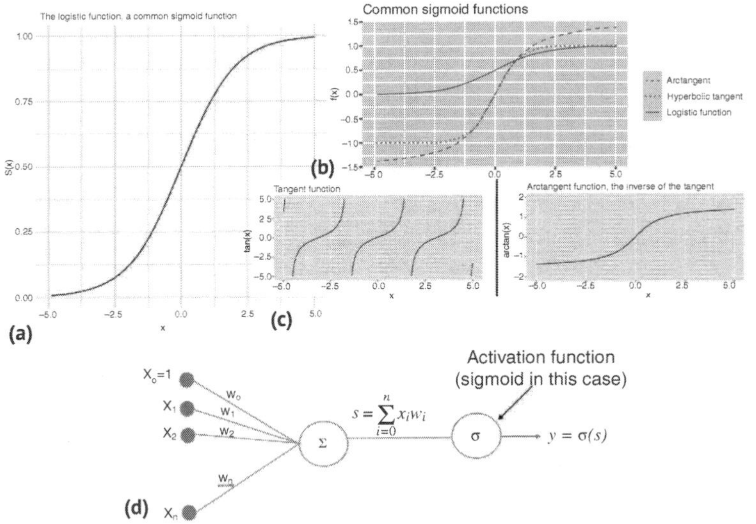

Figure 3.20 (a) The logistic sigmoid function, (b) common sigmoid functions, (c) tanhx, arctanhx, and (d) an artificial neural network.

machine learning terminology. The logistic sigmoid function is defined by the following formula:

Mathematical definition of logistic sigmoid function, an example of a typical sigmoid function. The logistic function accepts any real-valued input and gives an output between zero and one.

Example 1. Logisitic Sigmoid Function Calculation. Now we can compute the value of the function at a variety of points to gain insight into the shape of the function. At $x = 0$, the logistic sigmoid function evaluates to

$$S(0) = \frac{1}{1+e^{-0}}$$
$$= \frac{1}{1+1}$$
$$= 0.5$$

This is good for the explanation of the sigmoid as a probabilistic output from a logistic regression model, where a zero in will result in an output that is 0.5 with equal probabilities over both classes. At $x = 1$, we find a slightly larger value:

$$S(1) = \frac{1}{1+e^{-1}}$$
$$= \frac{1}{1+0.36}$$
$$= 0.73$$

and by $x = 5$, the value of the sigmoid function becomes very close to 1:

$$S(5) = \frac{1}{1 + e^{-5}}$$
$$= \frac{1}{1 + 0.0067}$$
$$= 0.993$$

In fact, if x is taken to infinity the sigmoid function approaches 1, while for negative infinity it approaches –1, but the derivative of the function never goes through zero. It has some quite useful properties: it approaches its limit, but always retains a nonzero gradient.

Hyperbolic Tangent Function Formula

Yet another very common sigmoid function is the hyperbolic function. It maps any real-valued input to the range between –1 and 1.

$$f(x) = \tanh(x)$$
$$= \frac{e^x - e^{-x}}{e^x + e^{-x}}$$

Arctangent Function Formula

A third alternative sigmoid function is the arctangent, which is the inverse of the tangent function.

$$f(x) = \arctan(x)$$

The arctangent function maps any real-valued input to the range $-\pi/2$ to $\pi/2$. In the graphs below, we can see both the tangent curve, a well-known trigonometric function, and the arctangent, its inverse:

Example 2. Calculation of hyperbolic tangent function. Similarly, we can calculate the value of the tanh function at these key points. Rather than being centered around 0.5, the tanh function is centered at 0.

$$\tanh(0) = \frac{e^0 - e^{-0}}{e^0 + e^{-0}}$$
$$= \frac{1 - 1}{1 + 1}$$
$$= 0$$

At 1, the tanh function has increased and has converged much more closely to 1, within 5 decimal places:

$$\tanh(1) = \frac{e^1 - e^{-1}}{e^1 + e^{-1}}$$
$$= \frac{2.718 - 0.368}{2.718 + 0.368}$$
$$= 0.762$$

$$\tanh(5) = \frac{e^5 - e^{-5}}{e^5 + e^{-5}}$$
$$= \frac{148 - 0.007}{148 + 0.007}$$
$$= 0.99991$$

Sigmoid as NN Activation Function: The sigmoid function is the activation function in use by NNs. To review briefly, recall how an activation function works within a single layer of an NN from the figure below. An activation function takes some weighted sum of inputs, and its output feeds into the next layer.

It is always ensured that the output of a neuron with a sigmoid activation function will fall between 0 and 1. Also, the output of this unit will be a non-linear function of the weighted sum of inputs because the sigmoid is a non-linear function. A sigmoid unit is a type of neuron that uses sigmoid as its activation function. Why do NNs require the sigmoid function?

An NN that has a linear activation function can only learn problems that are linearly separable. However, a nonlinearly separable problem can be readily learned by the NN with only one hidden layer and a sigmoid activation function in that hidden layer. The sigmoid function can be utilized in NNs to learn complex decision functions because using a nonlinear function results in nonlinear boundaries.

Only monotonically rising nonlinear functions can be employed as activation functions in NNs. For example, the functions sin(x) and cos(x) are not suitable for use as activation functions. Additionally, the activation function needs to be continuous and defined everywhere in the real number space. Additionally, the function must be differentiable across the whole real number space.

The usual back propagation algorithm uses gradient descent in order to learn the weights in an NN. In order to derive this algorithm, the derivative of the activation function needs to be obtained. The reason is that it is monotonic, continuous, and differentiable everywhere, coupled with the property that its derivative can be expressed in terms of itself, making it pretty straightforward to get the update equations when using back propagation algorithm in learning the weight in an NN.

Linear Versus Nonlinear Separability? Suppose we have a simple classification problem where we have a set of points in space, and each point has an assigned class label. Then, the problem is linearly separable if and only if there exists a straight line (or a hyperplane in an n-dimensional space) that can separate the two classes. Otherwise, if a straight line is not adequate for the division between the two classes, then we have a problem of nonlinear separability. The data in the two-dimensional (2D) space are plotted in figure below. For each point, a class label of either red or blue is assigned. In this example, the left figure describes a linearly separable problem that needs a straight-line boundary for distinction between the

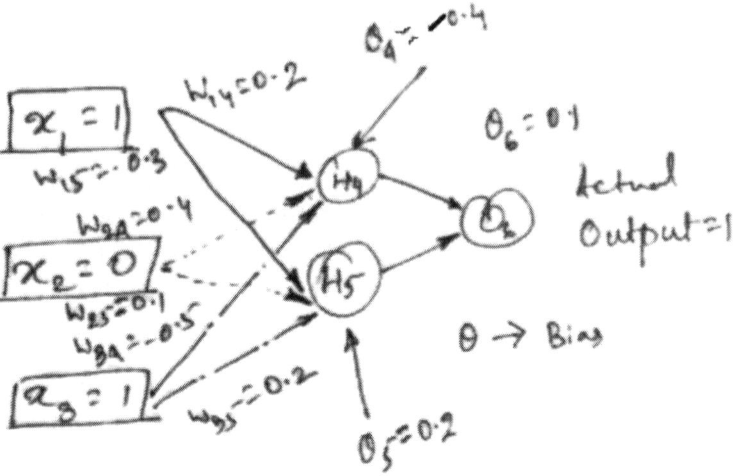

Figure 3.21 An ANN architecture with weights and biases.

classes. The right figure shows a non-linearly separable problem; in other words, its decision boundary is not linear. For three-dimensional space, a linear decision boundary can be represented by the equation of a plane. To extend it to an n-dimensional space, the equation of the hyperplane represents the linear decision boundary.

3.10.2 Solved Example A

Assume that the neurons have a sigmoid activation function and perform a forward pass and a backward pass on the network referenced in Figure 3.21. Assume that the actual output of y is 1 and the learning rate is 0.9, and therefore perform another forward pass.

Forward Pass: Compute output for y_4, y_5, and y_6.

$$a_j = \sum_j (w_{ij} \times x_i) \quad y_j = F(a_j) = \frac{1}{1 + e^{-a_j}}$$

$$a_4 = (w_{14} \times x_1) + (w_{24} \times x_2) + (w_{34} \times x_3) + \theta_4$$

$$= (0.2 \times 1) + (0.4 \times 0) + (-0.5 \times 1) + (-0.4)$$

$$= -0.7$$

$$O(H_4) = y_4 = f(a_4) = 1/\left(1 + e^{0.7}\right) = 0.332$$

$$\begin{aligned} a_5 &= (w_{15} \times x_1) + (w_{25} \times x_2) + (w_{35} \times x_5) + \theta_5 \\ &= (-0.3 \times 1) + (0.1 \times 0) + (0.2 \times 1) + 0.2 \\ &= 0.1 \end{aligned}$$

$$O(H_5) = y_5 = f(a_5) = 1/\left(1 + e^{-0.1}\right) = 0.525$$

$$\begin{aligned} a_6 &= (w_{46} \times H_4) + (w_{56} \times H_5) + \theta_6 \\ &= (-0.3 \times 0.332) + (-0.2 \times 0.525) + 0.1 \\ &= -0.105 \end{aligned}$$

$$O(O_6) = y_6 = f(a_6) = 1/\left(1 + e^{0.105}\right) = 0.474$$

$$\text{Error} = y_{\text{target}} - y_6 = 0.526$$

Each weight changed by amounts:

$$\begin{aligned} \delta_j &= O_j\left(1 - O_j\right)\left(t_j - O_j\right) && \text{if } j \text{ is an output} \\ \delta_j &= O_j\left(1 - O_j\right)\sum_k \delta_k w_{kj} && \text{if } j \text{ is a biden unit} \end{aligned}$$

where η is a constant called the learning rate, t_j is the correct teacher output for unit j, and δ_j is the error measure for unit j

Backward pass: Compute δ_4, δ_5, and δ_6

For output unit:

$$\begin{aligned} \delta_6 &= y_6(1 - y_6)\left(y_{target} - y_6\right) \\ &= 0.474 \times (1 - 0.474) \times (1 - 0.474) = 0.1311 \end{aligned}$$

For hidden unit:

$$\begin{aligned} \delta_5 &= y_5(1 - y_5)\, w_{56} \times \delta_6 \\ &= 0.525 \times (1 - 0.525) \times (-0.2 \times 0.1311) \\ &= -0.0065 \end{aligned}$$

$$\begin{aligned} \delta_4 &= y_4(1 - y_4)\, w_{46} \times \delta_6 \\ &= 0.332 \times (1 - 0.332) \times (-0.3 \times 0.1331) \\ &= -0.0087 \end{aligned}$$

Compute new weights.

$$\Delta w_{4j} = \eta \delta_j o_j$$

$$\Delta w_{46} = \eta \delta_6 y_4 = 0.9 \times 0.1311 \times 0.332 = 0.03917$$

$$w_{46}(\text{new}) = \Delta w_{46} + w_{46}(\text{old}) = 0.03917 + (-0.3)$$

$$W_{46}(\text{new}) = -0.261.1$$

$$\Delta w_{14} = \eta \delta_4 x_1 = 0.9 \times -0.0087 \times 1 = -0.0078$$

$$w_{14}(\text{new}) = \Delta w_{14} + w_{14}(\text{old}) = -0.007850.2$$

$$w_{14}(\text{new}) = 0.192$$

Similiarly, update all other weights

i	j	ω_j	δ_j	x_i	η	Updated W_{ij}
4	6	−0.3	0.1311	0.332	0.9	−0.261
5	6	−0.2	0.1311	0.525	0.9	−0.138
1	4	0.2	−0.0087	1	0.9	0.192
1	5	−0.3	−0.0065	1	0.9	−0.306
2	4	0.4	−0.0087	0	0.9	0.4
2	5	0.1	−0.0065	0	0.9	0.1
3	4	−0.5	−0.00087	1	0.9	−0.508
3	5	0.2	−0.0065	1	0.9	0.194

Similarly, update bias weights

θ_j	Previous θ_j	δ_j	η	Updated θ_j
θ_6	0.1	0.1311	0.9	0.218
θ_5	0.2	−0.0065	0.9	0.194
θ_4	−0.4	−0.0087	0.9	−0.408

-Forward Pass: Compute output for y_4, y_5, y_6

$$a_j = \sum_j (w_{ij} \times x_i) \quad y_j = F(a_j) = \frac{1}{(1+e^{-a_j})}$$

$$\begin{cases} a_4 = (w_{14} \times x_1) + (w_{24} \times x_2) + (w_{34} \times x_3) + \theta_4 \\ \quad = (0.192 \times 1) + (0.4 \times 0) + (-0.508 \times 1) + (-0.408) \\ \quad = -0.724 \\ O(H_4) = y_4 = f(a_4) = 1/\left(1 + e^{0.724}\right) = 0.327 \end{cases}$$

$$a_5 = (w_{15} \times x_1) + (w_{25} \times x_2) + (w_3 \times x_3) + \theta_5$$

$$= (-0.306 \times 1) + (0.1 \times 0) + (0.194 \times 1) + (0.194)$$

$$= 0.082$$

$$O(H_5) = 0.082$$

$$y_5 = f(a_5) = 1/\left(1 + e^{-0.082}\right) = 0.520$$

$$a_6 = (w_{46} \times H_4) + (w_{56} \times H_5) + \theta_6$$

$$= (-0.2611 \times 0.327) + (-0.138 \times 0.520) + (0.218)$$

$$= 0.061$$

$$O(O_6) = y_6 = f(a_6) = 1/\left(1 + e^{-0.061}\right) = 0.515$$

$$Error = y_{\text{target}} - y_6 = 0.485$$

$$= (1 - 0.515) = 0.485$$

Previously, it was 0.526. Now, it is 0.485. If it is okay for us, we can stop.

3.10.3 Solved Example B

Combining neurons into an NN. An example: Feed forward is represented in Figure 3.22.

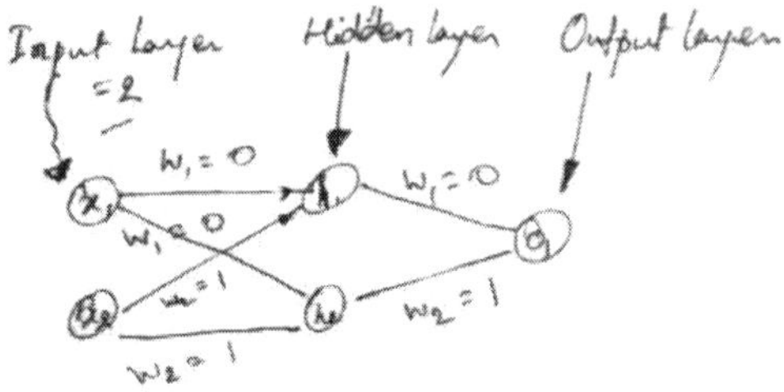

Figure 3.22 A feed forward NN.

$$w_2 = 1$$

$$\begin{cases} h_1 = h_2 = f(w,x+b) \\ = f(0 \times 2 + 1 \times 3 + 0) \end{cases} \tag{3.34}$$

$$= f(3)$$

$$= 0.9526$$

$$O_1 = f(w \cdot [h_1, h_2] + b)$$

$$= f((0 \times h_1) + (1 \times h_2) + 0)$$

$$= f(0.9526)$$

$$x = [2, 3]; w = [0, 1]$$

bias, $b = 0$, AF : sigmoid. Or,

$$O_1 = (x_1 \times \omega_1) + (x_2 \times \omega_2) + b$$

$$= (0.9526 \times 0) + (0.9526 \times 1) + 0 = 0.956$$

$$h_1 = (2 \times 0) + (3 \times 1) = 3$$

$$h_2 = (3 \times 1) + (2 \times 0) = 3$$

So we call detection an apple if the output layer has a value greater than 0.5, else orange.

3.10.4 Convolutional Neural Network

CNN, stands for convolutional neural network, which is a type of deep learning model especially developed for processing structural grid data like images. Through convolutional layers, CNNs automatically extract features from input data, so it's peculiarly targeted toward image recognition, object detection, and

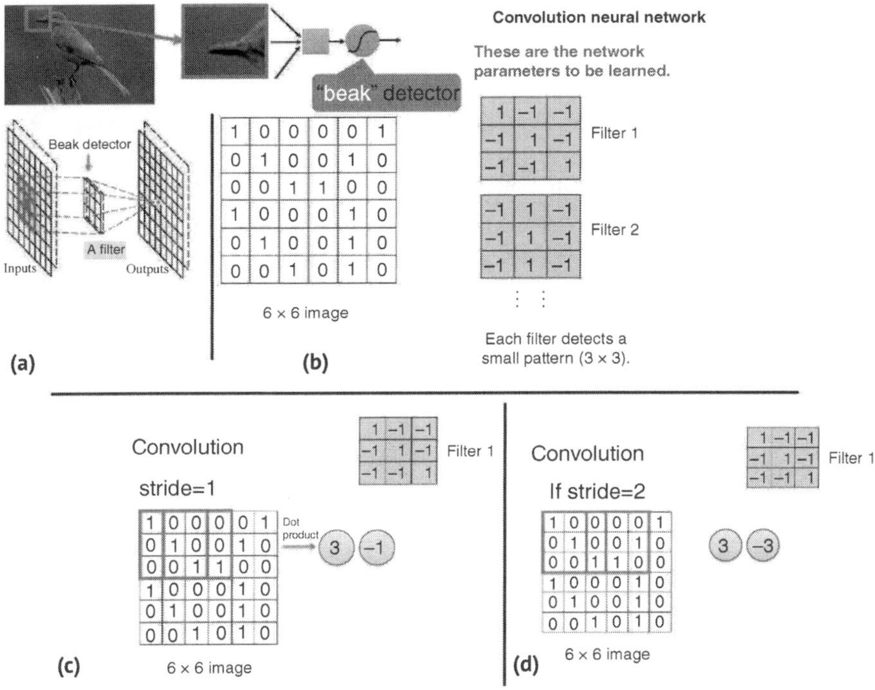

Figure 3.23 Sequential processes in a CNN.

classification. They comprise a number of layers, including convolutional layers, pooling layers, and also fully connected layers, where all work together to learn the hierarchical representation of the input data. A sequence of processes executed is represented in Figure 3.23.

3.11 Mapping, Navigation, and Obstacle Avoidance

Case Study: Autonomous Quadrotor Navigation
Autonomous Quadrotor Navigation with Vision-based Obstacle Avoidance and Path Planning Objective: Design a vision-guided autonomous navigation system for a quadrotor UAV that can also avoid and plan paths.
Autonomous Navigation System:
The system applies large object modeling, which includes virtual objects used to represent a virtual environment, for path planning. The ideal path between the waypoints is a straight line without interference by obstacles. This is a static type of path planning for terrain features and limitations.

The RRT algorithm is used for path planning and converts 3D waypoints into GPS coordinates for actual world navigation.

Obstacle Avoidance System:

The image sequence is pre-processed, which includes downsampling and conversion of them to grayscale levels, followed by Gaussian filtering. Dense optical flow techniques are used to calculate motion vectors and to track features that can be used to locate obstacles.

Motion vectors are combined through a sliding window approach to successfully detect the obstacles.

Implementation and Experiments:

A low-cost camera system is integrated to do vision-based obstacle avoidance.

Demonstrations outside the laboratory show how it is able to maneuver around obstacles.

Computer vision time is about 150–180 ms per frame. That means it achieves almost six frames per second, which is nearly real time.

Results

The researchers developed algorithms that can run on an embedded platform in real time. Our method for vision-guided quadrotor navigation is very feasible and practical. The system detects frontal obstacles and adjusts the trajectory of the quadrotor properly.

Future work involves advanced image-based approaches to handle moving obstacles.

Flow Sequence for Quadrotor UAV Navigation Based on Obstacle Image NavigationImage Acquisition:

The process begins with the onboard camera of the quadrotor UAV taking a sequence of images from a monocular view. The real-time environment perception and obstacle detection require the flow of images.

Image Pre-processing

All the captured images undergo the following pre-processing operations:

Downsampling: Diminishes the image processing computational load.

Convert Grayscale: Simplification of image data to make analysis easier.

Gaussian Filtering: Removal of noise in the images enhances the quality of data for detection algorithms.

Optical Flow Estimation

The dense optical flow technique is used to calculate the motion vectors between two consecutive frames of the images. The method will analyze pixel displacement to evaluate the dynamic change that has occurred in the scene.

Analysis of Motion Vectors

The optical flow field is analyzed to detect possible obstacles in a scene. Motion vectors are categorized based on noise reduction to determine what constitutes the foreground and background.

A threshold is established to determine which motion vectors indicate that the line of sight contains obstacles; those below the threshold are free of obstacles along the line of sight.

Obstacle Detection: The system segments the image to get independent motion regions related to detected objects. This segmentation assists in the validation that if a region corresponds to an obstacle, then it is determined by the optical flow field.

Path Adjustment: The quadrotor system adjusts in real time once obstacles are detected, adjusting its trajectory to navigate around the obstacles. The system periodically updates its path based on the latest image data and obstacle information (Lin and Peng, 2021).

Real-time Processing: In fact, the whole system is implemented on one single-board computer, allowing it to perform algorithms of real-time and online image processing. Therefore, the system works really efficiently: it processes images at a rate of about 150–180 ms per frame and acquires couple of six frames per second.

Navigation and Flight Guidance: This plan will guarantee safe movement in the environment since the quadcopter follows the planned path while dynamically avoiding obstacles. The integration of offline path planning and real-time obstacle avoidance enhances the overall effectiveness of the navigation system.

3.12 Fuzzy Logic Control Case Study in Food Processing

Blockchain Machine Learning-based Food Traceability System (BMLFTS)

Overview

The case study examines the practical feasibility of a BMLFTS to improve traceability in perishable food products in the supply chain, mainly because of the quality control advantage. It integrates blockchain technology with machine learning and fuzzy logic to address several issues arising from critical concerns around food safety and quality.

Problem Statement

The COVID-19 pandemic has complicated food safety issues and raised the consciousness of consumers regarding the quality and origin of food and shipping conditions.

Some of the difficulties facing the PFSC are in quality management and strict adherence to regulations.

Proposed Solution: BMLFTS

The solution integrates blockchain for secure data management and machine learning for predictive analytics and fuzzy logic for assessing quality.

This system will achieve an integrated traceability mechanism of food from production to its level of distribution with the utmost transparency and reliability.

Implementation Steps

Data Collection: Collect environmental conditions, the quality of the product, and supply chain processes.

Fuzzification: Convert crisp input data into fuzzy values by applying membership functions to assess quality parameters.

Decision-making Unit: The set of predefined fuzzy rules was used to interpret the data when it was fuzzified.

Defuzzification: The values obtained by the defuzzification process are converted from fuzzy outputs back into crisp values for actionable inputs and insights.

Smart Contract Integration

Here, three different smart contracts are created, namely product registration contract (PRC), batch additional contract (BAC), and transaction update contract (TUC). They handle data in relation to a product and trace its history. The contracts allow the easy tracking of batches and passing of information across, which helped boost consumer confidence.

Outcome and Advantages

Improved traceability of foodstuffs; this, in turn, meant enhanced quality control; and with these, consumers' confidence in perishable foodstuffs was boosted.

The system will track stakeholders from manufacturing to delivery of an entire food product life cycle, and this will go a long way in ensuring that the safety standards are complied with. Table 3.2 refers to AI/ML applications in the food sector (Lin and Peng, 2021).

3.13 Dijkstra's Algorithm for a Grid Map

An advanced version of the Dijkstra algorithm is proposed to meet the challenge of real-time navigation for mobile robots in large and dense environments, which is modeled using the grid approach. The approach begins by using a four-neighborhood search to develop a potential field based on Manhattan distance from the start to the entire grid within a linear time frame. It starts from the destination and performs an eight-neighborhood search towards the starting point, thus ensuring that it actually identifies a collision-free path, which is nearly optimal. A number of simulation tests performed in MATLAB confirm that this algorithm is capable of operating at a speed a hundred times faster than conventional Dijkstra and A-star algorithms with an error margin for the length of the path that can be considered acceptable compared to the shortest possible path. The grid method, first proposed by W.E. Howden in 1968, is a method through which the operational environment of a mobile robot is turned into a binary matrix. This is a grid of elements that can have a value of either 0 or 1, where the value of 0 represents open space for the navigation and the moving around of the robot, while a value of 1 represents blocked space with obstacles. Figure 3.25 in a MATLAB environment gives a clear realization of the actual representation of the working environment as presented by the grid method. In this, "s" is the starting point while "e" represents the destination.

To get an obstacle-free path from start(s) to end(e), some search rules have to be applied. In the algorithm used in this research work, a four-neighborhood search strategy is considered as illustrated in Figure 3.24, where the cost of moving in all directions from s is given as 1. After the entire area is covered, a search method is employed using an eight-neighborhood search to find the path. Here, a horizontal or vertical move costs 1; a diagonal move costs more at $\sqrt{2}$.

3.14 Path Planning with the A* Algorithm

A* Search algorithm motivation: To find the best possible shortest path in real-life applications such as maps and games, where there may be many obstacles. We can think of a 2D grid, which contains many obstacles, and we begin from a source cell (dark gray in Figure 3.25) to reach toward a goal cell (light gray).

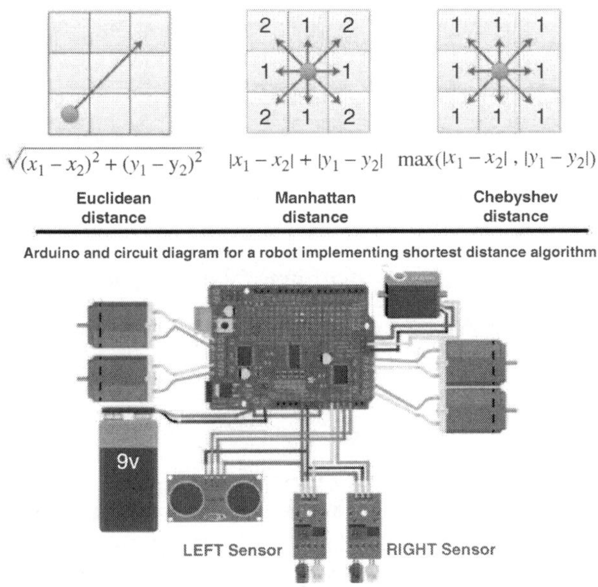

Figure 3.24 (a) Euclidean, (b) Manhattan, (c) Chebychev's distance, and (d) a mobile robot working environment.

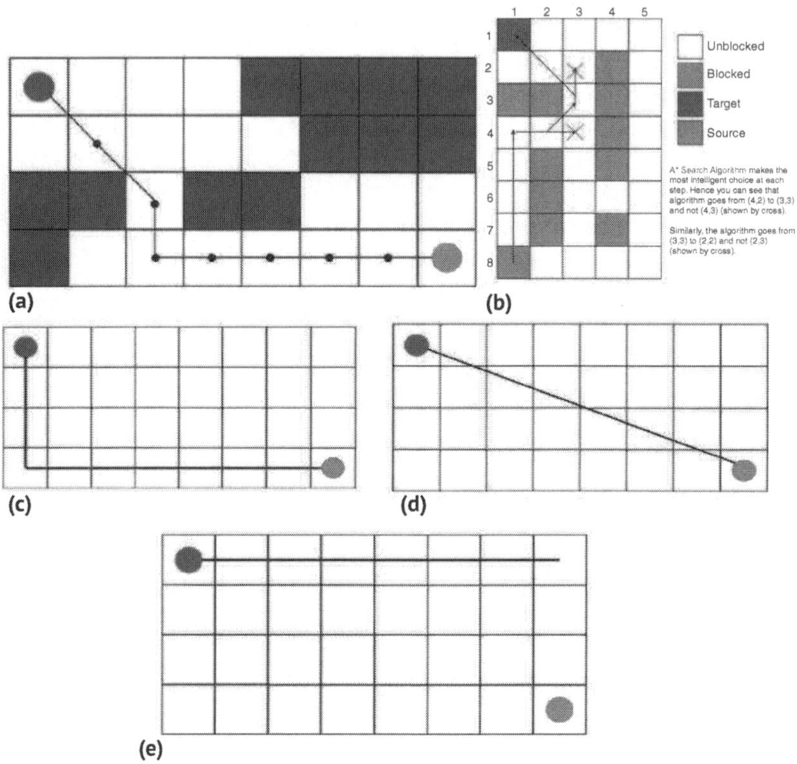

Figure 3.25 (a) 2D grid with obstacles, (b) A* search algorithm using Euclidean distance as a heuristics, (c) Manhattan distance heuristics (d) diagonal distance heuristics, and (e) Euclidean distance heuristics.

3.14.1 What Is A* Search Algorithm?

The A* Search Algorithm is one of the best algorithms when it comes to fast pathfinding and graph traversal.

3.14.2 Why Do We Use the A* Search Algorithm?

What makes A* unique compared to any other traversal algorithm is its intelligent design. Unlike all the others, A* performs at the highest complexity in terms of optimized capabilities, for which further details are brought below. What's equally important to note is the wide range of applications up to the present day—from video games and online mapping tools, where this algorithm has been favored in application for optimally finding routes. Figure 3.24 represents Euclidean, Manhattan, and Chebyshev distance.

3.14.3 Understanding the A* Search Algorithm

Suppose you would like to search through a 2D square grid filled with obstacles, starting from one point for the final target cell. In this case, the A* Search Algorithm can be the most efficient way to take this journey. The essential principle that drives its operation is node selection according to the value "*f*", which is the total cost produced by adding two key parameters: "g" and "*h*". The algorithm always chooses the node having the smallest "*f*" value to process.

The parameters 'g' and '*h*' are defined as: 'g' defines the cost of moving from the start point up to a specific square on the grid tracing the path it has covered; and "*h*" is a measure of the cost estimated to move from the current square to the destination. This is more or less an approximation or a heuristic, since that distance may be masked by many different factors. Now comes a variety of methods to estimate "*h*", and some of them will be covered later on.

Let's say we have a problem, as depicted in the Figure 3.25, trying to reach from a cell identified as the source cell to another cell—the target cell. On reaching this scenario, the A* Search algorithm would make an optimum choice of the path illustrated in the figure. Observe that this drawing assumes Euclidean distance for heuristic evaluation. Heuristics: Calculating g is obvious, but what on earth is *h*? We can do it two ways: (i) One can find the exact *h*, but it is very time-consuming. OR (ii) Estimating the value of *h* using different heuristics saves time. Let us see both the options:

1. Exact heuristics—finding the exact value of *h* is a very time-consuming process due to its exhaustive time complexity. Here are some ways to get the exact value of *h*.
   - In the previous stage, you can make a table of distances between two cells before running the A* search algorithm.
   - When no blocked cells or blocks exist in the situation, the exact *h* value can thus be computed directly using the distance formula or Euclidean distance beforehand.
2. Heuristics to approximate heuristics—the following are three heuristic methods generally used to estimate *h*:
   - Manhattan Distance

In this heuristic, the absolute differences between both the x and y coordinates of the goal cell and the current cell's x and y coordinates are added up.

h = abs (current cell.x – goal.x) + abs (current cell.y – goal.y)

When to use this heuristic?

When we are allowed to move in only four directions (right, left, top, and bottom). The Manhattan distance heuristic is depicted by Figure 3.25 (suppose the dark spot as the source cell and the light spot as the target cell).

- Diagonal Distance

It is nothing but the maximum of absolute values of differences in the goal's x and y coordinates and the current cell's x and y coordinates, respectively, i.e.

$$dx = \text{abs}(\text{current cell}.x - \text{goal.x})dy = \text{abs}(\text{current cell}.y - \text{goal.y})$$

$h = D \times (dx + dy) + (D2 - 2 \times D) \times \min(dx, dy)$ where D is the length of each node(usually = 1) and $D2$ is the diagonal distance between each node (usually = sqrt(2)).

When should you apply this heuristic?

It is applicable when movements are restricted to eight directions, akin to how a King moves in Chess. The diagonal distance heuristic is depicted in Figure 3.25, where the dark spot represents the starting point and the lighter one indicates the destination.

- Euclidean Distance

As it is clear from its name, it is nothing but the distance between the current cell and the goal cell using the distance formula.

$$h = \sqrt{((\text{current cell}.x - \text{goal.x})^2 + (\text{current cell}.y - \text{goal.y})^2)}$$

When to use the heuristic?

When we are free to travel in any direction, it is the ideal time to apply this heuristic. Figure 3.25 illustrates the Euclidean distance heuristics (assuming the dark spots represent the source cells and the light spots represent the target cells). Relation (differences and similarities) with other algorithms: Dijkstra is an A* search algorithm, a special case in which $h = 0$ for every node.

The A* search algorithm is as follows:

1. Begin by initializing the open list.
2. Also, initialize the closed list and put the starting node into the open list, whose f value can initially be zero.
3. Repeat as long as the open list is not empty:
   - **a)** Find a node in the open list with the lowest f (high priority in selection) and call it "q."
   - **b)** Remove q from the open list.
   - **c)** Create the 8 children of q, and make q their parent node.
   - **d)** For each of the children: (i) If one of the children is the goal, stop the search. (ii) Otherwise, compute the g and h for the child: the g is the cost of getting from q to the child plus the g of q. The h value is the distance from the successor to the goal. As distances can be computed in all kinds of heuristics, such as Manhattan, Diagonal, and Euclidean. The f value for the successor is the sum of its g and h values. (iii) If there exists another node in the OPEN list at the same position as the successor and it has a

lesser f value, then ignore this successor. (iv) If a node in the closed list is at the same position as the successor, but the f value of this node is smaller, ignore this successor as well. In all other cases, put the successor in the open list. Close the loop for every successor.

e) Put q in the closed list. Terminate the loop with an empty open list. Figure 3.25 shows some of the algorithms utilized in path planning and obstacle avoidance on a 2D grid map.

3.15 Path Following and Obstacle Avoidance

Path Following and Obstacle Avoidance for Autonomous Ships: Following the paper titled, "A collision avoidance decision-making system for autonomous ship based on modified velocity obstacle method," the authors discussed the most critical parts of path-following and avoiding obstacles by that strong decision-making system.

Collision Avoidance System (CAS): The paper points out how the collision avoidance system is the most critical component of the autonomous navigation system (ANS) in MASSs. The paper proposes a system that is efficient and robust, hence meeting the demands of actual applications.

Modified Velocity Obstacle Method: The collision avoidance model given in this paper is founded on an algorithm of the modified velocity obstacle. This enables the determination of safe velocities by which a ship can be kept from moving into the velocity obstacle regions of another vessel. The authors have proposed several different scenarios to validate this approach.

The decision-making mechanism should include various constraints, for example, ship maneuvers, multiple ship encounters, and also the COLREGS. This way, the system does not make poor decisions and the level of seamanship is excellent.

Path Following Constraints: Besides the above constraints, the system also considers the original navigation route of the ship. The decision course must not deviate much from the original path in avoiding unnatural detours, which means enhancing navigation efficiency.

Multiplicity in an Encounter Situation: This algorithm is constructed to face a situation of multiple ships. It would create its velocity obstacle regions regarding every target ship (TS) that could pose a collision. New decisions should avoid the selection of velocities in these regions to avoid new potential collision risks due to other ships that were not considered threats before. The architecture of ANS is shown in Figure 3.26.

Finite State Machine (FSM): The modified velocity obstacle method incorporates a Finite State Machine to study the interactive actions of other ships while navigating through a collision avoidance process. The analysis of this dynamic behavior is a requirement for effective path following and obstacle avoidance operations.

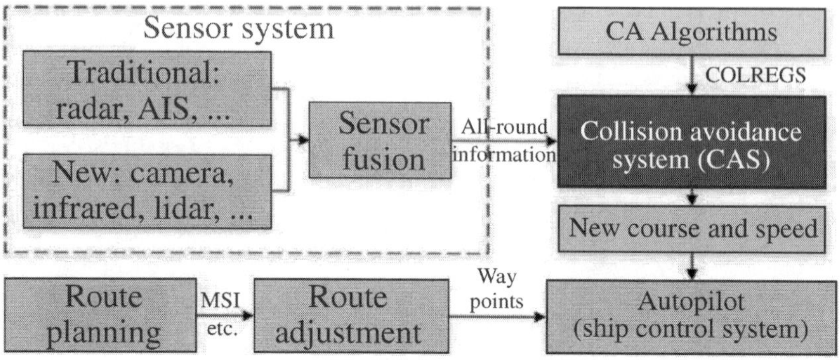

Figure 3.26 The structure of autonomous navigation system.

3.16 Fuzzy Logic Control

Things are not very clear and vague; this is what fuzzy refers to. Often, in the real world, we come across a situation that cannot be decided whether it's true or false. In such a way, there is so much freedom of fuzziness in reasoning with their logic. We can consider the inaccuracies and uncertainties of any situation in this way.

Fuzzy logic is a multivalued logic in which the values of the truth variables can be any real number between 0 and 1, not strictly two-valued like traditional logic. It is thus used to work with vague or uncertain information. Thus, it is considered a mathematical method for representing vagueness and uncertainty in decision-making processes.

Fuzzy logic is the methodology that insists that, on many occasions, the concept of true or false cannot be observed as, their does exist shades of gray between them. It allows partial truths; a statement is admitted partially to be true or false.

Fuzzy logic is applied in a variety of different scenarios, such as control systems, image processing, natural language processing, medical diagnoses, and artificial intelligence.

The key concept in fuzzy logic is the so-called membership function, representing to what degree an input value belongs to a certain set or category. A membership function maps an input value onto a membership degree between 0 and 1. Here, 0 means no membership at all, and 1 indicates full membership.

Fuzzy logic uses fuzzy rules working on if-then statements applied towards the relationship between input and output variables in a fuzzy manner. A fuzzy set is a set of membership degrees for every possible output value, hence the output of a Fuzzy logic system.

In short, fuzzy logic is a mathematical technique to represent vagueness and uncertainty in decision-making, as it accepts partial truths and is applied in a huge number of applications. It is based on the concept of the membership function, and the implementation is done using fuzzy rules. Figure 3.27 represents the Boolean and fuzzy logic representation.

The architecture consists of four main components:

1. **Rule Base**: This is a set of rules or some IF-THEN conditions developed by experts and is used to aid decisions made based on linguistic data. As a result of the development of fuzzy logic, highly effective design and adjustment methods for designing and adjusting fuzzy controllers have been developed—sometimes under that assumption,it makes the rule base less complex.
2. **Fuzzification**: It converts crisp input values or fuzzy numbers into fuzzy sets. These are crisp inputs—exact measurements obtained from sensors, for example, temperature, pressure, rpm etc.—which are fed to a control system for analysis.
3. **Inference Engine**: It determines the degree to which the real fuzzy input agrees with each rule and determines which of the rules to be activated depends upon the input. Then, there is a combination of those activated rules toward generating control actions.
4. **Defuzzification**: This is the process of converting fuzzy sets produced by the inference engine back into a precise value. Different techniques of defuzzification exist, and the most appropriate one to be used in a specific expert system is chosen in order to minimize errors.

A membership function is essentially a graph that determines how each point within a given input space is linked to a membership value that ranges from 0 to 1.

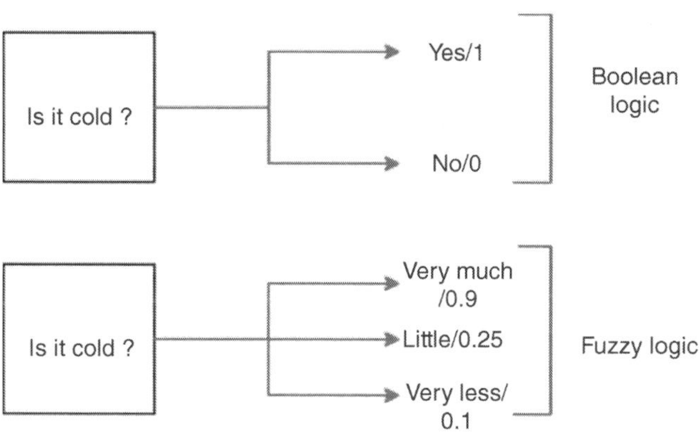

Figure 3.27 The Boolean logic and fuzzy logic representation.

This input space, commonly referred to as the universe of discourse or the universal set (u), encompasses all possible elements relevant to each specific application.

Within the domain of fuzzy logic, there are primarily three categories of fuzzifiers: (i) the Singleton fuzzifier, (ii) the Gaussian fuzzifier, and (iii) the Trapezoidal or triangular fuzzifier.

Fuzzy Control Exposed: Fuzzy control is an application that introduces human-like thinking to a controlling system. It is not based on rigid but soft reasoning, such as which an individual reasons from the premises of known facts for the inference of conclusions. It is very adaptable in uncertainties; it uses fuzzy logic in doing this.

3.16.1 Advantages of Fuzzy Logic Systems

1. They accept various types of inputs like fuzzy, noisy, or distorted data easily.
2. It is very easy to design and understand fuzzy logic systems.
3. Fuzzy logic deals with mathematical set theory principles, which provide simple processes of reasoning.
4. Even sophisticated problems of all such domains are dealt with effectively by imitating human reasoning and decision-making.
5. Algorithms require fewer data, which saves memory.

3.16.2 Problem with Fuzzy Logic Systems

1. The plurality of the problem-solving methodologies presented by the researchers can be an ambiguous solution, as a single method is not provided to solve the problems.
2. Sometimes, mathematical descriptions alone cannot confirm the properties of the systems.
3. As the methods involve precise as well as imprecise data while performing fuzzy logic, sometimes the precision gets compromised.

3.16.3 Applications

1. It is used for controlling altitude in spacecraft and satellites in aerospace.
2. It helps to control speed and traffic in the automobile industry.
3. Used in large companies for decision support systems and personnel evaluation.
4. The chemical industry uses it for controlling processes such as pH levels, drying, and distillation.
5. Now it is significant in natural language processing and most other AI-intensive applications.

6. Modern control systems and expert systems also make heavy use of fuzzy logic. With NNs, fuzzy logic is included in order to further improve the decision-making process, involving all the aggregation of data into more meaningful information in the formation of partial truths as fuzzy sets.

In the Boolean system, the absolute truth value was represented by 1.0, and the absolute false value represented as 0.0. In the fuzzy system, there is no logic or either an absolute truth or absolute false value present. But within fuzzy logic, a third value too present, which is partially true and partially false.

3.17 Robotic Motion and Odometry

Understanding robotic motion and odometry: robotic motion and odometry are important parts of the robotics field, especially for a case such as that of the four-wheel omnidirectional robot that is discussed in the paper. The following is a step-by-step explanation, based on the research:

3.17.1 Key Concepts in Robotic Motion Kinematic Modeling

Kinematic modeling is the study of the geometric structure of a robot as it influences its motion. The capability of controlling a robot's movement relies on the kinematic modeling of an omnidirectional robot, allowing it to move around in any direction properly.

The study emphasizes that the simplicity of the kinematic model facilitates better computation of motor direction and speed.

3.17.2 Motion Control

The system controls the movement by perfectly adjusting the motors of the robot. It presents the complexity of regulating four DC motors to achieve perfectly coordinated movement.

3.17.3 Implementation of Kinematics and Odometry Support

The incorporation of kinematics and odometry support ensures that the control of the robot's trajectories is fair and accurate, without causing errors in movement.

3.17.4 Significance of Odometry Definition

Odometry is the process of estimating changes in the time of a robot's position from sensor data, for instance, rotary encoders on the wheels.

3.17.5 Data Acquisition

The information gathered by the odometry system includes the coordinates, headings, and velocities of the robot that will guide the determination of movement and the position of the robot with respect to the initial place of deployment.

3.17.6 Performance

According to the study, movement accuracy greatly improves with the use of odometric sensors than with just kinematic control.

3.17.7 Future Work

Control Improvements: In the future, work will include the implementation of more advanced control methods such as PID or fuzzy logic, with more sensors, for instance, IMUs (inertial measurement units), for further improvement in movement control and accuracy.

Path Planning: The integration of kinematic modeling and odometry into path planning systems is very crucial since it helps ensure robots would always follow desired paths with great precision (Figure 3.28; Rijalusalam and Iswanto, 2021).

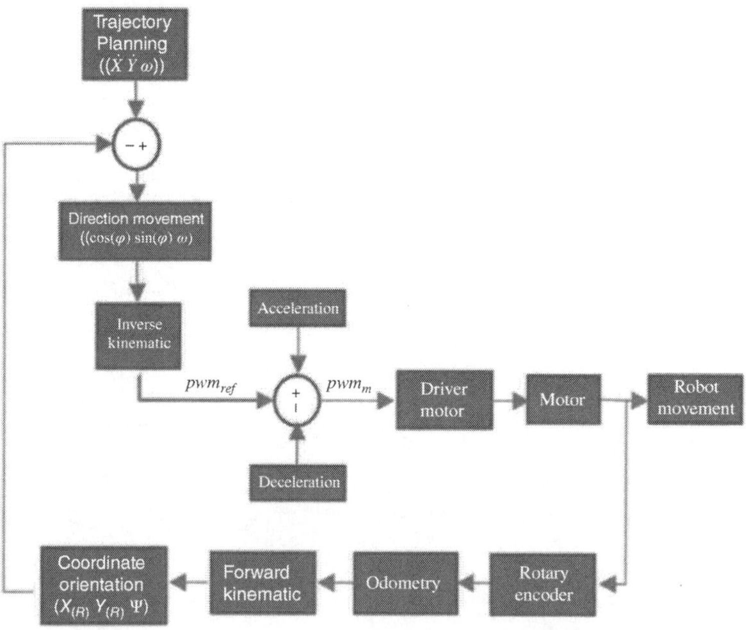

Figure 3.28 The block diagram of a control system (Rijalusalam and Iswanto [2021]/Journal of Robotics and Control (JRC)/CC BY SA).

3.18 Gripper: Its Various Types and Applications

Grippers are integral elements of the robotics that can be employed for grasping and manipulating objects. Some of the few types of grippers along with examples of a tree classification presented in Figure 3.35 are given below.

3.18.1 Based on the Motion of the Gripper Jaws

a) Based on primary motion of the gripper jaws
Parallel grippers: We have parallel grippers, where the jaws move in a parallel motion in relation to the gripper body. Accurate, simple, inexpensive, and widely used.
Angular grippers: Jaws are opened/closed around a central pivot point (sweeping/ arcing motion). Jaws need to move up and out of the way, in limited space.
Toggle gripper: The pivot point of jaw movement acts as an over-center toggle lock, provides high grip force/ weight ratio.

b) Based on the number of fingers: In this category we have concentric/self centering, individual servoed, and biomimic hand.

c) Based on the gripper type internal or external
Internal: part geometry allows locking well. Standard 2/3 fingered gripper may be used.
External: Common, closing force assists gripping, and works on adaptive encompassing.

d) Bernoulli and vacuum grippers

A Bernoulli gripper is a non-contact gripper that applies the Bernoulli principle, a fluid dynamics concept, using airflow to generate a grip. This principle dictates that when the velocity of a fluid is increased, its pressure will be reduced. In the context of a fluid such as air, this principle may result in a lifting or suction force. The force from the lifting or suction can, therefore, form the basis for a grip.

Air Flow and Suction: The gripper uses a high-speed jet of air directed at an object. When air moves rapidly over the surface of the object, the pressure above the object drops, creating a low-pressure zone. According to Bernoulli's principle, the faster-moving air (low-pressure) draws the object toward the gripper, as shown in Figure 3.29, because the surrounding (slower-moving) air exerts higher pressure on the object.

Grip Forming: The object is largely sucked into the gripper by the pressure difference created by the airflow, thus adhering to it. It is not pressed physically in any way by the gripper. The Bernoulli gripper does not physically touch the object to be gripped and acts on a variety of surfaces compared to vacuum grippers, which only require smooth surfaces and the object touches the gripper; this way, delicate or fragile things can be handled without being wrecked.

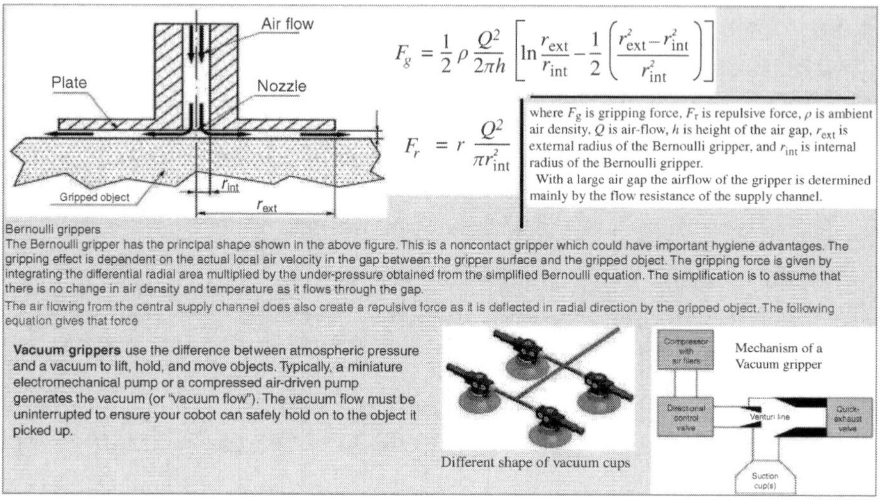

Figure 3.29 Bernoulli and vacuum grippers.

Keeping Grip: Provided that the air velocity is at a sufficient rate and the pressure differential is maintained, the object will remain in position. Once airflow stops or is modified, the object will fall since the pressure difference is lost.

A vacuum gripper, by definition, will make contact with the object, but it does this through the creation of a vacuum seal rather than through direct mechanical contact like a traditional gripper.

Creating the Vacuum: The vacuum gripper has a suction cup or pad that is placed against the object. A vacuum pump then creates a low-pressure area inside the suction cup, pulling air out and causing the pressure inside the cup to become much lower than the surrounding air pressure.

The gripping mechanism is based on the pressure difference: the higher pressure outside the cup and the lower pressure inside it make the suction cup adhere to the object. It is the force generated by this pressure difference that keeps the object fixed.

Contact with the Object: Even though the vacuum is "contactless" in the sense that it doesn't use mechanical gripping elements such as claws or fingers, the vacuum pad does have physical contact with the object surface to form the seal necessary for the suction force to work.

Parallel Gripper: The jaws move in a parallel motion in relation to the gripper body. These are accurate, simple, inexpensive, and widely used. Actuation is achieved by:

- Electrical solenoids
- Pneumatic pistons

- 4-bar (Grashoff mechanism)
- Twin rack and a pinion
- Wedge systems, etc.

Gripper Drives

- Pneumatic:
  Smaller units, quick assembly, high cycle rate, easy maintenance.
- Electric motors:
  High accuracy and repeatability, less floor space, low cost, easy maintenance.
- Hydraulic:
  High strength and high speed, large robots, takes floor space, mechanical simplicity, usually used for heavy payloads.
- Others:
  Hooks, adhesives, rough surface, rollers, etc.

Angular Gripper: Jaws are opened/closed around a central pivot point (sweeping/arcing motion). Jaws need to move up and out of the way in limited space.

Actuation: Electrical solenoids; pneumatic pistons; independent servos

Toggle Gripper: The pivot point of jaw movement acts as an over-center toggle lock and provides a high grip force/weight ratio.

Actuation: electrical solenoids and pneumatic pistons.

3.18.2 Gripper Type 2: Number of Fingers

Based on the number of fingers: In this category we have concentric/ self centering, individual servoed, and biomimic hand.

3.18.3 Gripper Type 3: Internal or External

Internal: Part geometry allows locking as well. Standard 2/3 fingered gripper may be used.
External: Common closing force assists gripping.

3.18.4 Other Types of Grippers

Magnetic gripper; Vacuum gripper; Multigripper; Vacuum gripper/Bernoulli/-Coanda gripper.

The forces impacting an object during varied acceleration scenarios depicted in Figure 3.30. Figure 3.31 represents calculation for the theoretical holding force, also Figure 3.32 represents various types of grippers.

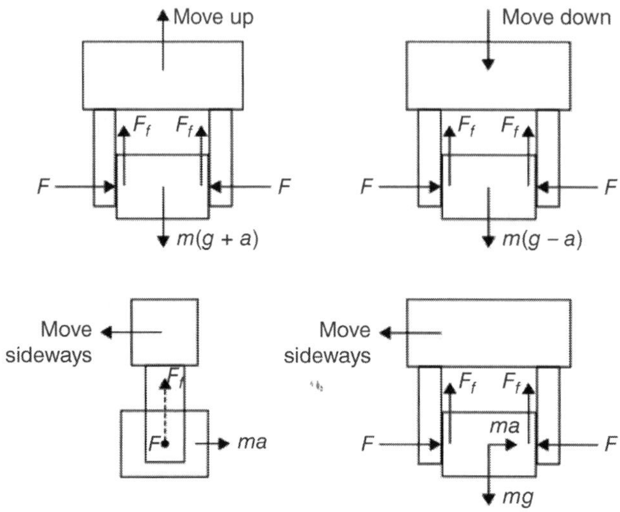

Figure 3.30 The forces impacting an object during varied acceleration scenarios include: "*m*" is representing the mass of the object being manipulated, "*a*" is indicating the object's acceleration caused by the robot's motion, and "*g*" is signifying the gravitational acceleration (Lien [2013]/Elsevier).

Vertical up:

$$F_f > m(g + a) \tag{3.35}$$

Vertical down:

$$F_f > m(g - a) \tag{3.36}$$

Horizontal parallel to finger face:

$$F_f > m\sqrt{g^2 + a^2} \tag{3.37}$$

Horizontal normal to finger face:

$$F_f > mg \tag{3.38}$$

Let us model a gripper–object interaction. The figure above shows that an arbitrary object placed between the gripper fingers, suffering by a constant disturbance force W, by a gripper force F_a, and by a frictional force $F_f = \mu F_a$, where μ is the coefficient of friction between the gripper and the object, then the dynamics of the slip velocity $v(t)$ can be represented as:

$$m\dot{v} = W - \mu F_a \tag{3.39}$$

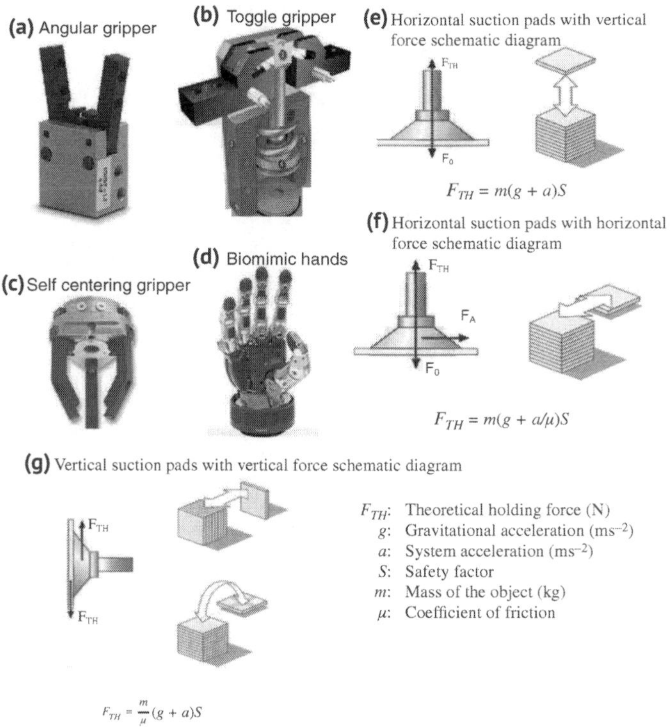

Figure 3.31 Various grippers and some calculation.

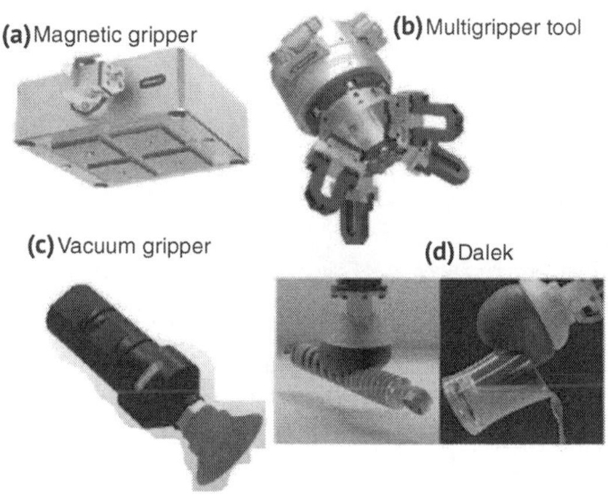

Figure 3.32 Some other types of grippers.

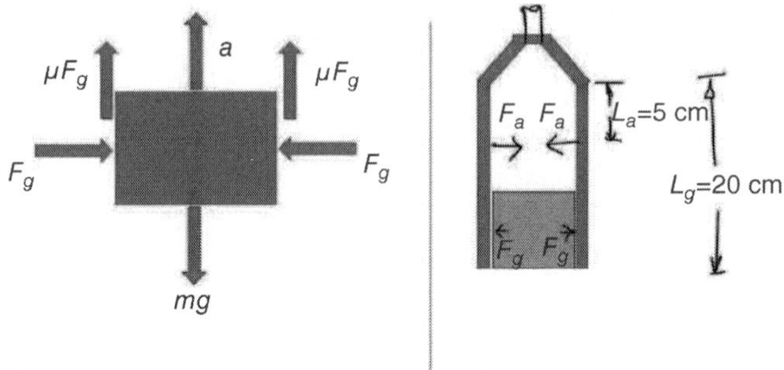

(a) Gripper (b) Pivot type gripper

Figure 3.33 Example numerical for grippers.

Example: A 5 kg rectangular block is gripped in the two finger gripper as shown in the Figure 3.33 and lifted vertically at a velocity 1 m/s. If acceleration is 27.5 m/s2 and the coefficient of friction between gripping pads and block is 0.48, calculate the minimum force that would prevent slippage. Solution:

mass $(m) = 5$ kg

acceleration of gravity (g) $= 9.8\ \text{m/s}^2$

acceleration of handling $(a) = 27.5\ \text{m/s}^2$

coefficient of friction $(\mu) = 0.48$

Considering the free body diagram,

$$2\mu F_g - mg = ma$$

$$F_g = m(a + g)/2\mu$$

$$F_g = 5(27.5 + 9.81)/2 \times 0.48 = 194.3\ \text{N}$$

This is the minimum amount of required force that would prevent slippage.

Example: A simple pivot-type gripper, as shown in Figure 3.33, is used to hold boxes. The gripping force (F_g) required is 20 kgf. The gripper is to be actuated by a piston device to apply an actuating force (F_a). What will be the value of actuating force (F_a)?

Solution: Gripping force $(F_g) = 20$ kgf; actuating force $(F_a) = ?$ Taking moments of the forces on one arm and summing them to zero, we get

$$F_g L_g = F_a L_a$$

$$F_a = F_g L_g / L_a$$

$$F_a = (20 \times 20)/5 = 80\ \text{kgf}$$

Therefore, the piston device would have to provide an actuating force of so kg to close the gripper with a force against a boxes of 20 kgf

Since it is not possible to directly control and apply the gripper force $F_a(t)$, we model the incremental displacement $x_g(t)$ of the gripper as proportional to the applied force such that

$$F_a \propto x_g \tag{3.40}$$

The time derivative of which can be related to the control input signal, the gripper velocity $v_g(t)$ is as follows:

$$\dot{F}_a = \kappa v_g \tag{3.41}$$

Thus, Eqs. (3.40) and (3.41) denote the overall system dynamics. Here, in deference to our problem statement earlier, W, μ, and κ are assumed to be unknown parameters, which will be adapted for during the control design process.

3.18.4.1 Vortex Gripper

The vortex gripper is another pneumatic type of handling device. It can pick up and hold a workpiece with zero contact. Its mechanism works by blowing compressed air tangentially through a nozzle into a vortex cup, which creates a high-velocity swirling flow of air. For this reason, negative pressure builds up in the center of the gripper, with the help of which the workpiece is lifted. The pressure over the upper surface of the workpiece is greater than that in the lower surface, thus picking up the workpiece. Simultaneously, the compressed air is released to the atmosphere from the clearance between the gripper's skirt and the workpiece. Provided that the supply of compressed air is continuous, there will be the levitation of the workpiece at some definite distance. Figure 3.34 represents the schematic of the vortex gripper.

The governing equations of fluid motion include the continuity equation, the Navier–Stokes equation, and the energy equation. The continuity equation is

$$\frac{\partial \rho}{\partial t} + \frac{\partial}{\partial x_i}(\rho u_i) = 0$$

The Reynolds-averaged Navier–Stokes equation is given as follows:

$$\frac{\partial \rho}{\partial t}(\rho u_i) + \frac{\partial \rho}{\partial x_j}(\rho u_i u_j)$$
$$= -\frac{\partial p}{\partial x_i} + \frac{\partial}{\partial x_j}\left[\mu\left(\frac{\partial u_i}{\partial x_j} + \frac{\partial u_j}{\partial x_i} - \frac{2}{3}\delta_{ij}\frac{\partial u_l}{\partial x_l}\right)\right]$$
$$+\frac{\partial}{\partial x_j}\left(-\rho\overline{u_i' u_j'}\right)$$

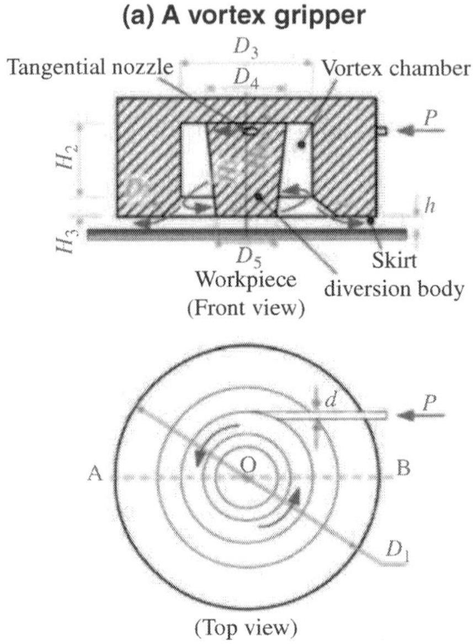

Figure 3.34 Schematic of a vortex gripper (Wu et al. [2012]/SAGE Publications).

To solve these equations, the Reynolds stress term $\overline{u_i'u_j'}$ must be properly represented. RSM uses differential transport equations to compute the individual Reynolds stress for the closure of Reynolds averaged Navier–Stokes equations. The Reynolds stress transfer equations can be written as

$$\frac{\partial}{\partial t}\left(\rho\overline{u_i'u_j'}\right) + \frac{\partial}{\partial}\left(\rho\overline{u_k}\overline{u_i'u_j'}\right) = D_{ij}^T + P_{ij} + \phi_{ij} - \varepsilon_{ij}$$

where the turbulent dissipation is ε_{ij}, the pressure–strain is ϕ_{ij}, the stress production is P_{ij}, and the stress diffusion is D_{ij}^T. While D_{ij}^T, ϕ_{ij}, and ε_{ij} must be modeled in order to finish the equations, P_{ij} does not require any modeling. This is how the generalized gradient-diffusion model can represent D_{ij}^T.

$$D_{ij}^T = -C_s\frac{\partial}{\partial x_k}\left[\rho\frac{k}{\varepsilon}u_k u_1\frac{\partial}{\partial x_1}(u_i u_j)\right]$$

The dissipation tensor is modeled as

$$\varepsilon_{ij} = \frac{2}{3}\delta_{ij\rho\varepsilon}$$

The pressure–strain ϕ_{ij} can be modeled using the following decomposition

$$\phi_{ij} = \phi_{ij.1} + \phi_{ij.2} + \phi_{ij.w}$$

$$\phi_{ij.1} = -C_1\rho\frac{\varepsilon}{k}\left(\overline{u_iu_j} - \frac{2}{3}\delta_{ij}k\right)$$

$$\phi_{ij\cdot 2} = -C_2\left[\left(P_{ij} - C_{ij}\right) - \frac{1}{3}\delta_{ij}\left(P_{kk} - C_{kk}\right)\right]$$

$$\phi_{ij.w} = C_1\frac{\varepsilon}{k}\left(\overline{u_ku_m}n_kn_m\delta_{ij} - \frac{3}{2}\overline{u_iu_k}n_jn_k\right.$$
$$\left.-\frac{3}{2}\overline{u_ju_k}n_in_k\right)\frac{k^{3/2}}{C_1\varepsilon d}$$
$$+C_2'\left(\phi_{km\cdot 2}n_kn_m\delta_{ij} - \frac{3}{2}\phi_{ik\cdot 2}n_jn_k\right.$$
$$\left.-\frac{3}{2}\phi_{jk.2}n_in_k\right)\frac{k^{3/2}}{C_1\varepsilon d}$$

where the slow pressure–strain, $\phi_{ij,1}$, is the turbulence–turbulence interaction and the rapid pressure–strain, $\phi_{ij,2}$, is the effect of mean flows on the turbulent field. The redistribution of normal stresses near the wall is due to the wall-reflection $\phi_{ij.w}$. The following transport equations give the turbulent kinetic energy, k, and its rate of dissipation, ε.

$$\frac{\partial}{\partial t}(\rho k) + \frac{\partial}{\partial x_i}(\rho k u_i) = \frac{\partial}{\partial x_j}\left[\left(\mu + \frac{\mu_t}{\sigma_k}\right)\frac{\partial k}{\partial x_j}\right] + \frac{1}{2}P_{ii} - \rho\varepsilon$$

$$\frac{\partial}{\partial t}(\rho\varepsilon) + \frac{\partial}{\partial x_i}(\rho\varepsilon u_i)$$
$$= \frac{\partial}{\partial x_j}\left[\left(\mu + \frac{\mu_t}{\sigma_\varepsilon}\right)\frac{\partial \varepsilon}{\partial x_j}\right] + C_{\varepsilon 1}\frac{1}{2}P_{ii}\frac{\varepsilon}{k} - C_{\varepsilon 2}\rho\frac{\varepsilon^2}{k}$$

where μ_t is the turbulence viscosity:

$$\mu_t = \rho C_\mu\frac{k^2}{\varepsilon}$$

$$C_1 = 1.8, C_2 = 0.6, C_1' = 0.5, C_2' = 0.3, C_{\varepsilon 1} = 1.44$$

$$C_{\varepsilon 2} = 1.92, C_\mu = 0.09, \text{ and } \sigma_k = 0.82, \sigma_\varepsilon = 1.0$$

1. Mechanical Grippers: In mechanical grippers, objects are gripped mechanically through jaws or fingers. This may be illustrated as a system that bakes robots inside the bakery. This system, in baking robots inside the bakery, utilizes mechanical grippers to take care of trays full of freshly baked products when packaging them.

2. Parallel Jaws Grippers: These are grippers which have more than one parallel jaw set that moves with the object in parallel to each other. For instance, in a fish fillet seafood processing plant, its jaws grippers are parallel in handling and sorting of the fillet before packaging. Those grippers are precise.
3. Vacuum Grippers: In vacuum grippers, vacuum pressure is employed for gripping purposes. Hence, they are useful when dealing with fragile food items—either pastries or fruits. For example, a confectionery factory uses vacuum grippers to lift and place fragile chocolates on trays.
4. Pneumatic Grippers: This is a gripper that operates based on compressed air. The pneumatic gripper provides very excellent gripping force and quickly responds. For instance, the processing line of vegetables utilizes pneumatic grippers in sorting fresh products into different containers.
5. Electric Grippers: An electric gripper is driven by an electric motor or actuator. It provides some degree of precision control and programmability. Applications include, as an example, in a dairy processing plant, gripping then packaging different sizes of blocks of cheese.
6. Finger or Jaws Grippers: These grippers that carry the fingers or jaws wherein the number of fingers or jaws is multitudinous and is adjustable according to the shape of the object to be held. For example, one of the meat processing plants uses finger grippers in order to portion meat for packing.
7. Actuators: Transducers that convert energy into movement to energize the gripper that may be hydraulic or electrical actuators, as in the illustration of a frozen food manufacturing line, that utilizes hydraulic grippers with gripper actuators in moving frozen food packages for sorting and palletizing.
8. Sensors: Grippers are fitted with integrated sensors to provide grip force feedback, object presence, or changing its position. For example, an assembly line of ready-to-eat meals uses grippers with sensors that place food containers and pack them out with correct geometries of packaging.
9. Bernoulli Grippers: These grippers work on the principle of the Bernoulli force, which really makes them very ideal for thin and porous food products like bread or biscuits. A bakery uses Bernoulli grippers when picking freshly baked loaves of bread from the baking oven to pack and box.
10. Multi-grippers Tools: These tools comprise gripper systems which have several arms or tools for picking up more than one food product in a single step. For instance, in a snack food manufacturing plant multi-gripper tools are applied to pick and place many snack bags in cartons to be dispatched.
11. Peltier Grippers: Peltier grippers are fitted with heating and cooling capabilities through the use of Peltier devices and can also be used in some areas where sensitive food to a change in temperature is involved. For frozen dessert production line, Peltier grippers have been used when handling ice cream containers without interrupting the integrity of the cold chain.

12. Magnetic Grippers: The magnetic grippers are used to pick the ferrous food items by magnetic fields. Thus, non-contact handling is introduced. For instance, magnetic grippers can be used in a canned food production plant in handling cans for filling and sealing operations.
13. Other Grippers: Innovative designs of grippers for specific food processing applications, like compliant grippers for delicate food products or bio-inspired grippers that mimic human handing skills by fingers. For example, there's a chocolate manufacturing plant using compliant grippers in handling its delicate chocolate pralines without breaking their shape.

Figure 3.35 represents various types of grippers in food industry, and Figure 3.36 represents various types of parallel grippers.

3.19 Robotics and Automation for Packaging and Handling

3.19.1 Mobile Robotics and Packaging Optimization Deliveries

Combining Mobile Robotics and Packing for Optimal Deliveries Exploring the Future of Warehousing in E-commerce: Meeting the customer expectation of rapid delivery discusses the potential achieved through the automation of packaging by mobile robots, exemplified by YouBot that can be applied for improving the efficiency in the order preparation and delivery processes.

Objective: An application that would enable the independent loading and stacking of pallets with assorted box sizes by a mobile robot. This would be an essential application for fast-paced environments like e-commerce houses.

Technology Used: IoT will be used to manage the YouBot operating system in the warehouse environment. Technology will be implemented through the use of RFID and indoor localization systems.

Optimization Algorithm: The KLS Optim's Bin Packing solver is of great importance in this project. It will output a JSON file containing the optimal positions, dimensions, and orientations of boxes for being packed on pallets to be input to the robotic application that coordinates the packing actions.

Feasibility Versus Optimality: It shows that a mobile robot needs to meet constraints versus practical feasibility, as opposed to optimization. While the developed packing algorithms may bring optimal solutions for computation, these solutions may not be feasible by the robot itself due to physics constraints.

Robotic Operations: The packing process involves guidelines indicated within the JSON file, as instructed by the YouBot to navigate the pallet and place items correctly. In turn, this process does not require much human oversight, so the amount of human error has been eliminated in most cases while efficiencies in the packing process have been improved.

Learning from Simulations: It incorporates real-world experiments and simulations in identifying potential failure cases such as instability or damage to items

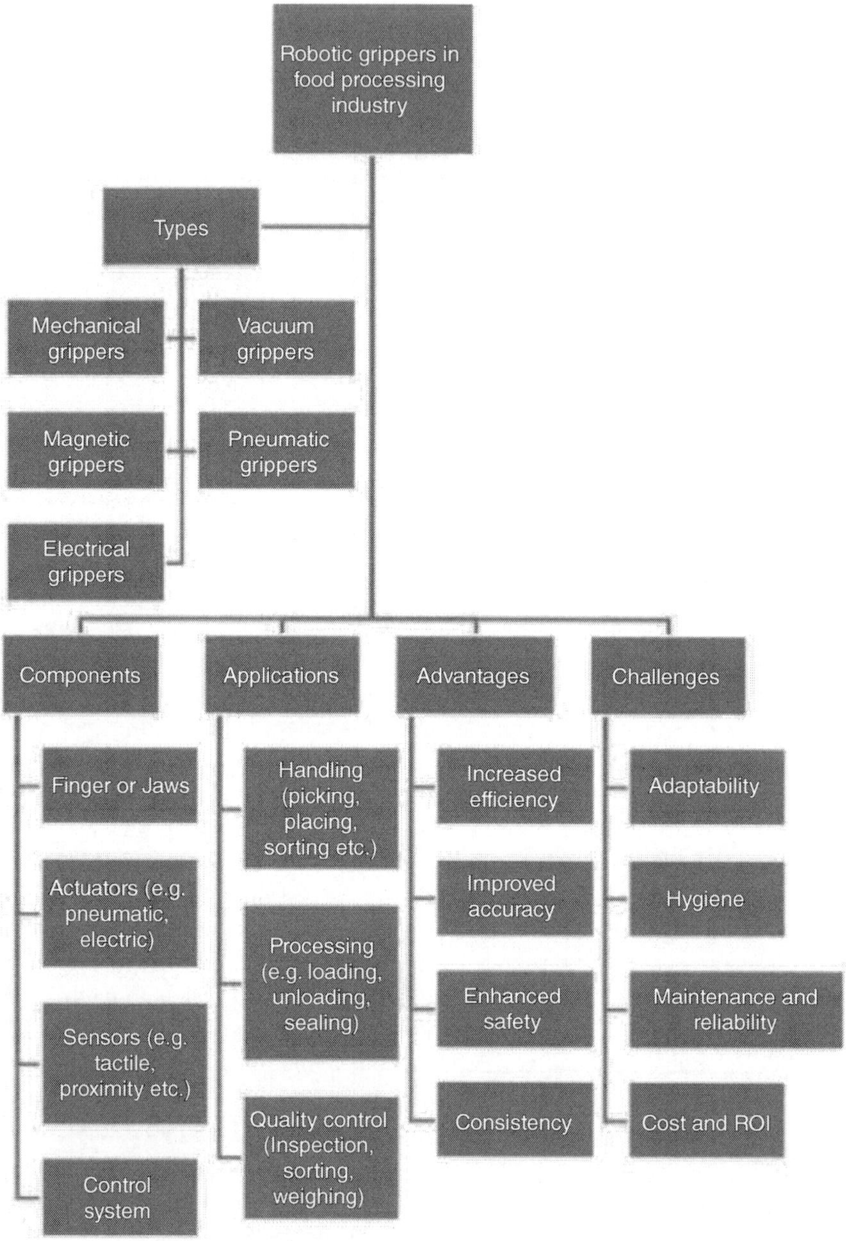

Figure 3.35 Various types of grippers in food industry.

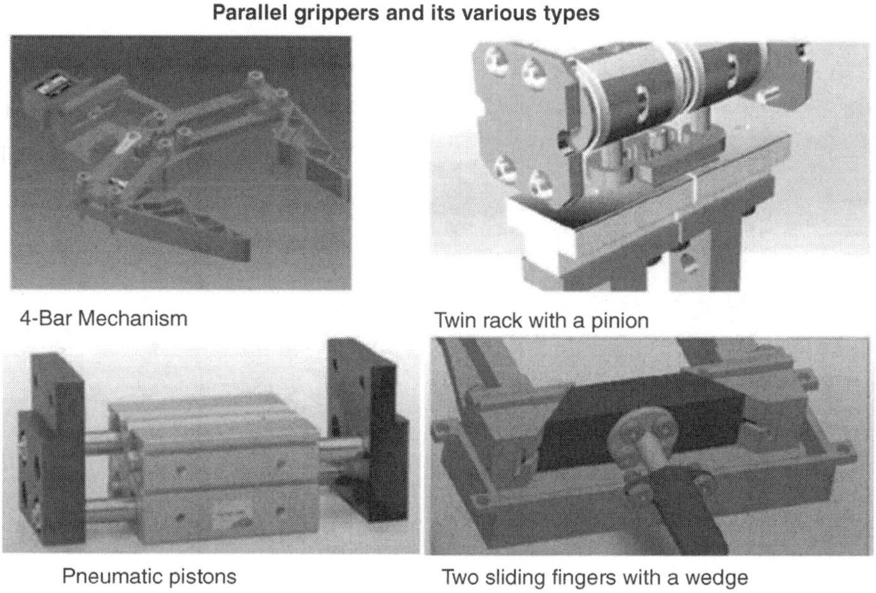

Figure 3.36 Various types of parallel grippers in food industry.

during packing. Such insights are used in the improvement of the strategies by which packing can be done, to ensure that the performance becomes better for the robot.

Integration of AI: We also introduce advanced heuristics and strategies in this proposal, which focus on modern techniques in artificial intelligence, such as those in deep learning. This will enhance an optimization solver to packing and may possibly yield results more applicable to autonomous mobile robots.

Navigation Challenges: It is through the ROS library, particularly the gmapping package of SLAM, that enables robot navigation within the warehouse environment. The technology enables the robot to make a map and also determine where it is at the same time.

This section also represents innovation in combining mobile robotics with optimization in packing so as to overcome the issues of modern warehousing and distribution. The work is aimed at trying to improve the efficiency and accuracy of packaging processes as automation contributes to the task and to fulfill the growing requirements of e-commerce delivery systems. While bringing together AI and advanced navigation techniques, this research puts it at the top rung of logistics technology.

3.19.2 Robotic Grasping and Control Strategies

Overview of Robotic Grasping Importance: Robotic grasping is critical for high-precision tasks in both industrial manufacturing and home service applications. Involves the achievement of high precision, reliability, speed, and flexibility at one go, which remains a very challenging goal.

Broadly classified into categories based on the use of sensing or constraints, with practical examples that cover the work done in the area of robotic grasping in hand.

Visual Sensors in Robotic Grasping: Computer Vision Algorithms: Methods that are applied in grasping include edge of the object finding, pose estimation as a graph-matching problem, among others.

Deep Learning Applications: Berczi et al. demonstrated in their study that through deep learning, robots learn to grab various objects present in daily life in a better adaptive way for the robots.

Task-parametrized Models: Duque et al. implemented a Gaussian mixture model for learning from robot motion supported by human demonstrations, allowing for flexible assembly operations without hard-coding.

Biomimic Approaches: Inspired by human grasping strategies, scientists like Kiguchi et al. studied how humans determine the amount of grasping force, and used this understanding to robotic systems.

Adaptive Learning: Rodriguez proposed the "abort and retry" mechanism for grasp learning, which effectively reduces training time by starting new experimental processes in case of failure.

New Trends in Robotic Manipulation Experience-based Control: In this category, humans learn to control their hands based on experience that can be mimicked in the robotic hands by using deep neural networks and reinforcement learning.

Minimizing Human Participation: Recent works are very much aimed at reducing human participation in robotic manipulation by using visual information as raw input for training robots in goal-oriented tasks.

3.20 Review Questions

Q1 What is a homogeneous transformation matrix? How is a point in 3D space represented using homogeneous coordinates?

Q2 What are the components of a homogeneous transformation matrix? How do you perform a translation using a homogeneous transformation matrix?

Q3 A camera is attached to a gripper in a robotic arm. The camera can focus on an object to determine its frame relative to the camera. Formulate the transformation matrix for the object relative to the robot arm using the matrices below.

$$^{Ccon}_{obj}T = \begin{matrix} 0 & 0 & 1 & 2 \\ 1 & 0 & 0 & 2 \\ 0 & 1 & 0 & 4 \\ 0 & 0 & 0 & 1 \end{matrix} \qquad ^{Grip}_{ccan}T = \begin{matrix} 0 & 0 & -1 & 3 \\ 0 & -1 & 0 & 0 \\ -1 & 0 & 0 & 5 \\ 0 & 0 & 0 & 1 \end{matrix} \qquad ^{Grip}_{Arm}T = \begin{matrix} 0 & -1 & 0 & 0 \\ 1 & 0 & 0 & 0 \\ 0 & 0 & 1 & 4 \\ 0 & 0 & 0 & 0 \end{matrix}$$

Q4 Solve for the missing elements of the following frame representation if the frame is attached to an object in space. Also, show the orientation and position of frame F wrt OXYZ?

?	0	−1	5	Hint 1.	(Cosθ)xu	(Cosθ)xv	(Cosθ)xw	*dx*
?	0	0	3		(Cosθ)yu	(Cosθ)yv	(Cosθ)yw	*dy*
?	−1	0	2		(Cosθ)zu	(Cosθ)zv	(Cosθ)zw	*dz*
0	0	0	1		0	0	0	1

*xu, zv, etc. are subscripts.

	u	v	w
x	?	90	180
y	?	90	90
z	?	180	90

Hint 2. (a) The matrice is required and (b) the xyz and uvw axes drawing are required?

Q5 Assume that the space frame {s} and the frame {*a*} are initially coincident and that the frame is instantly coupled to a little robot. Upon commencing its motion, the robot rotates 90°, then advances five units to arrive at the configuration {*b*}, rotates −90°, advances five units to arrive at the configuration {c}, and then rotates −90° and advances three units to arrive at the configuration {d} :. Determine the homogeneous transformation matrices for the robot's setup at every step:

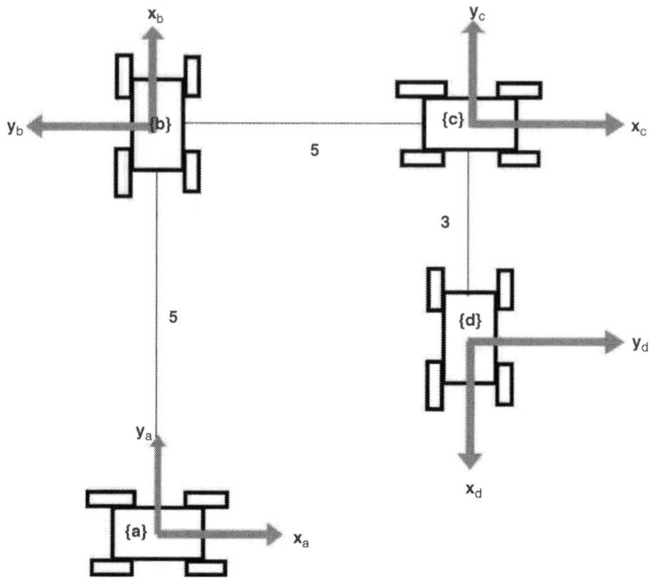

Bibliography

VH Benitez, R Symonds, and DE Elguezabal. Design of an affordable IoT open-source robot arm for online teaching of robotics courses during the pandemic contingency. *HardwareX*, 8:e00158, 2020.

T Brosnan and DW Sun. Improving quality inspection of food products by computer vision—a review. *Journal of Food Engineering*, 61(1):3–16, 2004.

M Elnour, H Haller, and M Martin. Life cycle assessment of a retail store aquaponic system in a cold-weather region. *Frontiers in Sustainability*, 3:1051091, 2023.

AS Jamaludin, MNM Razali, N Jasman, ANA Ghafar, and MA Hadi. Design of spline surface vacuum gripper for pick and place robotic arms. *Journal of Modern Manufacturing Systems and Technology*, 4(2):48–55, 2020.

A Khatamian. Solving kinematics problems of a 6-dof robot manipulator. In: *International Conference Computing*. 2015.

T Kondakci and W Zhou. Recent applications of advanced control techniques in food industry. *Food and Bioprocess Technology*, 10:522–542, 2017.

TK Lien. Gripper technologies for food industry robots. In: *Robotics and Automation in the Food Industry*. Elsevier. pages 143–170, 2013.

HY Lin and XZ Peng. Autonomous quadrotor navigation with vision based obstacle avoidance and path planning. *IEEE Access*, 9:102450–102459, 2021.

P McAllister, H Zheng, R Bond, and A Moorhead. Combining deep residual neural network features with supervised machine learning algorithms to classify diverse food image datasets. *Computers in Biology and Medicine*, 95:217–233, 2018.

DU Rijalusalam and I Iswanto. Implementation kinematics modeling and odometry of four omni wheel mobile robot on the trajectory planning and motion control based microcontroller. *Journal of Robotics and Control (JRC)*, 2(5):448–455, 2021.

S Rokhva, B Teimourpour, and AH Soltani. Computer vision in the food industry: accurate, real-time, and automatic food recognition with pretrained mobilenetv2. *arXiv preprint arXiv:2405.11621*, 2024.

Q Wu, Q Ye, and GX Meng. Experimental and numerical study of vortex gripper with a diversion body. *Proceedings of the Institution of Mechanical Engineers, Part C: Journal of Mechanical Engineering Science*, 226(6):1526–1534, 2012.

Y Zhang, S Wang, G Ji, and P Phillips. Fruit classification using computer vision and feedforward neural network. *Journal of Food Engineering*, 143:167–177, 2014.

4

AI/ML and Robotics Applications in Food Safety, Quality, and Challenges

After reading this chapter, you should be able to understand:

- The food processing industry now stands at the threshold of a revolutionary period in its history, which will be under the significant influence of the so-called fourth industrial revolution: Industry 4.0. This is very popularly known as Food Processing 4.0. It combines into foods at production as well as the supply chain level modern sophisticated digital technologies such as Internet of Things (IoT), artificial intelligence (AI), big data analytics, robotics, and blockchain. These technologies include precision, efficiency, and quality control through real-time monitoring, automatic decision-making, and streamlining traceability.
- At present, the Food Processing 4.0 context is represented by the incorporation of innovations concerning automated quality assurance, elevated safety measures, and sustainable techniques in accordance with consumer desires for greater clarity, food safety standards, and more eco-friendly goods.
- Future advances in Food Processing 4.0 will revolutionize production through smart machinery, adaptive manufacturing systems, and autonomous processing lines. As such, change also entails a shift in the workforce, with an emphasis on new skills to manage highly interconnected, intelligent systems–ones that have digital literacy and technical knowledge. Moreover, with the uptake of such technologies in industries, there will be moral and cybersecurity considerations that will require proactive frameworks to further protect data integrity and foster equal access to technology along the global food sector chain.

Food Engineering Automation with Robotics and AI, First Edition. Abir Chakravorty.

- This section discusses its current applications, potential challenges, and future opportunities concerning Food Processing 4.0, which should be appreciated in terms of its added potential to bring about further operational efficiency in food production as well as sustainability.

This section discusses how AI and machine learning (ML) should shape the food quality control and safety assessment sectors. By doing this, the use of AI and ML fills significant gaps in food safety, yet with higher compliance with regulations. The various ML applications are represented in Figure 4.1, whereas Figure 4.2 categorizes supervised and unsupervised learning, and Figure 4.3 illustrates the subset of AI. Table 4.2 represents various studies applying AI/ML in food sector.

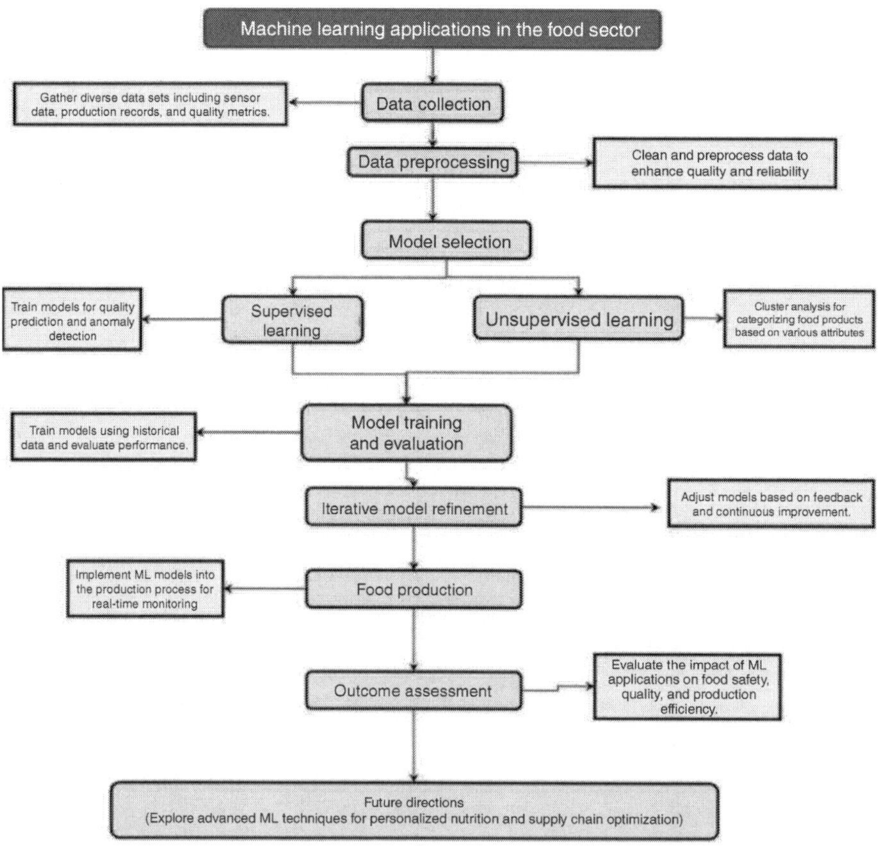

Figure 4.1 ML applications in the food sector.

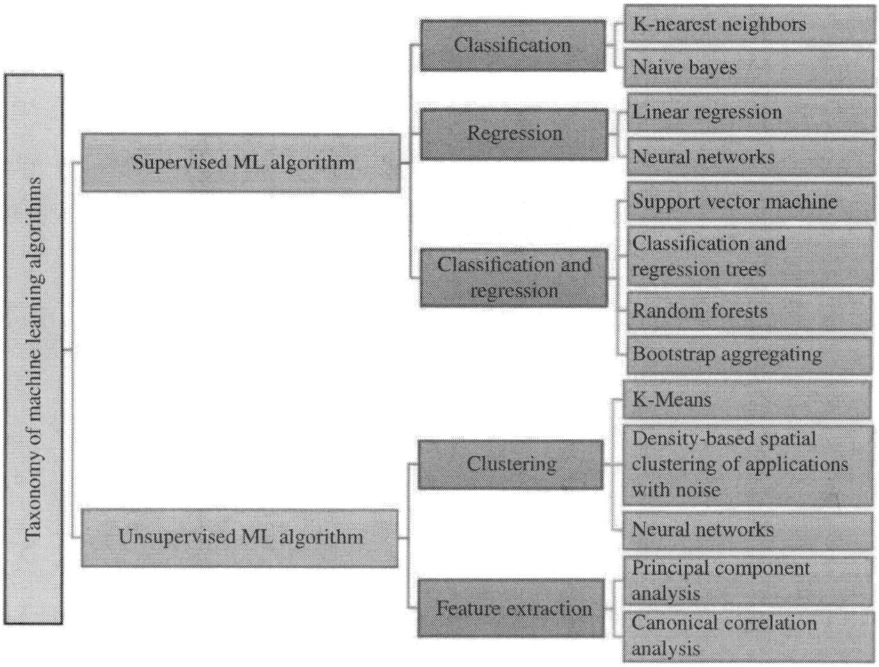

Figure 4.2 Categorizing supervised and unsupervised learning (Alsharif et al. [2020]/MDPI/CC BY 4.0).

4.1 Food Safety and Quality Detection Using Robotics and AI/ML

In food quality detection, several ML algorithms are commonly used to analyze data and make predictions, which are as follows:

- Support Vector Machines (SVM): Effective for classification tasks, SVM can be used to differentiate between high and low-quality food samples.
- Decision Trees: Useful for creating interpretable models that can predict food quality based on various features (e.g. color, texture).
- Random Forest: An ensemble method that improves prediction accuracy by combining multiple decision trees, effective in handling complex datasets.
- K-Nearest Neighbors (KNN): A simple algorithm that classifies food samples based on the characteristics of their nearest neighbors.
- Neural Networks: Particularly, deep learning (DL) models can analyze complex patterns in image data (e.g. detecting spoilage through visual inspection).
- Logistic Regression: Often used for binary classification problems, such as determining whether food is safe to consume.

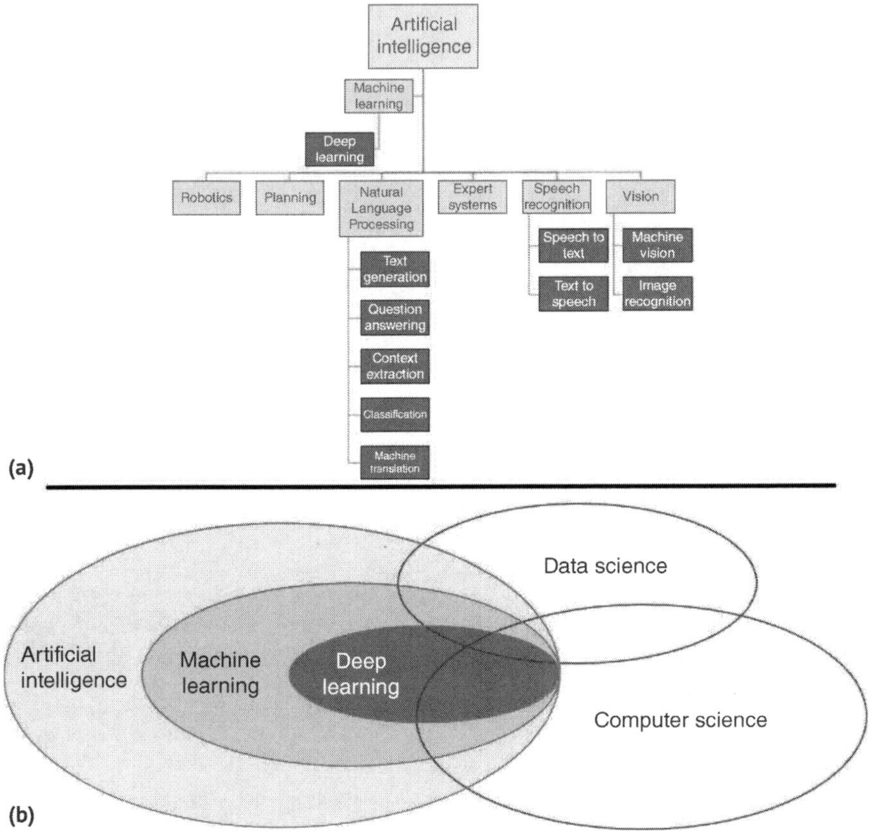

Figure 4.3 The subsets of AI.

- Naive Bayes: Suitable for classification tasks, particularly with categorical data, and can be used for detecting contaminants.
- Principal Component Analysis (PCA): While not a classification algorithm, PCA is used for dimensionality reduction and feature extraction, helping improve the performance of other algorithms.
- Gradient Boosting Machines (GBM): This includes methods like XGBoost and LightGBM, which are powerful for regression and classification tasks due to their ability to model complex relationships.
- Convolutional Neural Networks (CNN): Specifically for image data, CNNs can analyze images of food for quality detection based on visual features.
- Applications: *Sensor Data Analysis*: Algorithms can process data from sensors measuring temperature, pH, and -other factors. *Visual Inspection*: DL models can analyze images to detect spoilage or quality issues. *Predictive Modeling*: ML can forecast food quality based on historical data.

Presented below is Figure 4.1 for the work on ML applications in the food sector by Chhetri (2024).

4.1.1 Explanation of the Workflow

- Robotic Arm: Collects food samples and communicates with sensors.
- Sensors: Receive primary information about the quality of the food stuff such as temperature and pH.
- Data Preprocessing: Cleans and gets ready the raw data to be further analyzed.
- Feature Extraction: Determines the features from the processed data.
- ML Algorithms: Several algorithms are used, such as SVM, decision trees, random forest, neural networks, and gradient boosting to make predictions on the quality of food stuff. In so doing, it shall give conclusions in the form of possible flagged issues based on the algorithmic outputs.
- Database: The system preserves history and results for further analysis.
- User Interface: This interface warns and provides operators with choices based on analysis and control of the robotic arm.
- SVM Applicability: Suitable for classifying food quality based on multiple features like chemical composition and physical characteristics. The various ML algorithms can be applied to evaluate different food products and assess their freshness. Example: Distinguishing between fresh and spoiled fruits based on measurements from sensors (e.g. firmness, sugar content).
- Decision Trees Applicability: Useful for interpreting the criteria that determine food quality. Example: Assessing vegetable freshness by evaluating features like color, moisture content, and odor in a step-by-step manner.
- Random Forest Applicability: Handles complex interactions among features well, reducing overfitting. Example: Classifying meat freshness based on a combination of temperature, pH, and other sensory data from various samples.
- Neural Networks Applicability: Ideal for complex pattern recognition, especially in high-dimensional data such as images. Example: Evaluating the freshness of fish by analyzing images to detect color changes or spoilage signs (e.g. discoloration, texture).
- Gradient Boosting Machines (GBM) Applicability: Strong performance in predictive tasks with structured data. Example: Predicting the shelf life of packaged foods by analyzing past sales data and environmental conditions during storage.
- K-nearest Neighbors (KNN) Applicability: Simple classification algorithm based on similarity. Example: Classifying dairy products as fresh or expired based on attributes such as smell, color, and texture compared to known samples.

- Logistic Regression Applicability: Useful for binary classification tasks. Example: Determining whether a food product meets quality standards (e.g. acceptable versus unacceptable freshness).
- Naive Bayes Applicability: Effective for classifying categorical data. Example: Evaluating processed foods based on ingredient analysis, predicting freshness based on the presence of preservatives and other factors.
- PCA Applicability: Helps in reducing dimensionality while retaining essential features. Example: Visualizing the quality of various food products by plotting them based on their key features, making it easier to identify outliers or spoilage.
- CNN Applicability: Primarily used for image recognition tasks. Example: Analyzing images of fruits and vegetables to detect bruising, mold, or other signs of spoilage.

4.2 Automation of Fruit and Vegetable Sorting, Grading

In the section "Automated Sorting And Grading of Vegetables Using Image Processing," a computer vision-based system is proposed to automate grading and sorting of agricultural products—strawberries, brinjals, and similar items. The prime goal of this system is to replace human grading and sorting methods with a machine-based system that will obtain consistency, uniformity, and accuracy for fruits and vegetables in sorting. However, the present challenges in AI/ML applications are depicted in Table 4.1.

The system applies several techniques of image processing, among which:

Preprocessing: Image preprocessing improves the quality of images by filtering out noisy contents with median, Wiener, and adaptive filters. Median will be used here because it is efficient to give a noise-reduction effect while preserving edges (Usha, 2017).

Color Detection: It brings out the spotted areas of fruits that are essential for grading.

Segmentation: K-means clustering is used for segmenting the images. With the help of segmentation, it is easy to classify the fruits according to maturity levels.

Feature Extraction: The system extracts features like entropy, standard deviation, and mean so that the maturity level of the strawberries as well as brinjals can be judged. These are some important features for classification.

The system proposed is developed and simulated through MATLAB; the GUIDE of MATLAB is used for the purpose of creating the desktop application, so a friendly user interface will be there for the sorting and grading process.

Table 4.1 Challenges in AI/ML applications.

Challenge	AI/ML Applications Affected	Elaboration of application-specific challenges	References
High computational requirements	Computer vision	Applications of computer vision face difficulties owing to the significant computational power needed. This is essential for jobs like automated inspection and swift pollutant identification, requiring sophisticated hardware capabilities.	Botero et al. (2021)
Overfitting	Computer vision	Overfitting poses a challenge in computer vision applications, since models may become excessively adapted to training data, hence undermining their generalizability. Maintaining an equilibrium in model complexity is essential for optimal performance.	Mahadevkar et al. (2022)
Data security concerns	IoT-enabled sensors	IoT-enabled sensors encounter issues with data security, especially in real-time monitoring. Ensuring data integrity and confidentiality is essential, necessitating strong encryption and authentication protocols.	Kaur et al. (2024)
Integration challenges	IoT-enabled sensors	The incorporation of IoT-enabled sensors into pre-existing systems presents difficulties. Compatibility challenges and smooth integration for ongoing monitoring necessitate careful preparation to guarantee effective risk mitigation strategies.	Parekh et al. (2024)
Privacy concerns	Blockchain integration	Blockchain integration raises privacy issues due to its intrinsic immutable and decentralized data storage. Achieving equilibrium between transparency and personal privacy is an essential aspect.	Politou et al. (2019)
Regulatory acceptance	Blockchain integration	Securing regulatory approval for blockchain integration in the food sector is a complex challenge. Surmounting legal and regulatory obstacles is essential for the extensive implementation of this technology.	André et al. (2021)

(Continued)

Table 4.1 (Continued)

Challenge	AI/ML Applications Affected	Elaboration of application-specific challenges	References
Interpretability challenges	Predictive modeling	Predictive modeling faces issues with interpretability. Comprehending and elucidating the decisions rendered by these models is essential for fostering trust and guaranteeing the practical applicability of data-driven decision-making.	Eloranta and Boman (2022)
Dependence on data	Predictive modeling	Predictive modeling encounters difficulties due to its reliance on data. Guaranteeing the accessibility of high-quality and diverse data is essential for the reliability and efficacy of models in enhancing food processing.	(Schmitt et al., 2020)

The section depicts the outcome of the preprocessing techniques used in the strawberry and brinjal images; it shows that the methods selected are effective in terms of quality enhancement of the images and accurate grading. The table represented elaborates the application-specific challenges to AI/ML applications in food processing.

4.3 Prediction Models in Measuring Color and Food Quality

This section discusses how diffusion-based time-temperature indicators (TTIs) can be used as cheap alternatives in assessing the quality of stored fruits. TTIs were, for this purpose, used to predict the condition of three fruit types: kiwifruit, strawberry, and mango during different storage temperatures of 5, 10, 15, and 20 °C. We established the relationship between color changes in TTIs and several measurements of fruit quality, including weight loss, soluble solids content, vitamin C content, titratable acidity, and antioxidant capacity. This was achieved through the calculation of Ea.

We demonstrated the impact of storage time and storage temperature on color changes of the TTIs and, consequently, on fruits. However, with time and increasing temperatures, we found the RGB – Red, Green, Blue is an additive color model in which colors are created by combining different intensities of red, green, and blue light, widely used in digital imaging, displays, cameras, and computer – of the TTIs to continuously drop. Further, fruit quality deterioration was accelerated at higher storage temperatures. Most importantly, Ea for color changes in the TTIs versus changes in fruit quality was not over 25 kJ/mol. This similarity brings forth an argument that TTIs might predict the quality of such fruits.

Based on these results, we developed predictive models using TTIs' RGB values. This aspect, therefore, presents the possibility of developing cost-effective quality monitoring solutions that can be applied in different areas of predicting food quality.

4.3.1 TTI Fruit Quality Model

Accordingly, we attempted to evaluate the predictions of the predictive models. We attached TTIs to the surface of fruits in three different conditions: 3 °C for 10 days, 12 °C for 8 days, and 20 °C for 10 days. We then evaluated the color change of the TTIs as well as fruit quality metrics. We used the following formula to calculate the discrepancies between the predictions:

$$\%\text{ error in prediction} = 100 \times |\text{ predicted values } - \text{ actual outcomes }| \\ /\text{actual outcomes}$$

4.4 Data Management and Its Importance

Data management is a key feature of food quality and safety, especially through the implementation of AI and ML technology. These technologies naturally provide real-time overviews and evaluations of food processing, thereby enhancing decision-making and risk management in food quality management through the use of AI and ML.

AI and ML are used to extract insights from the enormous amounts of data produced through IoT devices to enhance food quality estimations using spectral analysis and sensor fusion ("IoT, Big Data, and Artificial Intelligence in Agriculture and Food Industry," 2022).

Thus, big data analytics has the potential to analyze trends and factors that affect food quality beyond the traditional experimental method (Pollard et al., 2015).

Big data inclusion in food safety management allows dynamic risk assessment as it uses the inputs of various stages of the food supply chain to effectively manage foodborne diseases.

Blockchain technology benefits food safety by ensuring that data are kept secure and unalterable, which is required for preserving consumer trust.

4.5 Ethics and Safety

In recent years, AI and ML have been transferred to food quality and safety. This step promises much but presents an ethical challenge. As AI technologies improve, they will build on monitoring and managing food safety issues. However, this needs very careful attention to ethical concerns and public trust in it.

4.5.1 AI in Food Safety Surveillance

AI systems may continuously monitor food attributes, for example, temperature, to provide calculative safety and quality indices, thus increasing the handling of perishable commodities.

Big data and complex algorithms can be used to make food supply chains safer, ensuring safer food delivery and reduction of waste.

4.5.2 Ethical Issues

Decision-making by AI should be transparent; stakeholders must understand the driving algorithms and the implications that they have for food safety.

Responsible innovation of culture will reduce potential risks associated with AI, thus forming a need for ethical governance in AI implementation.

While there is much hope in AI for changing food safety, there is a need for ethical imperative and public trust to ensure its effective integration.

This revolution is to change the food process engineering industry in terms of its efficiency and safety plus better quality products. Its foundation is laid in innovation by robots, biotechnology, and control systems.

These enhance effectiveness and cut costs due to labor.

4.6 Adoption of Automation in Food Process Engineering

Industrial robots replace humans in the meat-processing parts, such as slaughtering and cutting, reducing muscle-straining injuries among workers, according to Kim et al. (2023).

Including sensors in the production for self-regulated process control, ensuring quality and safety of food products (Caldwell, 2013). SCADA systems assist in real-time monitoring of food processes and administer them appropriately.

4.6.1 Effect on Food Processing

Process Efficiency: Automation increases the rate at which food products are processed and significantly reduces food wastage—an aspect that greatly impacts the food processing industry with trillions wide.

Quality Support: Increased automation promotes better adherence to health and safety regulation requirements in food manufacturing.

Although automation offers a number of benefits, major barriers such as significant investments and the need for specific labor remain major deterrents to the widespread adoption of automation.

4.7 Automation, Economic Return, Food Manufacturing Sector, Increased Safety Assurance, and Quality of Products

The automation of the food manufacturing industry boosts the economic returns, safety assurance, and quality of products. Installation of modern technological equipment, such as robots, sensors, and controlled systems, ensures smooth operation, minimal labor costs, and little or no human error. Thus, productivity and profitability levels are increased (Caldwell, 2013).

Financial Benefits: Automation lowers overhead costs since resource use is maximized and waste is minimized. Higher production by automated systems will have higher yield and, consequently, higher revenues generated.

Safety Assured: Mechanized systems ensure uniform adherence to all safety guidelines for minimizing contamination risks.

Machine vision and online spectroscopy are technologies that enable instant quality and safety analyses.

Product Quality: Automation ensures that strict quality control measures are implemented, ensuring that products attain safety and quality standards.

Implementation of total quality management systems ensures that the reliability of the product is improved and that consumers have confidence in the particular product.

While bringing several advantages, automation raises significant issues of job loss and the need for experienced human personnel to monitor sophisticated technologies. Thus, it is very important to balance these factors in order to achieve sustainable growth in the food manufacturing sector.

4.8 AI, Robotics, and Automation for UN Sustainable Development Goals

Great opportunities and challenges will be created by AI, robotics, and automation to realize the UN Sustainable Development Goals (SDGs). The technologies will be able to augment data collection, optimize resource management, and provide for social fairness, together working toward how progress can be made.

4.8.1 Function of AI in Monitoring and Evaluation

Earth observation technology with AI will deliver dependable data on tracking the success of SDGs such as crop detection utilizing complex ML models.

Artificial neural networks could use fewer numbers of variables as input for the computation of SDG indices, thereby optimizing the process of appraisal.

4.8.2 Overcoming Global Challenges

AI4SG initiatives are those that try to come up with solutions to other issues of global concern, like poverty and climate change. They aim to achieve SDGs.

On doing a SWOT analysis, AI has been seen to help in the achievement of SDGs but is itself also fraught with some threats and weaknesses facing mitigation.

4.8.3 Ethical Considerations

The ethics of AI applications must be evaluated to ensure that they contribute positively toward sustainable development.

Robotic and automation technologies imply several possibilities for the good and progressive attainment of the UN SDGs through the application of AI. A few of the most relevant ways in which these technologies can contribute are:

1. No Poverty (SDG 1)
   Microfinance and Credit Scoring: AI can analyze data to assess creditworthiness for underserved populations, facilitating access to loans.
   Job Creation: Robotics can enhance productivity in agriculture and manufacturing, creating jobs in emerging sectors.
2. Zero Hunger (SDG 2)
   Precision Agriculture: AI-driven analytics can optimize crop yields by analyzing soil health, weather patterns, and pest outbreaks.
   Supply Chain Optimization: Automation can streamline food distribution, reducing waste and improving access to food in remote areas.
3. Good Health and Well-being (SDG 3)
   Telemedicine and Diagnostics: AI can assist in remote diagnostics and patient monitoring, improving healthcare access.
   Robotic Surgery: Robotics can enhance precision in surgical procedures, reducing recovery times and improving outcomes.
4. Quality Education (SDG 4)
   Personalized Learning: AI can tailor educational experiences to individual learning styles and paces, improving student engagement and outcomes.
   Access to Resources: Automation can help create and distribute educational materials, especially in remote areas.
5. Gender Equality (SDG 5)
   Empowering Women: AI tools can provide resources and information for women entrepreneurs, improving their business outcomes.

Addressing Bias: AI can be designed to identify and reduce gender biases in hiring and promotion processes.

6. Clean Water and Sanitation (SDG 6)
   Smart Water Management: AI can analyze usage patterns to optimize water distribution and detect leaks in real-time.
   Pollution Monitoring: Robotics can monitor water quality in various bodies, ensuring safe drinking water.
7. Affordable and Clean Energy (SDG 7)
   Energy Efficiency: AI can optimize energy use in buildings and industries, reducing consumption.
   Renewable Energy Integration: Robotics can automate the maintenance of renewable energy sources, enhancing their efficiency.
8. Decent Work and Economic Growth (SDG 8)
   Skills Training: AI can identify skill gaps and provide tailored training programs, helping workers adapt to automation.
   Entrepreneurship: Automation can lower barriers to entry for new businesses by streamlining processes.
9. Industry, Innovation, and Infrastructure (SDG 9)
   Smart Manufacturing: AI and robotics can improve production efficiency and sustainability through automation.
   Innovation Hubs: AI can facilitate collaboration between industries and research institutions, fostering innovation.
10. Sustainable Cities and Communities (SDG 11)
    Smart Transportation: AI can optimize traffic flow and public transport, reducing congestion and pollution.
    Urban Planning: Robotics can assist in the construction and maintenance of smart infrastructure.
11. Responsible Consumption and Production (SDG 12)
    Waste Management: AI can improve recycling processes and reduce waste through better sorting and analysis.
    Supply Chain Transparency: Automation can track product life cycles, promoting sustainable practices.
12. Climate Action (SDG 13)
    Climate Modeling: AI can enhance climate models, predict impacts, and help in policymaking.
    Disaster Response: Robotics can assist in disaster recovery efforts, providing support in hard-to-reach areas.
13. Partnerships for the Goals (SDG 17)
    Collaboration Platforms: AI can facilitate partnerships by matching organizations with similar goals and resources.

Data Sharing: Automation can streamline data collection and sharing, improving transparency and collaboration.

4.9 AI in Food Supply Chain, Tracing/ Warehouse/ Distribution

The integration of AI in the food supply chain significantly enhances tracing, warehousing, and distribution processes. AI technologies streamline operations, improve efficiency, and ensure food safety from production to consumption.

4.9.1 AI Applications in Food Supply Chain Management

1. Predictive Analytics: AI models, such as DL architectures, predict plant growth and optimize yield, aligning production with market demand and minimizing waste.
2. Logistics Optimization: AI facilitates real-time monitoring and management of logistics, addressing challenges like spoilage and inefficiencies in storage and transportation.
3. Digital Supply Chain Surveillance: AI enhances the tracking and analysis of supply chain data, improving risk management and supplier performance through frameworks like SDAR (Surveillance, Detection, Action, Response) (Bintrup et al., 2022).

4.9.2 Challenges and Considerations

Despite the benefits, implementing AI in food supply chains poses challenges, including data management complexities and ethical concerns related to surveillance and bias in AI algorithms (Rahmadhaningsih, 2022). Balancing technology with human oversight remains crucial for effective logistics planning.

4.10 Data Analysis, Discrete and Continuous Data on Classification and Regression

Classification and regression techniques are used in data analysis, and they describe discrete and continuous data methods. Classification techniques, like a decision tree, Bayesian network, and SVM are very useful and helpful for the prediction of group memberships from datasets. Regression methods focus more on modeling the kind of relationship between variables, which often require continuous data.

4.10.1 Classification Techniques Decision Trees

It represents a decision with possible outcomes graphically, which is intuitive for classification tasks (Phyu, 2009). Bayesian Methods: It makes use of pre-existing probabilities while classifying the data. These types of methods of classification apply to both types of variables—dichotomous and continuous variables. Continuous Data Handling: Continuous variables, when categorized, are sometimes less informative and lossy. Continuous data can be easily converted into categorical forms that may remove or blur hidden patterns; hence, regression models may lose some robustness. Although classification capabilities extend to many categories, certain types of data are best analyzed with regression methods. Hence, there is a symbiotic relationship between these two techniques for data analysis.

4.11 Case Studies

Detecting Food Adulteration Using AI

4.11.1 Introduction

Food adulteration stands for a practice where the quality of food is compromised for economic gain, and thus this production is hazardous to the health and safety of consumers. This section emphasizes the detection of adulterants to protect consumers and assure food integrity.

4.11.2 Objective

In the current section, an attempt has been made to systematically review the role of AI in detecting food adulteration in relation to different ML and DL techniques that were recently used for this purpose.

4.11.3 Methodology

The section encompasses 112 research articles published over the past 10 years that discuss several AI techniques and their applications in the detection of food adulteration. Only studies that followed a defined review protocol, such as specific research questions and inclusion criteria, were selected.

4.11.4 Summary of Key Findings

Methods: According to the section, 75% of the studies utilized ML and DL approaches to indicate their helpfulness in detecting food adulteration (Goyal et al., 2022).

Food Categories: Major food products analyzed include dairy items, fruits, and meat, with a focus on parameters that impact food quality.

Research Trends: The review reveals a rising trend in the publications of research in this area, mostly in conference proceedings, pointing towards an active interest in food safety technologies.

4.11.5 Challenges

The section claims that in the available literature, very scanty work is there on ML-based techniques specially for detecting food adulteration, hence calling for more exhaustive literature surveys to be done in the near future in this area.

It further elaborates the unavailability of annotated datasets to the public, thereby hindering the development of reliable detection systems.

The section concludes that AI has vast potential for food safety because it can efficiently work on the adulteration of the produced food. Figure 4.4 relates the number of methods involved in detecting the adulteration of food products. The article suggests the need for more research related to DL techniques, building mobile applications for real-time detection systems, and the incorporation of AI with sensor technologies to monitor food quality.

A summary of annotated datasets for food quality detection (Goyal et al., 2022) is presented in Table 4.2.

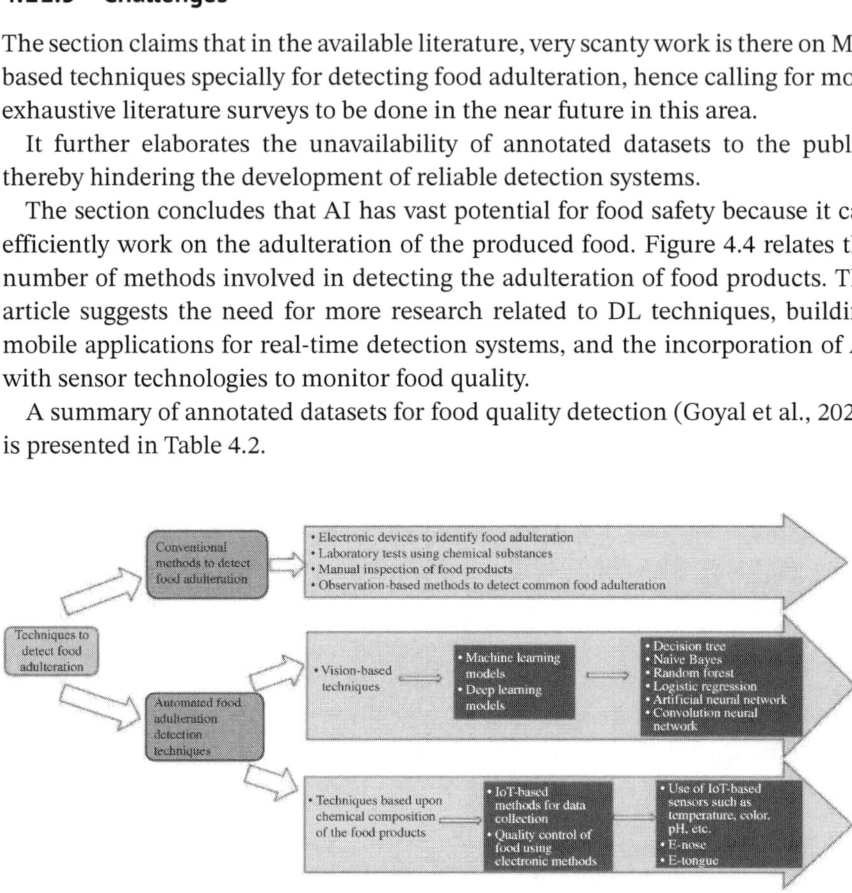

Figure 4.4 Various methods to detect food adulteration (Goyal et al. [2022]/Springer Nature).

Table 4.2 Various annotated datasets for food quality detection.

Dataset	Dataset details	Model and references	Link
Food-101	101 food categories, with 1000 images per category	Author defined, CNN-FOOD, FoodNet, DeepFood, Inception Module ResNet, ResNet-50, Inception V3, wide-slice residual networks (WISeR)	https://www.kaggle.com/dansbecker/food-101/home

Dataset	Dataset details	Model and references	Link
UECFood-256	256 Japanese food image Categories	DeepFood, DeepFoodCam, CNN-FOOD, ResNet, ResNet-50, Inception V3, Inception-v3 with FP-CNN, WISeR	http://foodcam.mobi/dataset256.html
UEC Food-100	Images of 100 kinds of Japanese food with at least 100 images per category	DeepFoodCam, DeepFood, CNN-FOOD, ResNet, Inception V3, MultiTaskCNN, Inception-v3 combined with FP-CNN, WISeR	http://foodcam.mobi/dataset100.html
Fruit-360	90,483 images of 131 categories of fruits and vegetables	CNN EfficientNet, CNN	https://www.kaggle.com/moltean/fruits
FIDS30	971 diverse images of 30 fruit categories with around 32 images in every class	CNN, CNN LeNet	https://www.vicos.si/Downloads/FIDS30

4.12 Review Questions

Q1 What is an activation function?

Q2 What is the difference between forward propagation and backward propagation?

Q3 Explain the working of perceptron?

Q4 How is CNN used in NLP?

Q5 How to choose the features for a neural network?

Bibliography

MH Alsharif, AH Kelechi, K Yahya, and SA Chaudhry. Machine learning algorithms for smart data analysis in internet of things environment: taxonomies and research trends. *Symmetry*, 12(1):88, 2020.

M André, J Margarida, H Garcia, and A Dante. Complexities of blockchain technology and distributed ledger technologies: a detailed inspection. *Fusion of Multidisciplinary Research, An International Journal*, 2(1):164–177, 2021.

UJ Botero, R Wilson, H Lu, MT Rahman, MA Mallaiyan, F Ganji, N Asadizanjani, MM Tehranipoor, DL Woodard, and D Forte. Hardware trust and assurance through reverse engineering: a tutorial and outlook from image analysis and machine learning perspectives. *ACM Journal on Emerging Technologies in Computing Systems (JETC)*, 17(4):1–53, 2021.

DG Caldwell (ed.). *Robotics and Automation in the Food Industry: Current and Future Technologies*. Elseiver. 2013.

KB Chhetri. Applications of artificial intelligence and machine learning in food quality control and safety assessment. *Food Engineering Reviews*, 16(1):1–21, 2024.

S Eloranta and M Boman. Predictive models for clinical decision making: deep dives in practical machine learning. *Journal of Internal Medicine*, 292(2):278–295, 2022.

K Goyal, P Kumar, and K Verma. Food adulteration detection using artificial intelligence: a systematic review. *Archives of Computational Methods in Engineering*, 29(1):397–426, 2022.

K Kaur, A Kaur, Y Gulzar, and V Gandhi. Unveiling the core of iot: comprehensive review on data security challenges and mitigation strategies. *Frontiers in Computer Science*, 6:1420680, 2024.

SV Mahadevkar, B Khemani, S Patil, K Kotecha, DR Vora, A Abraham, and LA Gabralla. A review on machine learning styles in computer vision—techniques and future directions. *IEEE Access*, 10:107293–107329, 2022.

R Parekh, B Sedhom, SK Padmanaban, and AA Eladl. A review of iot-enabled smart energy hub systems: rising, applications, challenges, and future prospects, 2024.

E Politou, F Casino, E Alepis, and C Patsakis. Patsakis Constantinos. Blockchain mutability: challenges and proposed solutions. *IEEE Transactions on Emerging Topics in Computing*, 9(4):1972–1986, 2019.

S Pollard, H Namazi, and R Khaksar. Big data applications in food safety and quality, 2015. https://doi.org/10.1016/B978-0-08-100596-5.21839-8.

J Schmitt, J Bönig, T Borggräfe, G Beitinger, and J Deuse. Predictive model-based quality inspection using machine learning and edge cloud computing. *Advanced Engineering Informatics*, 45:101101, 2020.

5

Application of Robotics in Processing Meat, Fish, and Poultry

After reading this chapter, you should be able to:

- Robotics and Computer Vision: Integration of fish processing to improve efficiency is being done by robotics together with computer vision. Sophisticated algorithms for quality checking while avoiding human errors will be accomplished by sorting and inspection of fishes.
- Robotic Manipulators with End-effector: Robotic manipulators, along with the specifically developed end-effector, improve speed in the process of handling and cutting meat, poultry, and fish, whereas they also encourage packaging within them. Thus, all the processes aid to accomplish efficiency and standard quality.
- Predictive Maintenance and Smart Technologies: Predictive maintenance is integrated with artificial intelligence (AI) and Internet of Things (IoT) to optimize food processing operations and improve the reliability of equipment. Case studies reflect significant improvements in efficiency of production and streamlined processes.

5.1 Introduction

An effector is any robot or robotic device that has an effect on the environment.

There are two fundamental categories of effectors based on their effects on the environment:

1. Locomotion: This occurs when the effector itself causes motion in the environment. This category of effector is called mobile robot.

Food Engineering Automation with Robotics and AI, First Edition. Abir Chakravorty.

2. Manipulation: This occurs when the effector causes motion in objects around it.

A mobile manipulator is a combination of a fixed manipulator and a mobile robot. A robotic manipulator is a software-controlled mechanical device that employs sensors to guide one or more end effectors in accomplishing programmed motion in a given workspace to manipulate physical objects.

Manipulators can also be classified as: (a) parallel manipulator (b) serial manipulator.

In a serial manipulator, all the bodies are arranged in an open chain configuration to form a shape, and the end effector is connected to the base in an open chain configuration.

A parallel manipulator is similar to a parallel open chain manipulator, which, when two or more open chain manipulators are connected, forms a closed loop chain.

5.2 Application of Robotics in Processing Meat, Fish, and Poultry

For instance, Aly et al. (2023) and Romanov et al. (2022) in their paper explain an advanced automated meat processing system with three key stages for such tasks of deboning. It begins with the perception phase, where sensory inputs are received by the system in its evaluation of the shape and texture of the meat before it moves into judgment based on comparison of the initial state and the desired state according to accreted knowledge. Cutting with proper technique and tools. The final step is cutting, utilizing the proper technique and tools. The abstract also focuses on the state and development of automation in meat processing, having elaborated on the available technologies to guide the cutting process and through which these systems adjust to the variability and unpredictability of meat.

In like manner, the processing of pork meat has become significantly automated. Of course, the robing of pork meat, one of the most important actions in any type of meat processing, can be summarized with RoBUTCHER, a concept subsidized by European funds, which develops autonomous robotic cells known as meat factory cells, as Pame et al. (2023) report. The MFC concept is supposed to replace the linear configuration of the traditional production line with a more efficient cell-based setting.

The MFC approach therefore proposes a different workflow from the historical traditional slaughtering lines; typically in slaughterhouses (Haidegger et al., 2024).

The process begins at the slaughtering stage, which is followed by dehairing, evisceration, and subsequently halving of the carcass. Later, it is segmented into primary and tertiary cuts with a procedure for chilling. The MFC method proposes a change in this workflow since it suggests that the primary cuts should undergo hot boning directly after dehairing and followed by removal of internal organs, according to Sødring and others, in 2022.

The MFC system of Mason et al. is based on two robotic arms, one for manipulation and grasping and the other for cutting. CHU secures and positions the carcass during this operation. Variance in carcasses is adapted by integrating a combination of CT data with real-time 3D visualization and expert-derived cutting data into neural networks to create accurate cutting trajectories.

Using the RGB-D camera technology, imaging can define more precisely the different parts of the carcass and their critical features. De Mederios Esper et al. and Mason et al. explain that this graphical data in 2021 will be ML algorithms that try to calculate the ideal grip points as well as cutting routes. Along with these lines, various sensing techniques are discussed, informing the "smart knife" cutting path in areas where there is inadequate visual support, including electrical impedance, force sensing, optical and electromagnetic spectroscopy, according to Mason (Romanov et al., 2022).

Of these, only optical and electromagnetic spectroscopy were found suitable for further use in the automation of meat processing. Further experiments with electromagnetic (EM) spectroscopy yielded promising results in the direction of smart knives having almost nil contact and depth detection errors to move potentially toward development in the improvement of the automation of meat processing. Besides, significant progress in terms of autonomous robotics for meat processing has been introduced with the development of the MFC concept. It has brought with it various innovative technologies and methods that could probably revolutionize the industry.

Augmented reality (AR) and virtual reality (VR), two different visual technologies, may be used separately or in combination as a command interface for cooperative robots, or cobots. Overlaid digital information augments human sensory experiences or provides additional data. The most common method of accessing AR is with a smartphone, headset, or projecting digital images onto real-world settings. AR systems typically connect to sensors or cameras to supply the data that will be displayed. However, in the last few years, the promise of AR technology has been highly spectacular in meat processing. For example, the research paper published by Christensen & Engell-Nørregård in 2016 showed the potential to facilitate a better pork belly trimming and cutting process. The premise of the experiment was to develop three varieties of pork belly, which were differentiated from different raw materials and varying weights, as well as percentages of tissues. Each raw material was CT-scanned to provide color-coded maps: these maps segmented tissue according to density (fat, meat, bone) and showed the thickness of fat in other colors. Staff were helped by an overlay of fat coverage in color, warnings of the recipe ID and the changes needed and marks of incision lines. Though there are some challenges, the application of AR technology yields a better quality product compared to the more conventional manual method. This new practice gives operators a kind of "x-ray vision" to the carcass as to minimize costly errors that abound in the field.

On the other hand, VR presents a fully immersive digital experience, literally shifting users to a custom virtual landscape from a headset. This high technology has a major role in testing the machinery and carrying out the operation safely. It has been used in the red meat processing industry for training purposes. VR enables workers to develop their safety measurements and responses to potential danger virtually by simulating the risky environments of slaughtering rooms; no exposure is made to actual danger. This further empowers workers to sharpen their ability to cut little details individually, with minimal ongoing supervision and the wastage of traditional ways of training.

In this paper, the deformation-measuring estimation method proposed depends on the estimation of the clamping force. Ribbed belts - steel cord-reinforced polyurethane belts, width 16 mm on one side of each finger and 10 mm on the other side of each finger, 260 mm long, T5 profile. It may be assumed that the bulk of the transmitted force in the belts is borne by the cords, so the stress–strain curves of such belts should approximate the mechanical behavior of the underlying steel wires, which is, to be approximately linear. The tensile testing of the belts using a tensile testing machine (H & P WPM Retrofit 400 KN, by Quantify, Tienen, Belgium, 5 independent measurements) confirmed the linearity theory with $R > 0.99$ for all belts tested in the elastic zone. Assuming that the polyurethane "matrix material" and the cords constitute a valid composite, the deformation of the belts must be proportional directly to the tensile stress applied to them, according to Hooke's law.

$$\sigma = E \cdot \epsilon, \tag{5.1}$$

where σ is tensile stress, and E represents the elastic modulus of the wires in steel, and ε represents mechanical strain. Hooke's law may be restated in terms of the tension pull (F) in terms of the extensional stretch of the belts (Δl) by using the effective cross-area (i.e. the sum of the cord cross-areas, ΣA_{cords}) as well as the original effective length of the belts-again, here denoted simply as l, which denotes the distance from the shaft:

$$F = \sum A_{\text{cords}} \cdot E \cdot 1/l \cdot \Delta l. \tag{5.2}$$

As the gripper's actuating mechanism closes to pick up an item, the side of the belt drive that is taut, or "active," sees the center of the gripper assembly. This error, although present, can be fully compensated for by changes within the controller software. The limitation is more fundamental and due to the way in which the strain gauges are connected. These are usually precision devices for metal elongation, but they do not respond much beyond a deformation threshold of about 5% and are therefore less useful than in materials that undergo higher deformation rates such as polymers. However, this restriction is transcended in the case of steel cord-reinforced polyurethane belts because the polyurethane undergoes large strains, which thus enhances the elongation potential, such that strain gauges can be applied directly on the polyurethane matrix for correct elongation measurements.

In the measurement setup, we used four similar strain gauges arranged in a Wheatstone bridge configuration. This arrangement allows considerable signal amplification and thermal compensation. We use the "full bridge stain gauges" type, which consists of four strain gauges arranged on a single plane in a Wheatstone bridge layout whereby two gauges are aligned perpendicular to the other two. This arrangement means that whereas two gauges of the deformation are parallel to belt movement, perpendicular gauges supply mainly compensate against thermal effects and suffer no deformation. Preliminary testing indicated large amounts of noise relative to the signal; thus, it seemed that the precision of the approach was going to be acutely sensitive to how well fixed the gauges might be. This led to the development of an alternative force estimation method for clamping forces, based on measuring the angular positions of the shafts within the mechanism. Like the previous approach above, this further approach also follows the linear elastic model of the belts, given by Eq. (5.1). In this further approach, however, elongation of the belt is calculated using the angular difference between the pair of belt-gears associated with each finger. When energized, the rotating motor causes the input gears of the belt drive assembly to rotate; simultaneously, the output gears will rotate as well. However, due to the elastic nature of the belts, the rotational velocity of the output gears is offset by a decrease that is proportionate to the elongation of the belts.

$$\begin{aligned} \phi_2 &= \phi_1 - \Delta l/r \\ \Delta l &= (\phi_1 - \phi_2) * r \end{aligned} \tag{5.3}$$

where Δl is the elongation, r is the radius of the gears (20 mm), and ϕ_1 and ϕ_2 are the angular locations of the input and output gears, respectively. (3) implies that the extension, and hence the clamping force according to (2), is directly proportional to the difference between the two angular positions of the gears driven by a belt, and depends neither on their angular velocity nor on the actual angular position. The angular positions of the shafts are measured using magnetic encoders (As5601) whose magnets are attached to the shafts. The 12-bit contact-less sensor has about 0.0015 rad of resolution, translating to a stretch of 30 μm in the belt. Another application of magnetic encoders is as a means for determining the angles of the grasping fingers, which can report abnormal grasping or even be reported as defective if they detect a heavy clamping force before the "closed" finger position is actually reached.

5.2.1 Utilizing Computer Vision and Robotics for Fish Processing

A new method for detecting fish cutting points in fish processing has been developed, focusing on target weight with AI machine vision. First, a structure of the fish is created as a 3D model using image processing techniques, then 3D feature accumulation is performed using partitioned random sample consensus. From this initial point, it then evolves into designing a neural network model that predicts where, on the fish, cutting would be the best to attain an instructed target weight. The training of this model would be based on the extracted 3D features and the actual weight data. One of the key advantages of this approach is that it would be possible to cut fish precisely based on the predictions of the model directly. This method was benchmarked on the difference in actual weight of cut fish segments versus the target weight, with an average error of around 3%.

To calculate the weight of a fish, the volume is simply multiplied by its density. However, as outlined above, there are differences in densities throughout various parts of the fish, meaning that there would not be a uniform density value that could be used for the whole fish. This problem is addressed by the proposed method through its adoption of a machine-learning method for weight forecasting of fish. This is basically carried out by making an analysis of the relationship between the obtained 3D data, which embraces volume, length, the longest axis, and the shortest axis from images, with the actual measurement of weight (see Figures 5.1 and 5.2).

5.2.1.1 Preprocessing Phase

Extracting 3D forms requires visual data from at least two angles. For this purpose, two cameras are placed in the preprocessing stage to capture images of the fish from both a top-down and a front perspective at the same time. Then, these images

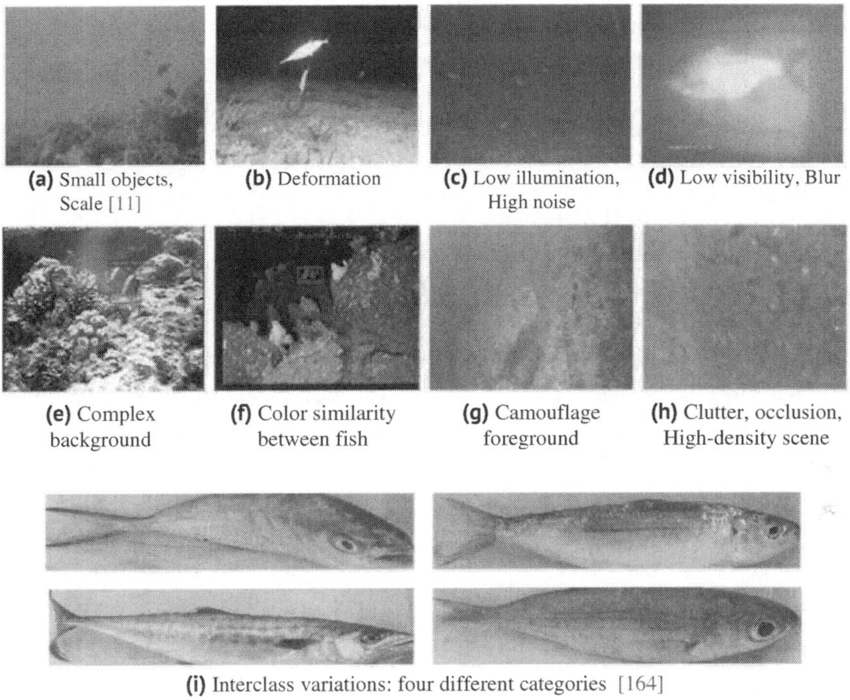

Figure 5.1 Various categories of challenges for fish detection. Most images are from Fish4 and NOAA (Yang et al. [2021]/Springer Nature).

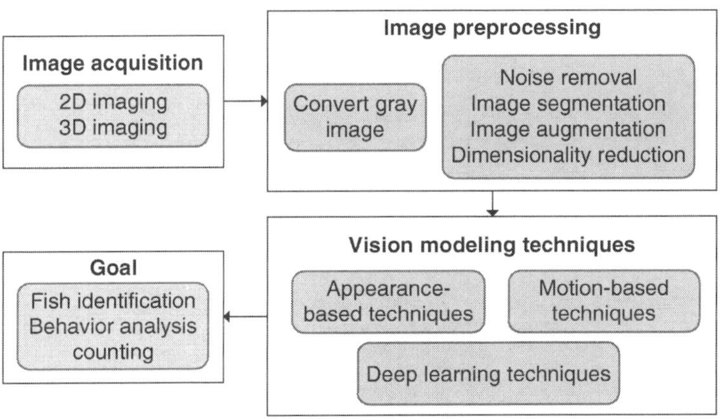

Figure 5.2 Block diagram of computer vision in fish detection (Jang and Seo [2023]/Elsevier).

undergo numerous filtering techniques that refine the visual information. These involve the conversion of images to grayscale for reduced computation, the use of high pass filters to guarantee an accurate representation of edges, and the utilization of masking, Gaussian blur, and binarization techniques to provide noise reduction (Chuang et al., 2014; Tharwat et al., 2018; Hernández-Serna et al., 2014). All these processes significantly enhance the accuracy of detecting the contours of the fish.

5.2.1.2 2D Modeling Phase

The process was to create a silhouette of a fish using images taken from both top-down and frontal perspectives. RANSAC, which is fast computing and effective for the robust handling of noise, was used in outlining the fish. To create a 2D silhouette, the first step was to select three random points from the image of the fish that has gone through processing. A quadratic model was formulated to connect these points. The number of points within a distance "t" of the model was counted to deduce its adequacy to be used as an outline of representation. The more such a count, the nearer the model was to the actual outline. The process continues with choosing points and determining models, of which the model giving the maximum number of proximity counts turns out to be the final outline.

Equation (5.4) shows the probability of m randomly selected dots by RANSAC being inliers for an appropriate model.

$$p = 1 - (1 - a^m)^N \tag{5.4}$$

In this context, "m" represents the number of dots selected randomly, "a" represents the ratio of inliers to the total number of dots, and "p" represents the probability that all "m" randomly chosen dots through the random dot selection process repeated N times are inliers. Equation (5.5) is used to calculate the minimum number of repetitions necessary to select "an appropriate model" through the random dot selection process.

$$N = \frac{\log(1 - p)}{\log(1 - a^m)} \tag{5.5}$$

$$\begin{aligned} \text{Area}_p &= a_p b_p \pi \\ \text{Vol}(l) &= \int_0^{l-1} \text{Area}_x dx \end{aligned} \tag{5.6}$$

It is considered that the artificial neural network (ANN) is the training method for an ML-based approach to correlation between 3D feature data and weight. The pretrained algorithm predicts the cutting point of the fish based on the first reception of the input fish's 3D feature information. This step is further lengthened for l as from minimum tail direction up to maximum head direction till when the computed weight pw = tw. Upon equaling of pw = tw, then l1 will act as the initial point for cutting off. Provided pw = 2.tw then l2 will act as a secondary point for cutting. The length of the nth cutting point is computed whenever pw equals n.tw. For each fish cutting point under the condition of tw held at 30 g.

One hundred fifty Pacific sauries are used to test the validity of the proposed approach. an example of 30 fish cut into different lengths. The rest 120 Pacific sauries are used for the method test. Of these, 60 were in one test focusing on a fish weight of 40 g, and the other 60 in a separate test focusing on a fish weight of 60 g. The experiment started by placing the target Pacific saury and its target weight inside a specially designed capture box, which always places the fish in a fixed position within the box. It is equipped with two cameras and a light source. These will give the capture of two perspectives of the fish, from above and from the frontal view. After the image of the fish, an initial processing and modeling phase gives a 3D representation of the Pacific saury, which, due to the complex 3D data, allows the trained model to forecast and depict it in detail.

The Mean Absolute Error (MAE), Mean Relative Error (MRE), and Maximum Error were used to test the effectiveness of the proposed method by analyzing the data. MAE, MRE, and the Maximum Error were calculated. In this case, *haty* is the target weight, y_k is the measured weight of the k-th sample, and n is the number of samples.

$$MAE = \frac{1}{n}\sum_{k=1}^{n} |\hat{y} - y_k|$$

$$MRE = \frac{MAE}{\hat{y}} \times 100 \tag{5.7}$$

$$\text{MaxError} = \text{Max}\,(|\hat{y} - y_k|)$$

The text presents the design and development of the method for fish cutting points based on weight. This method is divided into four distinguishable stages: preprocessing, generation of the 2D representation of the fish, development of 3D representation generation, and machine learning (ML)-based prediction for the identification of the cutting points from the weight of the fish. A general overview of this method is depicted in Figure 5.1. The first operation done is that of an image-processing filter to better prepare and optimize images from the fish so that they can be represented in 2D for the construction of a quadratic equation model. By the 2D outline of the fish, its 3D shape is then constructed with all dimensions concerning volume, length, and thickness. It finally predicts cutting points from its 3D data. Figure 5.2 represents the flowchart for image acquisition, image preprocessing, and vision modeling techniques for fish identification.

The phase is incorporated into the cutting point prediction model, which utilizes ML and is influenced by the desired weight. This model is designed to forecast each cutting point based on that specific weight target. To achieve this, the model was previously trained with target weight data, taking into account the fish's volume, length, and thickness.

5.2.1.2.1 Appearance-based Techniques

The size, shape, and color change with the growth of the fish. Methods that are appearance based have an advantage over motion-based in environments where the movements of the fish are minor. This means that appearance-based methods would work really well but require lots of pre-existing knowledge. As can be seen in the figure below, these all fit into three broad categories of processes: generating candidate regions of interest, extracting distinguishing features, and classifying or regressing those features. For singles, the process reduces to just extracting and classifying features. However, since there are several fish that lie within a scene, it first detects features from candidate regions developed by sliding a window technique and then employs a classifier in order to classify those regions. The subsequent section puts attention to major aspects such as feature extraction and classification because they are critical to the correctness of fish detection. Our elaboration mainly includes analysis based on crucial characteristics: low-level as well as local features. These include color, texture, and shape, among others, and techniques include scale invariant feature transform (SIFT), speeded up robust features (SURF), Haar-like features, edges, and binarized normed gradients (BNG). All of them have advantages and disadvantages in each, and when a set of features is used, even the power to differentiate more objects or patterns becomes strengthened.

Color and Texture Feature: Feature extraction based on color and texture is one of the fundamental techniques in image analysis. Many studies have been conducted on color feature extraction techniques and texture feature extraction techniques. The uniqueness of species makes it challenging to identify the fish from the images, but requires diversity in video qualities and lighting conditions. To overcome the drawbacks of the dependency on colors alone, it has been seen that the combination of color and texture features is effective for overcoming this limitation. The covariance-based model is, in fact, a very sophisticated method that allows the feature integration and establishment of structural relationships. However, it has major drawbacks when trying to separate the fish from the background—especially if camouflage causes a mix of both textures and color similarities.

Texture Identification with the Help of Local Binary Pattern: The local binary pattern (LBP) approach has gained widespread popularity for its effectiveness in texture description, outshining other techniques in its robustness against pose and illumination changes. Typically, the LBP characteristic is effectively paired with the HOG feature, enhancing performance in fish detection by mitigating issues like occlusions. However, a notable drawback of the LBP method is its vulnerability to noise.

Normed Gradients (NG) Feature: Swift and precise fish localization is crucial for effective fish detection. This technique is designed to efficiently identify all potential fish locations, thereby speeding up the evaluation process and decreasing spatial complexity while maintaining high accuracy and detection rates. Research indicates that BNG's performance is 2000 times quicker than traditional objectness measurement methods and selective search strategies.

Haar-like Feature: Haar-like characteristics, edge are simple and inexpensive computationally. They capture the variations in gray levels within an image. These are, in principle, designed to detect changes in intensity in horizontal, vertical, and diagonal directions. Their effectiveness, however, heavily depends on the orientation information of the object being detected. Although they offer real-time detection capabilities, the training stage for these algorithms is much more time-consuming compared to the LBP and HOG detector algorithms.

SIFT and SURF Feature: The SIFT is designed to be resistant to changes in rotation, scale, and illumination levels, making it a highly stable feature for local analysis. By integrating SIFT with SURF, enhanced localized data extraction was achieved in the context of fish classification.

HOG Feature: The Histogram of Oriented Gradients (HOG) feature functions as a descriptor in object detection tasks. It generates features by assessing the distribution of gradient directions and intensities within a specific region of the image. Recognized as an excellent local feature, it excels at capturing objects' edge or shape details. Consequently, it has demonstrated outstanding effectiveness in detecting both single objects and multiple targets.

Weber's Local Descriptor (WLD): WLD offers notable robustness against noise and variations in lighting when compared to the SIFT feature, and operates at a quicker pace due to its singular adjustable parameter. Even though incorporating WLD features with color moments post-normalization can enhance the accuracy of the corresponding model, it's noteworthy that the WLD feature's reliability gets compromised by alterations in image scale, rotation, and shifting.

DPM Detector: The ability to track multiple entities in an unstructured underwater environment with a high detection rate while keeping computational complexity low is crucial. The DPM detector, specifically the proposed DMK, takes the camera's ego-motion into account and has demonstrated reliable performance. Where possible, appearance-based DPM models have commonly been used along with tracking techniques to address the issues of detection reliability due to changes in fish appearance. The employment of tracking algorithms based on adaptive multiappearance models and strategies enables alterations in the varied changes in fish appearance, thus yielding commendable results. In contrast, the performance of this approach may fluctuate with changes in the environment.

Fish Head Parts Detector: During the studies based on tracking, it has already been proven that the fish head is the most resistant part of the body because of its specific color and shape. For these reasons, ellipse fitting was often used to outline the area of the fish head in many studies, after which information about the features of this area can be obtained via detection and tracking processes. Another important technique developed in order to combat issues about position updating of the fish location is trajectory linking. Nevertheless, issues related to tracking problems arise where other fish cover the view of the head of the fish, thus loss of data about the position of the fish.

Classification/Regression: A classifier that is trained to distinguish fish and non-fish based on features extracted from image sequences is widely used for the classification of the appearances of fish with high accuracy. Up till now, numerous classification approaches, such as SVM, kNN, ANN, AdaBoost, RF, and SRC, have been commonly applied to classification problems. Among these algorithms, SVM and ANN were respectively utilized for regression to predict the count of fish fry.

Support Vector Machine (SVM): This system is widely acknowledged for its ability to solve the problem of binary classification. Using a range of kernel functions, it maps nonlinear problems into the linear domain and hence their application in multi-class problems and over-fitting. Most recent studies have highlighted its superiority in the identification of multiclass fish types in both small and large data set cases. However, determination of suitable kernel function is highly expertise-constrained and directly proportional to the volume of data that increases computational complexity.

Artificial Neural Network (ANN): The ANN with its complex neuron-based network architecture computes output through multiple learning and training cycles with backpropagation assisting in adjusting the weights to minimize the error rates. ANN surpasses all the classical approaches as it can handle fuzzy, random, and nonlinear data. ANNs have been utilized for species identification and fish count estimates when appearance characteristics are concerned. Conversely, due to fewer layers of ANNs, they may be less effective than other algorithms.

K-nearest Neighbors (KNN): KNN is one of the simplest classification algorithms which is commonly employed for fish species classification. It categorizes the test data based on the distance between feature points and a specific category centroid. Distances between several feature points are calculated. KNN is very effective for small-sized datasets with moderate accuracy. This approach can be expensive with high sizes in terms of computational cost, since the value of k needs to be chosen to achieve good performance.

Naive Bayes: The Naive Bayes classifier is mostly overlooked, and its algorithms are not much prioritized due to being very complex. It has been used only to classify nine fish species from 282 images for recognition, therefore, making its applicability confined within broader or more complex classification tasks.

AdaBoost Classifier: AdaBoost algorithm is an efficient ensemble technique that combines multiple weak classifiers to generate a strong classifier. The performance of AdaBoost algorithm has been well proven in a number of conditions by updating the weights for both the classifiers and the training samples. In conclusion, AdaBoost has poor performance when the number of individual classifiers used is less. Hidden Markov Model (HMM). HMM is a probabilistic model that well infers unobserved states from observed sequences. This has been so successful in the analysis of fish behavior, with especially the detection and tracking of their movements in underwater videos. The HMM, however, becomes computationally expensive and difficult to use since it is specifically dependent on parameter and the length of data trajectories.

Other Classifiers: Besides the above classifiers, there are many other algorithms used for fish classification. These classifiers include Random Forest, SRC, which is sparse representation-based classification, Copula Theory, Hierarchical Partial, Fisher Discriminant, Balance Enforced Optimized Tree with Reject Option, Short-time Fourier Transform and Discriminant Analysis, and Graph Embedding Discriminant.

Motion-based Techniques: Techniques for tracking in which detection and location are considered in video sequences have been developed to analyze fish behavior. Motion-based methods have been developed and widely used for the study of both individual and collective fish behavior in order to create a dynamic background model reflective of fish movements. Though they differ between them, all three primary categories for these methods are generally frame differencing, background subtraction, and optical flow.

Frame Differencing: In that case, it separates background from foreground by subtracting the pixel values across the consecutive frames in the video. It's typically used for the detection of moving fish but is really hard because of quite a lot of problems: its sensitivity to lighting changes and difficulty in the immediate updating of the background image.

Background Subtraction: Background subtraction differs the current image from the background to separate regions in motion. Among these algorithms for foreground target extraction of such fish to track and analyze their behavior, we have the Gaussian Mixture Model (GMM), Kernel Density Estimation (KDE), and Visual Background Extractors (ViBe). GMM can fail to capture slow-moving objects and demands considerable computational power. Almost all the papers use the GMM method for fish detection using Kalman filter for tracking analysis of behavior. However, the extraction of pure background frames from freely available underwater videos is much harder, as there is a potential chance that fish will appear at any instance. To overcome these challenges, scientists combined GMM with other techniques for better detection efficiency. For example, Moving Average Detection, combined with GMM, achieves a relatively good balance between processing speed and accuracy. A Speed-up Background Modeling technique has also been added using a nonparametric histogram concept in order to enhance the efficiencies of GMM and, consequently, decrease tracking speeds.

Combining Gaussian mixture models with pixel-wise posteriors addressed the issue of detecting stationary or semi-mobile fish. It evades the weaknesses of GMM as well as the addition of an opening operation. The Vibe detection algorithm has the highest-reported performance in Fish-4Knowledge with an F-score of 84.3%. Its background updating mechanism makes it computationally quick and noise-resilient. To detect fish in underwater video characterized by noise, murkiness, low visibility, and changing illumination, Shevchenko et al. used the following techniques of background subtraction: ViBe, adaptive GMM, and KDE. Each technique has different advantages on a set of data. According to their authors, adaptive GMM is highly skilled in motion detection if there is commonality with the background to be detected. If sensitivity to noise is required, the highest level is obtained by ViBe. KDE balances both motion detection and noise resistance.

But there are a few inherent limitations with the motion-based methods, as they depend on continuous movement of the fish, motion data, and stationary cameras that are adversely affected by lighting changes, dynamic backgrounds, camera stability, and obstructions. Besides this, detection and monitoring of fish with mobile cameras in challenging terrains are particularly challenging owing to the inefficiency of background modeling techniques.

Optical flow has been studied as a way of capturing motion information, such as the velocity vectors of frame pixels, in order to monitor fish group behavior without having to track individual fish. Zhao et al. have applied an optical flow technique upgraded with a particle advection scheme and a kinetic energy model for detecting comprehensive fish behavior, that is, gathering and scattering in recirculating aquaculture systems (RAS). Despite its merits, this approach had some disadvantages, such as higher missed detection rates and high susceptibility to disturbances by water surface fluctuations, especially during strong emergent behaviors. Optical flow techniques are also notorious for being quite sensitive to small changes in hyperparameters, which may induce network instability.

Deep learning techniques seem promising as well. Traditional ML algorithms coupled with appearance-based features can be used for small datasets, where it highly involves human effort in developing the features. Thus, there is a bit of uncertainty. Deep learning, on the other hand, drags out both basic and intricate features from a large number of datasets effectively, thus allowing features of fish to be accurately identified from images, which works spot on even when variations are on lighting, positioning, and orientation, thus making the application thoroughly suitable for computer vision modeling. In addition, deep learning appears to be able to process the issues in object localization tasks and, therefore, provides answers to overcome the previous methods that were confined in their capacity due to the limitations of motion data. Among the best known are convolutional neural network (CNN), generative adversarial network, and long short-term memory networks, which have been widely used for fish recognition tasks. More precisely, they have been outstandingly adapted to the classification of various fish species as well as associated tasks of detection and localization. The quality and efficiency with which GANs produce output make them less susceptible to the limitations associated with the limited availability of labeled data. The transfer of learning with CNN has been found useful to enrich samples of data for training. CNNs on fish tracking enable various target tracking inside an enclosure incorporating head characteristics from the fish to address problems in variance of appearance, intricate movements, and frequent obstructions. Despite these advancements, such technologies are not commonly used in fish counting and behavioral analysis; there are a few instances where CNNs have been employed for counting purposes, along with extracting behavioral patterns from movement data as well as the usage of LSTMs in identifying unusual behaviors.

Object Detection Framework

If the image contains multiple objects of interest, then traditional detection methods that rely on exhaustive searching over many window configurations result in an enormous computational requirement due to the proliferation of unnecessary windows. Hence, there is a need for an efficient detection framework to minimize computational overheads. In this case, frameworks for fish detection via deep learning are discussed, distinguishing between two-stage (region-based) and one-stage (regression-based) models, as they form the most important aspects for the improvement of efficiency in the detection process.

The region-based model uses selective search in region proposal generation, efficiently solving the problem with windows that prove to be superfluous. It means that this approach improves the process by reducing the time complexity and windows in opposition to the traditional sliding window techniques. Those regions are then applied in conjunction with a feature-extracting CNN and an SVM classifier to make the final classification of which species is within an image. The R-CNN architecture built up to Fast R-CNN introduces an optimization step. In fast R-CNN, a CNN goes through the entire image and produces a feature map from an image. Region proposals are then obtained by the identification of corresponding patches from this feature map to derive patches from it and shorten the overall detection process.

Generation of region proposals is a time-consuming critical bottleneck in accelerating the process and handicaps the efficiency of the detection speed of an algorithm. Unlike traditional methods using selective search for region proposal and SVM classifiers, the novel approach adopted by Faster R-CNN uses Region Proposal Network and CNN and allows end-to-end detection. It is still quite challenging to reach the speed levels for real-time object detection in video.

Region-based methods are now being utilized for the video-based application of detection in underwater scenes of fish. Jäger et al. used a high-end R-CNN algorithm whose method of object proposal generation does not apply selective search but rather background subtraction to detect the fish. However, this method still had large false detections with an average precision of 40.66%. While the Faster R-CNN implementations heavily reported a Mean Average Precision improvement over DPM, R-CNN, and Fast R-CNN, tripling the detection speed of Fast R-CNN, efforts to counterbalance the tedious nature of hand-labeling huge video datasets encompass combining object proposals with optical flow and a segment-based unsupervised detection technique. This approach minimized overlapping windows on luminance-only underwater fish videos by NOAA and increased mean precision of detection by 10% with respect to nonfusion methods. It was also at this point that significant computational and time requirements for training CNNs led to using pretrained models in order to avoid sample size constraints and optimize speed. The performance that the pretrained PVANet on the SeaCLEF 2016 dataset attained stands at a mean average precision (mAP) of 89.95% (Li et al., 2017). It, therefore, outperforms faster R-CNN. Similarly, a model of a Faster R-CNN fitted with three VGG16 classification networks set up to perform small, medium, and large size outperformed others with an 82.4% mAP on fish data from the Remote Underwater Video Stations (RUVS), establishing a new standard on a surf species dataset developed by the researchers. Additionally, a cascaded ensemble structure in which components are connected using an LSTM network with an attention mechanism proved to yield better accuracy, correct prediction errors, and be resistant to distortion of images. However, the algorithms detected in the series of R-CNN algorithms take a relatively longer time to process compared to real-time detection.

YOLO algorithm improved detection speed by introducing a one-stage, regression-based algorithm that splits input images into grids. This process produces a sequence of grids that are subjected to box detection and classification, making the process simple. YOLO is faster than two-stage methods because of its lightweight backbone network and few candidate regions for predictions. The accuracy metrics of YOLO surpass the combination of HOG features with SVM classifiers: sensitivity is 93%, specificity is 62%, IOU is 0.634, and it takes 16.7 frames per second. The proposed approach can outperform other methods in scenarios that are challenging due to noise, dim light, or haze.

The overall performance of YOLOv3 is promising by obtaining a mean average precision of 53.9 in the MHK and hydropower underwater fish videos dataset. In this way, this result establishes the model to be effective in discriminating between fish and other nontarget features, such as bubbles and clutter. Class imbalance is one of the major drawbacks of the one-stage detection approach of YOLO, where candidates generated are highly dominated by bac.

5.2.1.3 Design of the Robotic Fish with Specific Details About the Structure and Components

Computer vision applications are more recently implemented in aquaculture for the execution of various inspection tasks. This paper discusses the development of three main activities involved in the creation of a biomimetic robotic platform with onboard intelligence allowing the autonomous execution of different tasks. This proposed robot mimics swimming activity in carangiform fishes and performs various paths during its way with the use of a magnetic propulsion system driven by one single motor. This type of propulsion can be optimized by research in fluid dynamics, making the energy consumption better. Modular and scalable by nature, the platform might easily be designed to incorporate various elements, such as a vision system. The vision-based approach discussed here has many promising applications in marine environments and can be adapted to identify many species of marine life. This novel fish-like robot platform looks set to play a vital role in the sustainable oversight and maintenance of aquaculture operations.

In ideal design, the robotic fish has its head with all electronics sealed in the compartment. Thus, it guarantees perfect operation in water. The segments are three linked and attached in a mechanism of an oscillating arm, with the caudal fin at the other end. The rotation is conducted from the head to the oscillating arm and further through the links via a wire mechanism with a motor specially designed within the head. This spinning creates a disc furnished with four permanent magnets strategically situated to produce distinct poles on the disc. The oscillating arm itself is made up of two magnets in contact with the disc by alignment to its polarity. This magnetic attraction and repulsion allow for a contact-less movement mechanism. In this innovative setup, the robotic fish is able to move through water by simulating the natural oscillation of a fish's tail fin; thus, the motor does not directly interact with the oscillating links.

Figure 5.3 callout shows the outline of the robotic fish, which can be used to describe the complex system of its parts. It uses a hall effect sensor to provide and convert the accurate measurements developed from the magnetic fields from magnets attached to the tail into position measurements from the tail end.

The whole model is quite large, as it measures 40 cm from head to tail, and weighs around 875 g. Design and production were made with the help of SolidWorks-the leader in CAD-software-and finally, the model was printed with the help of a Zortrax M200 3D printer using FDM technology. Its building material is high-strength, ABS-based plastic, Z-ULTRAT, wherein superior mechanical properties have been chosen over standard ABS.

It comes with a high torque 6 V brushed DC motor, which is further upgraded with a metal spur gearbox in order to achieve maximum force and control. A quadrature encoder attached directly to the shaft of the motor provides speed and position feedback. A microcontroller board ESP32 serves as the "brain" of the system and comes integrated with a single-channel motor driver. The operation requires neodymium magnetic discs, every disc measuring 12 mm in diameter and 10 mm high, with a maximum load capacity of 5.3 kg, and the wires binding it to the tail are 0.3 mm in diameter, which can endure 20 kg, thus making it very robust and reliable.

5.2.1.4 Using Computational Fluid Dynamics for Future Design Enhancements

Movement of a flexible body in fluid is often indicated by changing the configuration of a 2D flexible hydrofoil to show the evolution of form as it travels through a thick, nonsquashable liquid. Of particular use is the analogy of living organisms that travel in a wavy fashion, using their bodies or fins to induce the wave they travel in. The in-plane sinuous motion of the fish is represented by a harmonic function multiplied by a function A(x) describing how the size of this side-to-side motion changes with x. The midline can similarly be described as the wave motion it experiences as it moves in the same direction of the flow:

$$y = A(x)\cos(2\pi(x/\lambda - ft) + \phi) + d, 0 \leq x \leq 1$$

where $A(x)$ denotes amplitude, λ denotes the wavelength, f denotes phase speed (frequency) and the wavephase angle; ϕ and d refer to the mid-line lateral position of the foil when placed at a balanced position. Such an expression is obtained for carangiforms by correlating information gotten from observing directly the data emanating from swimming fish.

$$A(x) = A_0\left[1 - 0.825(x-1) + 1.625\left(x^2 - 1\right)\right]$$

A_0 represents the peak amplitude of the tail-beat. After determining this deformation, the fish is described as engaging its entire body in an undulating manner. This allows every part of the fish to contribute to the creation of thrust, which, by transferring momentum to the surrounding fluid, initiates its movement through the fluid. The incompressible Navier–Stokes equations, which are pertinent in this context, can be articulated in the following way.

$$\nabla \cdot \mathbf{u} = \mathbf{0}$$

$$\frac{\partial \mathbf{u}}{\partial t} + (\mathbf{u} \cdot \nabla)\mathbf{u} = -\frac{1}{\rho}\nabla \mathbf{p} + \nu \nabla^2 \mathbf{u}$$

Here, u it is a velocity field, and p is the pressure field, whereas symbols ρ ν denote the fluid density and kinematic viscosity, respectively. Again, another complexity added is by solving body dynamics and even in moderately simple cases like 2D motion. It introduces many problems. Thus, by Newton's laws, the force acting on the body would then be equal to the rate of change of its momentum as it passes through the water. The fish motion is decomposed into a rigid body motion and deformation. The rigid motion is described by its translational and rotational velocities, v_b and ω_b associated with the body's center of mass. Meanwhile, the deformation motion, expressed as v_{def}, can be calculated using the swimming style adopted. Because of this, the developed formulation encompasses the dynamic motion of the foil.

$$\mathbf{v} = \mathbf{v_b} + \omega_\mathbf{b} \times (\mathbf{x} - \mathbf{x_{cm}}) + \mathbf{v_{def}}$$

where (x − $\mathbf{x}_{\text{cm}}$) **r** is the position vector of the body relative to the center of mass, $\mathbf{v_b}$ the symbol represents a 2D vector (v_b^x, v_b^y), while the angular velocity vector $\omega_\mathbf{b}$ has a singular component, ω_b. We then introduce the generalized velocity **V**, the force vector **F**, and the generalized mass matrix $\mathbb{M}$.

$$\mathbf{V} = \begin{bmatrix} v_b^x \\ v_b^y \\ \omega_b \end{bmatrix}, \mathbf{F} = \begin{bmatrix} F_x \\ F_y \\ M_z \end{bmatrix}, \mathbb{M} = \begin{bmatrix} m & 0 & 0 \\ 0 & m & 0 \\ 0 & 0 & I_b \end{bmatrix}$$

where m is the mass of the body which has density $\rho_b = \rho$ and I_b its moment of inertia with respect to the center of mass. The net forces and moments on the body are defined as

$$F_x = \oint_{\partial\Omega_b} f_x ds, F_y = \oint_{\partial\Omega_b} f_y ds$$

$$M_z = \oint_{\partial\Omega_b} [(\mathbf{x} - \mathbf{x}_{cm}) \times \mathbf{f}]_z \, ds$$

where $\mathbf{f} = (f_x, f_y)$ are the hydrodynamic (pressure+viscous) loads on the fish and $\partial\Omega_b$ is the surface of the foil. The motion of the body is governed by

$$\frac{d}{dt}(\mathbb{M}\mathbf{V}) = \mathbf{F}$$

The motion equations were integrated over a particular time-step using a standard viscous solver that was validated against a referenced study. Our research introduces a vision-based model that instantly detects and categorizes various underwater creatures by using a robotic fish. Kinds of animals which we focus on are sharks, fish, rays, and jellyfish. Distinguishing and accurately classifying these creatures prove difficult because of their physical similarities.

We used two public datasets available online. We combined them: Roboflow Aquarium Dataset and Google's Open Images Dataset V7. The combination would bring a rich set of images that was helpful in having a balanced representation of the animal classes under study. Several criteria were applied to the choice of images selected for usage. Their targets were those variedly posed and differently illuminated images with exclusion of duplicates or images with very high similarities. I had processed the images on the Roboflow platform, where collection, inspection, labeling, and resizing to a uniform dimension of 640 x 640 pixels were accomplished. The dataset was then further optimized, with the addition of augmentations on the images for further expansion of the images. Even though the images were prelabeled, more validation and enhancements were carried out, and the images ended up being divided into training, validation, and testing sets at the ratio of 82:12:6, respectively.

Since the objective of the project was to classify four classes of underwater fauna appropriately while ensuring a model feasible for real-time execution on the embedded hardware of the robotic fish, we chose the YOLOv8s model. The YOLOv8s algorithm consists of three primary parts, including backbone, neck, and head. It applies the variant version of CSPDarknet53 to the use of the backbone and also helps the neck to fuse multiscale features from images. This head is designed specifically for decoupled detection and classification processes. A pre-trained YOLOv8s model, in terms of applying the COCO dataset, was fine-tuned through our collected data.

This enhanced version was tested by processing the footage taken from the Lisbon Oceanarium from October to November 2022. A pair of test videos, 22 and 36 s long with a frame rate of 30 fps and a resolution of 1080 x 1920, was filmed through a 12 MP camera with f/2.8 video capabilities.

5.2.1.5 Computer Vision Algorithms in Food Processing

The above is because of increased global demand for quality foodstuffs and, consequently, growing interest in the use of less destructive and more rapid detection methods than are currently found in food industries. The analysis of data obtained by such technologies is complex and time-consuming, requiring highly skilled personnel. In addition, the success of several statistical and chemometric approaches appears to be significantly affected by a variety of factors such as noise, sample variability, and the complexity of data under various testing scenarios. In this context, recognition is being given to machine learning (ML) techniques pertaining to their adaptability in the food sector especially in improving nondestructive technologies and equipment, which are incapable of highlighting irrelevant details, identifying key features, and building accurate models.

This review provides an introductory comparison between TML and DL approaches. In addition, this paper critically reviews more emerging nondestructive technologies such as acoustic analysis, machine vision, electronic noses (e-noses), and spectral imaging technology. These technologies are discussed with a variety of advanced ML methods, thereby strengthening their applicability to important tasks for food quality management, which include type differentiation and classification, safety verification, and processing oversight. The review will also present challenges in this field and ways forward.

The section shows the possibility of real-time food quality and safety monitoring using the integration of advanced ML strategies with techniques for nondestructive testing. Figure 5.3 represents the swimming tests conducted in a pool with a fish-inspired robot. Each of the ML algorithms has unique benefits and a preferred area of application, whereas deep learning definitely seems to become one of the most effective tools in the feature-learning domain, indicating the need for further research of its full potential in the food industry.

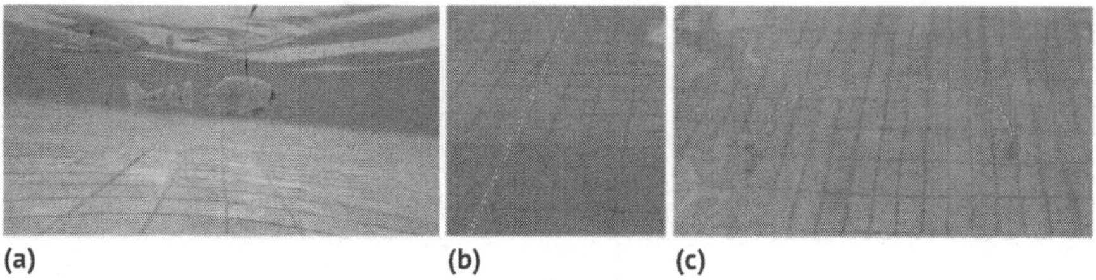

Figure 5.3 Swimming tests conducted in a pool using the fish-inspired robot: Picture of the fish-inspired robot (a). Image created by overlayed frames showing how the robot driven by one motor could create different paths, linear (b) and circular (c) (Jang and Seo [2023]/Elsevier).

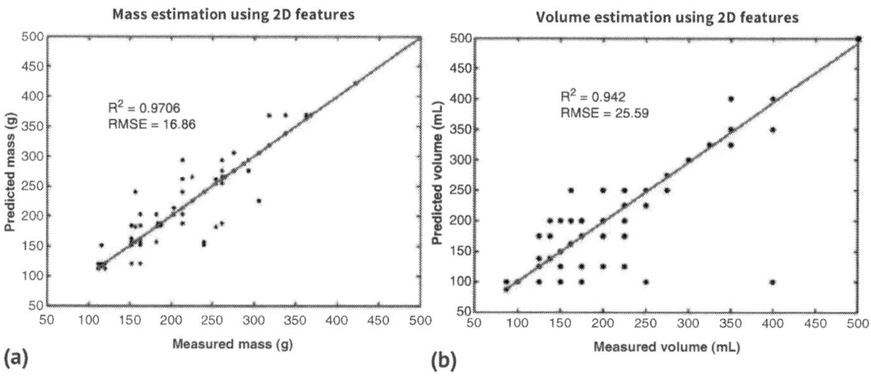

Figure 5.4 Scatter plot of the correlation between predicted V and M with the measured V and M of the three models (M1, M2, and M3) with the RBF-SVM (Nyalala et al. [2019]/ Elsevier).

5.2.1.5.1 Estimating the Volume and Mass of Cherry Tomatoes Through the Application of ML and Computer Vision Techniques

This developed a system that predicts the mass and volume of cherry tomatoes using a computer vision system and some ML algorithms. The study then sets up a formula that correctly describes how to relate the mass to the volume of the tomatoes. Figure 5.4 represents the scatter plot of the correlation between predicted V and M with the measured V and M of the three models (M1, M2, and M3). In 2019, Nyalala accurately computed the mass of tomatoes $M = 1.312V(0.9551)$ in a test set by using this model with an R^2 score of 0.9824 and an RMSE of 15.84 g (Nyalala et al., 2019). It captures depth images of tomatoes in various poses through image processing and determines relevant features. Then, five regression prediction models were built based on both 2D and 3D image data. With this group of models, the best performance was observed in the radial basis function support vector machine (RBF-SVM) with 2D features, which resulted in an accuracy of 0.9706, and combined features for estimating mass with an accuracy of 0.9694, and volume as well, respectively. The predictions from this model could then be further used to apply the derived relationship between mass-volume. The developed system uses inline sorting and grading with mass, volume, or density values, noninvasively, accurately, and consistently for cherry tomatoes.

In an acquisition from a local Pukou district market, 300 ripe cherry tomatoes of assorted sizes were procured. This variety is notably popular in eastern China and was abundantly accessible throughout the time of the investigation, the experimental setup and image acquisition system (Manduca et al., 2023). These tomatoes were carefully selected to be free of any visible damage or imperfections. For the purposes of this study, each tomato underwent labeling and was precisely weighed using a calibrated electronic scale (model YP10001, manufactured by Shanghai Instruments Company) with an accuracy of ± 0.1 g. Additionally, the volume of each tomato was calculated using the water displacement method (WDM), as described by Mohsenin (1970). In terms of image capture, the setup included a 3D Kinect sensor (Model Kinect 2.0, produced by Microsoft in Seattle, WA, USA), which was positioned atop a metal stand at a height of 0.7 m.

Figure 5.5 represents the scatter plot of the correlation between predicted *V* and *M* with the measured *V* and *M* of the three models with Bayesian ANN. A system to predict the mass and volume of cherry tomatoes has been developed employing the computer vision system along with ML algorithms. This research establishes a formula that could reliably relate mass to the volume of the tomatoes: $M = 1.312\ V(0.9551)$. This formula correctly utilized the test dataset and was able to give appropriate mass estimation with an R^2 score of 0.9824 and a root-mean-square error (RMSE) of 15.84 g, according to Nyalala (2019). The system

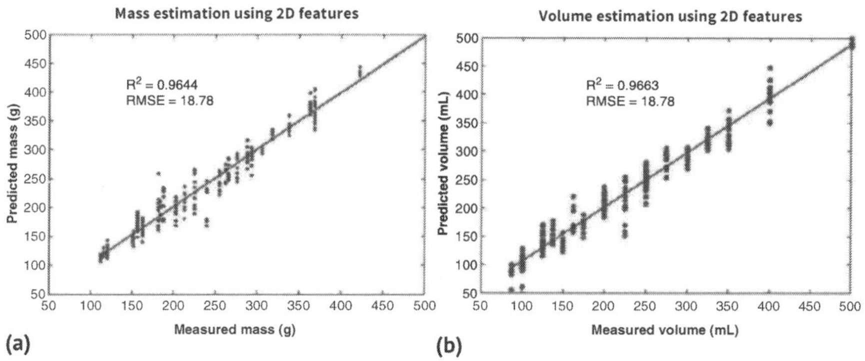

Figure 5.5 Scatter plot of the correlation between predicted V and M with the measured V and M of the three models (M1, M2, and M3) with Bayesian-ANN (Nyalala et al. [2019]/ Elsevier).

took images of the tomatoes at different positions and used techniques of image processing to gain relevant features on it. It then came up with five regression prediction models based on both 2D and 3D image data. Considering these models, the best performance would be achieved using RBF-SVM with a prediction accuracy of 0.9706 for estimating the mass and 0.9694 for estimating volume using the respective 2D and combined feature sets. The predictions with this model can then be used in the derived mass-volume relationship. This new system provides an option for inline sorting and grading of cherry tomatoes by mass, volume, or density in a noninvasive, precise, and consistent way.

5.2.1.5.2 Image Preprocessing

The processing of raw depth image data was executed following this procedure: Background removal was executed with the image subtraction approach, as demonstrated in the following equation:

$$g(x,y) = \begin{cases} 0, & \text{if } |f(x,y) - b(x-y)| \leq T \\ f(x,y) & \text{otherwise} \end{cases}$$

where $g(x,y)$ denotes the output final picture after removing the background, $f(x,y)$ is the original image, $b(x,y)$ is the image of the background, and T represents the threshold. An area of interest is defined by establishing maximum and minimum thresholds based on the distance intensities of the depth image in order to carry out distance thresholding. The collected depth image was smoothed by applying a Gaussian kernel filter with 15 × 15 pixels, zero mean, and subsequently processed by morphological opening with a disk-shaped structural element of 9 pixels in order to remove small holes and to get the refined depth image. Subsequently, the depth image was binarized to convert it into a binary image using Otsu's approach as illustrated below (Nyalala et al., 2021):

$$h(x,y) = m[g(x,y)]$$

where $h(x,y) = h$ is the resultant-filtered image, and $m = \left(\frac{1}{2\pi\sigma^2} e^{-\frac{1}{2}\left(\frac{e^2+f^2}{\sigma^2}\right)}\right)$ is the Gaussian filter kernel.

h. $Tn = (g \oplus Tn)\Theta Tn = \cup\{((Tn)_p \mid (Tn)_p)\}$ where Tn is the structural element, $(Tn)_p$ is the translation of Tn by a point p.

5.2.1.5.3 2D Features

2D features have been used, such as projected area (A), perimeter (P), eccentricity (E), major-axis length (λ_1), minor-axis length (λ_2), and radial distance (D). Fruits' form has been used as a parameter for mass and volume estimations. A is the total number of pixels that the object occupies or the area that the tomato covers when projected onto the picture plane. The following equation defines the variable A (Nyalala et al., 2019):

$$A = \frac{1}{2} \sum_{n-1}^{N-1} (y_{n+1}x_n - x_{n+1}y_n) \tag{5.5}$$

This formula has been derived from Green's theorem, which relates a 2D integral $\iint_{\text{ROI}}$ on a ROI with a line integral $\oint_{\partial\text{ROI}}$ on the closed boundary of the ROI (Schulz et al., 2018).

P is the distance around the boundary of the object; it was computed by counting the number of pixels in the closed contour. Consider an image shape having N vertices: $C = \{(x_n, y_n)\}$, with $n = 1, \ldots, N$, and $(x_1, y_1) = (x_N, y_N)$, then P is defined as Eq. (5.6).

$$P = \sum_{n-1}^{N-1} (x_{n+1}, y_{n+1}) - (x_n, y_n) \tag{5.6}$$

E is a ratio of λ_1 and λ_2 when an ellipse is fitted on an image; a more circular object shape will have a lower eccentricity. E can be computed using the principal axes method according to Eq. (5.7).

$$E = \frac{\lambda_1}{\lambda_2} \tag{5.7}$$

Equation (5.7) provides λ_1, which is the pixel distance between the major-axis endpoints.

$$\lambda_1 = \sqrt{(X_2 - X_1)^2 + (Y_2 - Y_1)^2} \tag{5.8}$$

where the major-axis endpoint coordinates are described by (X_1, Y_1) and (X_2, Y_2). The result is the Euclidian distance between the two points.

Equation (5.9) provides λ_2, which is the pixel distance between the minor-axis endpoints.

$$\lambda_2 = \sqrt{(M_2 - M_1)^2 + (N_2 - N_1)^2} \tag{5.9}$$

where the minor-axis endpoint coordinates are described by (M_2, N_2) and (M_1, N_1). The result is the Euclidian distance between the two points.

The average distance between an image's center of gravity and its boundary points is denoted by the D. According to Eq. (5.10), the radial distance is calculated from the central point (centroid) in the object of each pixel $x(n), y(n)$ on the boundary.

$$d(n) = \sqrt{[X(n) - \bar{X}]^2 + [Y(n) - \bar{Y}]^2} \quad n = 0, 1, \ldots, N-1 \tag{5.10}$$

5.2.1.5.4 3D Features

As 3D features the volume (V) and surface area (S) were used. These features were related to the volume of the tomato. These features that describe the 3D shape of the tomato were observed in the depth images that were captured. S was estimated using 3D Delaunay triangulation. A set of triangles is generated in three dimensions by the Delaunay triangulation. Equation (5.11) represents the overall surface area. The areas of the triangles were approximated by Eq. (5.11):

$$S = \sum_{i=1}^{T} \frac{1}{2} |x_i \times y_i| \tag{5.11}$$

where T is the total number of triangles, and x_i and y_i are the two vectors that span the ith triangle.

V was estimated using numerical integration of the depth pixels using Eq. (5.11).

$$V = \sum_{i=1}^{N} (b - c_i) \cdot A_i$$

where b is the tomato's average contour depth, c_i is the depth value, A_i is the approximated pixel area of the ith pixel, V is the estimated volume, and N is the total number of pixels.

In our experiment, we present the investigation of regression models for the mass and volume prediction of tomatoes by using different input features. Models tested include SVMs with different kernels, linear, quadratic, cubic, and radial basis functions (RBFs), along with Bayesian Artificial Neural Network (Bayesian-ANN). Among these, SVMs were especially known for their high precision on training data, for being interpretable geometrically, and for their very efficient mathematical processing. Another major advantage of SVMs is that they do not overfit the data during training, even with relatively little training data. Bayesian optimization was used to fine-tune the SVM classifiers. We used MATLAB's version, namely, R2019a release, applying its Neural Network Toolbox in performing the implementation of the RBF-ANN model. Table 5.1 specifies the architecture of the models with various input, hidden, and output layer combinations. Bayesian regularization training method was adopted in this work for the best performance of models. Training the model continuously was continued until the error between the predicted output and actual target is minimized. A 10-fold cross-validation approach on the training set was adopted, and further performance metrics were evaluated on a separate test set. Hence, we thus suggest three models for either mass or volume prediction, depending on the type of training features. The first model employed 2D features (M1). The second made use of 3D features (M2). In the third model, we combined both 2D and 3D features (M3), whose specifications are clearly presented in Table 5.1 represents the comparison of different classification techniques which deepen the understanding of the SVM and various other models. Wu and Zhou (2006) and Benmouna et al. (2022) had 958 samples, where they used 671 samples, about 70%, for training purposes. Then they kept the rest, which is 287 samples, or about 30%, for testing and verification purposes. The investigators hand-sampled the samples to ensure that the tomatoes didn't touch the edges of the images because they tend to move on the conveyor belt. During the training phase, the parameters of the SVM with RBF kernel are fitted with the help of 10-fold cross-validation to balance the two models' capacity to be accurate and prevent overfitting without causing any underfitting and thus ensure generalizability. The analysis made during the development process of the Bayesian-ANN model indicated that 10 hidden layers is effective, whereas the number of neurons, although having a negative relation with percent error during the validation, increases the computation times.

5.2.1.6 Plant Leaf Disease Detection Using Computer Vision and Machine Learning Algorithms

Agriculture is very crucial in ensuring food security because the population is growing so fast. Detection of diseases in crops at initial stages ensures the supply of food. However, the early detection of diseases in crops is somewhat hard to trace. The objective of the paper, therefore, is educating farmers on newer available technological developments that would help in curbing the effects of plant diseases in general and tomato plants, widely consumed as a source of nutrition. The research employs ML and image processing techniques with a high-resolution algorithm in the detection of diseases in the tomato plant's leaves. In this case, it addresses the tomato leaf samples that present with symptoms of the disease. Analysis of these samples of the diseased leaves brings an early identification of possible problems based on the symptoms for farmers. The process begins with resizing all the leaf samples into 256×256 pixels and then applying Histogram Equalization to the obtained images in order to enhance the quality. Then, it further divides the data space into several independent regions using K-means clustering and outlines the borders for the leaf samples by contour tracing. Several methods are used to extract meaningful features from the samples, such as Discrete Wavelet Transform, Principal Component Analysis, and Grey Level Co-occurrence Matrix. Once these features are extracted, they are classified using a variety of ML approaches by SVM, Convolutional Neural Network (CNN), KNN, etc. The proposed model is validated for its efficacy through the accuracy rates obtained with SVM at 88%, KNN at 97%, and CNN at 99.6% on the diseased tomato leaf samples.

5.2.1.6.1 Proposed Model

This part of the section describes the IP and ML techniques that were used in developing the model. Figure 5.6 depicts the proposed model (DWT+PCA+

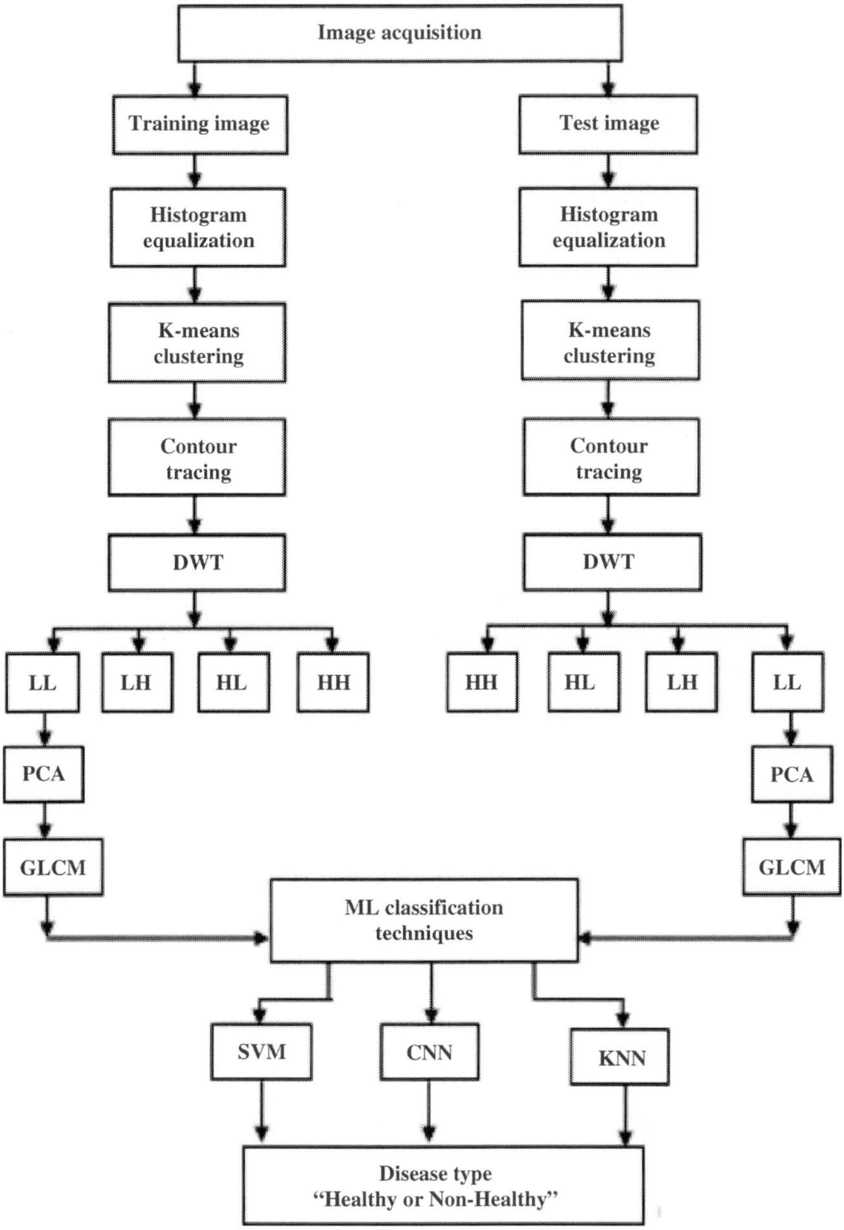

Figure 5.6 Proposed model for leaf disease detection identification (Nyalala et al. [2019]/Elsevier).

GLCM+CNN) for leaf disease detection using computer vision and ML techniques. Tomato samples considering six problems are used in order to check the accuracy of this image and determine whether the disease found on a leaf is healthy or unhealthy. The picture processing increases the tomato samples to 256×256 pixels in order to make them of the same size for all the experiments.

The employment of Histogram Equalization (HE) and K-means clustering techniques is pivotal in enhancing the segmentation quality of leaf samples. This segmentation process allows for the early prediction of whether a leaf is diseased or not, based on the outcomes of the K-means clustering analysis. Additionally, the technique of contour tracing is utilized to delineate the perimeters of the leaf samples effectively.

For the purpose of feature extraction from these samples, methodologies such as Discrete Wavelet Transform (DWT), Principal Component Analysis (PCA), and Gray Level Co-occurrence Matrix (GLCM) have been adopted. These methods aid in identifying the informative regions or features within the samples. Subsequently, in the phase involving ML strategies, tools like SVM, KNN, and CNN are applied to categorize these features. The efficacy of the model is meticulously documented following the classification process.

5.2.1.6.2 Dataset

The research utilizes a specialized dataset, specifically the village database of tomato leaves, which encompasses plants having been affected by various diseases. This dataset contains images of tomato leaves exhibiting six distinct disorders and serves as the foundation for the conducted experiments aimed at leaf disease detection. Illustrations of leaf image samples from this dataset are presented.

5.2.1.6.3 Preprocessing

The preprocessing stage identifies the infected regions through leaf images and is done using the K-means clustering technique. This includes finding the center of each data image, forming the clusters, and the actual distance from the center of one cluster to another. After applying the K-means clustering algorithm, the samples are presented in Figure 5.6. Besides this, contour tracing also takes place on the digital samples of leaves so that all essential shape information can be extracted, which will be of the utmost importance in the analysis of contour characteristics and subsequent pattern classification, thus playing a crucial role in the entire process of effective feature extraction. Images generated in the contour tracing process are presented.

For features extraction purposes from improved samples of tomatoes, the DWT method has been employed. DWT breaks down the data into sub-bands of lower (LL, LH, HL) and higher frequency (HH). It was observed that the LL component of DWT carries most of the information content compared with higher frequency components. Further, we analyze the selected optimal features using PCA. This is predominantly for the reason that the GLCM method applied disperses higher orders of gray levels based on a relevant proximity criterion. This technique has significant importance in the feature extraction from leaves and applies measures like Homogeneity, Auto-correlation, Dissimilarity, Entropy, Sum of Squares, Average, Variance, and Entropy. The compilation of features obtained from DWT, GLCM, and PCA into a single feature vector enables the classification of image samples using various classifiers.

5.2.1.6.4 Image Classification Approaches

Classification of the sample uses various techniques, like SVMs, KNN, and CNNs. It's one among ANNs designed to process data through layers, mainly including input (IL), output (OL), and more than one hidden layer (HL). These hidden layers comprise convolutional layers, a RELU layer which acts as an activation function, along with pooling, fully connected, and normalization layers. Mathematically, it has been shown that the CNN structure is more aligned to cross-correlation than classical convolution, thus explaining its relevance in interpreting matrix indices (Adhikari et al., 2024; Kadam et al., 2024). A typical standard three-layer network setup is illustrated in Figures 5.7 and 5.8 representing a typical CNN algorithm.

The confusion matrix for a CNN that analyzes both output and target classes is depicted in Figure 5.9. Figure 5.10 illustrates the advancement of training samples of leaf characteristics, which have been classified by the CNN to determine their accuracy and error rates.

5.2.1.6.5 Evaluation of Leaf Disease

The parameters such as Precession, Recall, and F-measure for the proposed model are calculated and are given in the following Eq. (5.12). The precision, recall, and F1-scores for some ANN models are as follows:

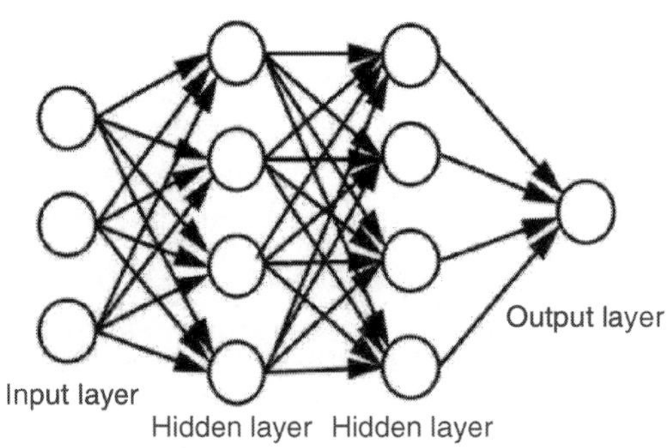

Figure 5.7 A three-layer neural network (Harakannanavar et al. [2022]/Elsevier).

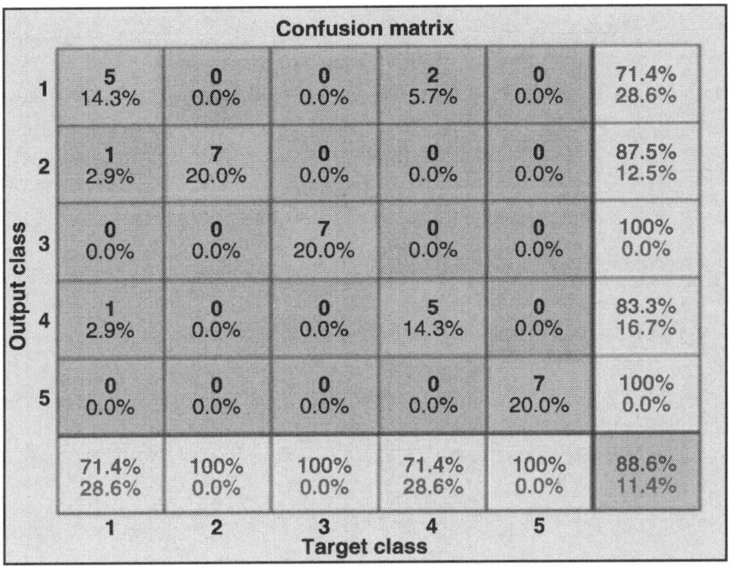

Figure 5.8 A CNN algorithm (Harakannanavar et al. [2022]/Elsevier).

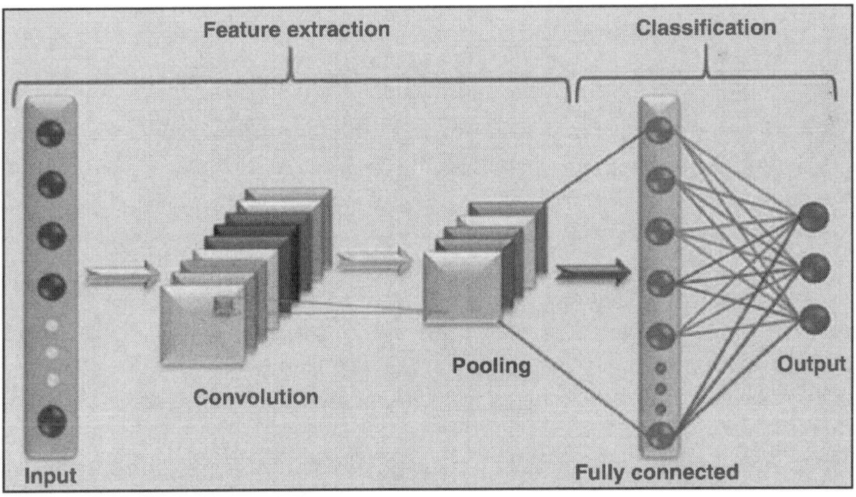

Figure 5.9 Performance of a CNN classifier (Taner et al. (2021)/MDPI/CC BY 4.0).

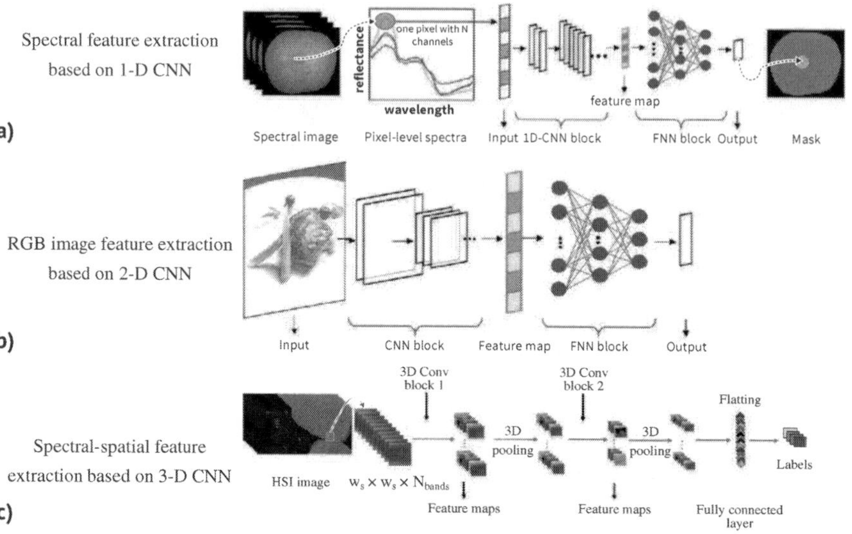

Figure 5.10 Comparison of the proposed model (Liu et al. [2021]/Elsevier).

$$
\begin{aligned}
\text{Precision Measure (\%)} &= \frac{\text{True Positives}}{\text{True Positives} + \text{False Positives}} \times 100 \\
\text{Recall Measure (\%)} &= \frac{\text{True Positives}}{\text{True Positives} + \text{False Negatives}} \times 100 \qquad (5.12) \\
\text{F-measure} &= \frac{2 * \text{Precision} * \text{Recall}}{\text{Precision} + \text{Recall}} \times 100
\end{aligned}
$$

A dataset on samples of tomato leaves was used in testing a new model during the experiment. Hundred samples of healthy leaves were subjected to analysis and the model accurately classified 99, hence an accuracy rating of 99%. The model also classified all the samples infected by Mosaic virus with an accuracy rating of 100%. It achieved the same 100% accuracy rating as the leaf mold. In testing 100 yellow curl leaf samples, it recorded a 99% accuracy rate for the model. Moreover, in Spotted spider mite and Target Spot diseases, it has gained 99 and 100% accuracy rates correspondingly. The total samples tested were 600, and the proposed model achieved an average accuracy of 99.5%, the performance of a CNN classifier is represented in Figure 5.9.

There are samples for training and testing datasets verified by the approach.

Recommended hardware and software to be used: The recommended hardware and software for the task comprise: Operating System: Windows 10, Processor: NVIDIA GPU, Programming Language: Python, and Libraries: Image Data Generator, OpenCV, Keras, TensorFlow, NumPy, and Matplotlib.

Table 5.1 Comparison of different classification techniques.

Techniques	Disorders	Precision	Recall	F1 Score	Accuracy (%)
K mean clustering + GLCM + SVM (Wu and Jiang (2022))	Healthy	0.983	0.834	0.881	81.9
	MV	0.913	0.722	0.821	86.8
	LM	0.895	0.817	0.896	80.2
	YC	0.811	0.889	0.992	86.3
	SM	0.977	0.861	0.985	85.9
LBP+PCA + K mean clustering + KNN (Chen et al. 2022)	TS	0.987	0.898	0.994	87.9
	Overall	0.967	0.978	0.915	8 9
	Healthy	0.858	0.998	0.905	95.9
	MV	0.834	0.852	0.932	92.8
	LM	0.722	0.844	0.916	89.6
	YC	0.817	0.985	0.892	87.7
	SM	0.889	0.876	0.915	88.9
	TS	0.861	0.928	0.958	94.8
	Overall	0.978	0.975	0.952	97.3
	Healthy	0.985	0.984	0.979	98.9
	MV	0.949	0.945	0.962	95.3
	LM	0.932	0.934	0.956	95.5
	YC	0.938	0.968	0.979	96.4
	SM	0.953	0.919	0.899	97.2
	TS	0.967	0.978	0.898	95.5
	Overall	0 .995	0 .995	0 .988	99.09

These experiments have been conducted on the dataset Plant Village dataset on samples of Tomato with six disorders. This model is evaluated based on the key metrics like Precision, Recall, and F1 score for the model considered. The model considered was tested on a dataset of 600 tomato leaf samples. Figure 5.10 compares this model with previous models. Table 5.1 is a comparison of precision, recall, and F1-scores for SVM, KNN, and CNN. The current model DWT+PCA+GLCM+CNN shows a substantially high accuracy since it has obtained 99.09% accuracy rate as against the previous models. This model, which integrates computer vision with ML classification techniques, is further compared with other methodologies reported by Hossain et al. (2021) and the comparison is provided in Table 5.1. The accuracy obtained is remarkable and the proposed methodology yields improved results compared to existing ones.

5.3 Robotic Manipulator in Food Processing

Let's take an onion's or artichoke feeding robotic system, adapted to cleaning and cutting machinery. First, they feed the veggies in the hopper, then they are meant to feed them into the processing line through manually transferred workers. The most basic problem with the system, as it stands now, is that the number of workers that can be accommodated is limited, and they can only process about 100 units per minute. The objective of the project was to design and develop a robotic solution that could be easily integrated with existing machinery and meet the following specified requirements: (i) A work envelope dimension of 620 mm in the X-axis, (ii) a pick and place frequency of between 80 and 120 cycles per minute, (iii) the maximum handling weight for the onions/artichokes of 1 kg, and (iv) 4 or 5 degrees of freedom, with 3 for positioning and one or two for aligning the vegetable. The proposed paper was to consist of: Section 2 introduced the selection and design of the proposed robotic solution. The analysis of kinematics and the workspace of the parallel manipulator were dealt with in Section 3, while Section 4 treated the computer vision system. This section explains the hardware implementation of the prototype and the early results. The final section of the paper is a description of the current outcome of the projects undertaken and possible future research ventures.

Design and Development of the Robotic Solution: This chapter describes the design phase of the solution, which, starting with the evaluation of different options, continues into the detailed design of the robotic solution using CAD tools and presents the conception of the vacuum suction cup feature.

5.3.1 Review of Different Options

Before the feasibility study, a review was carried out of commercial alternatives for use. Included in the review was both serial and parallel kinematic structures for robotic arms. Summarized in the findings of this review, various serial and parallel robotic systems, respectively. Data presented in these tables show that from all serial robots, SCARA configuration stands closest to our requirements, although it does not hit all of our marks perfectly, with one major drawback-no serial robot exceeds 35 cycles per minute. To achieve a target of 120 cycles per minute, this would need four units of SCARA robots. On the other hand, in relation to the usage of parallel robots, there is a growing trend in its usage over the years. For example, as demonstrated by Das and Dülger (2005), it indicates the widespread use of applications using parallel robots for versatility within their varying scope.

Although the machines are very fast and accelerated for a seamless performance with maximum efficiency in the food packaging industry, it still gets much more utilization in manufacturing. The models in commerce are usually large in size and have overly large workspaces compared to what is needed, complicating their use in agricultural equipment because of excessive dimensions.

5.3.2 Innovative Design of Robots

In response to these challenges, a new form of robotic arm, using a delta kinematic structure, was initiated to aid the problem. To be able to ensure the feasibility of such an innovative approach, a number of simulations were conducted to premultiply evaluate several different delta architecture-based designs for the robotic arms as depicted in Figure 5.11. The tests were performed attached to the 0.5 kg payload on the robot end-effector, at which point the motors were mounted on the upper platform of the robot. The horizontal operational workspace of the robot is 800 mm (±400 mm) and obtained a cycle speed of 60 cycles per minute in the simulation. This would mean that if the operational conditions are to be met, then two units of the developed robot would be required. Bottom half of Figure 5.13. Torque demands from the actuators of a standard trajectory. Smaller actuators than 2 Nm

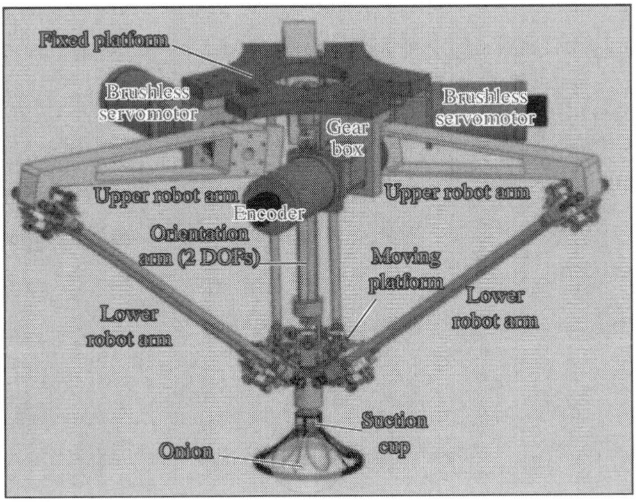

Figure 5.11 Description of the final design robotic solution over a CAD image (Harakannanavar et al., 2022).

are indeed enough to fulfill the continuous torque demands. This makes it feasible to use compact brushless motors and a low-gear coupling system. However, peak torque demands of up to 10 N m were made during the first movement from a static position (emphasized in simulation without path planning). Figure 5.11 shows a graphical representation of the robotic system with the solution as integrated on the overall agricultural machinery. It is shown that two robot units are mounted on the hopper and together perform 120 cycles. The image further goes on to prove that the robot is actually able to pick up the vegetables and drop it into a carry conveyance line in such a manner that it does not interfere with the hopper or the neighbor's robot unit.

CAD Tools for Designing: The robotic solution's design was executed utilizing Autodesk Inventor. Autodesk Inventor is a 3D CAD application providing a straightforward suite of tools for 3D mechanical design and documentation. The robot consists of two platforms as illustrated in Figure 5.13—one static platform on the top and another platform on the bottom that can move. Additionally, it includes three sections, Figures 5.12 and 5.13, representing a designed vacuum suction cup and the geometrical model of the robot.

The most efficient method to manipulate delicate objects such as vegetables and fruit is through the creation of suction using an air technique known as a suction-cup technique. It applies the negative air pressure developed for suctioning to lift and move an object. There has been a specialized suction cup designed for onions and artichokes. This particular cup is constructed so that it can automatically suit the shapes of onions and artichokes. Then, it is made of a very flexible material, such as silicon. One important feature of this particular cup is the adaptive side concavities. It is a pivotal point where these concavities of the cup shape will change to fit the different shapes of the vegetable, thereby sealing all the gaps to efficiently create a vacuum. Initially, the design of the cup hadn't included reinforcements across these concavities that bent the sides of the cup upward as sometimes to interfere with adjacent vegetables.

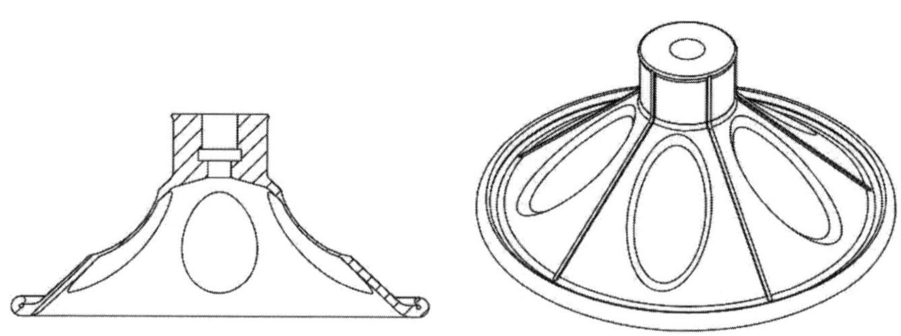

Figure 5.12 Designed vacuum suction cup (Harakannanavar et al., 2022).

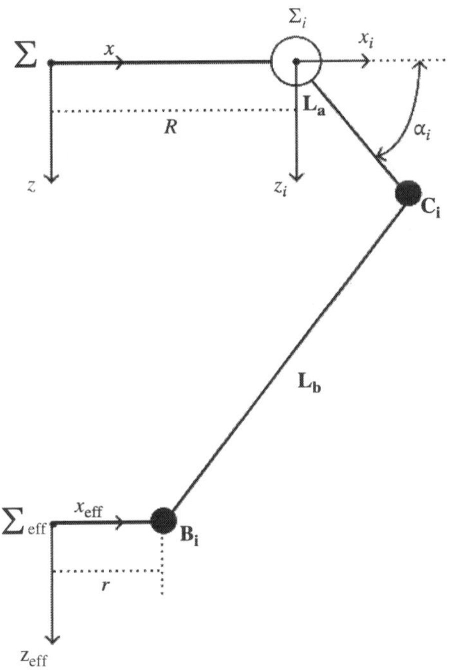

Figure 5.13 Geometrical model of the robot (Harakannanavar et al., 2022).

5.3.3 Kinematic Analysis and Workspace

Kinematic Analysis: After the kinematic analysis of the delta robot, the required equations for the control of the system are then written (see Eq. (5.13)). The geometrical model that is used is the one in Figure 5.13. In the middle of the top (fixed) platform is the global reference system $\sum(X, Y, Z)$, with the x-axis perpendicular to the axis of the actuator 1 and the z-axis perpendicular and pointing downward. At the bottom of the moving platform, there is a second reference system $\sum_{eff}(X_{eff}, Y_{eff}, Z_{eff})$ which shows the current position of the robot's end-effector in relation to the system. The robot's symmetry meant that the forearm and kinematic chain bar could be treated separately. Arm and forearm lengths are L_a and L_b, respectively. Each arm has a reference system $\sum_i (X_{if}, Y_i, Z_i)$, which has been rotated by an angle $\theta_i\,(0°, 120°)\,.(i = 1, 2, 3$ legs) and 240°), at a distance R from the global reference system $\sum$.

The joint positions are denoted by α_i and the $\sum_i$ system has the active joints there. They are passive joints at $\mathbf{B}_i$ and $\mathbf{C}_i$. The lower platform's radius is r while the top one is R. The geometrical parameters of the robot are ($i = 1, 2, 3$ legs):

$$\left\|\sum_i \sum_i\right\| = R; \quad \left\|\sum_{\text{eff}} \mathbf{B_i}\right\| = \mathbf{r};$$
$$\|\mathbf{B_i C_i}\| = \mathbf{L_b}; \quad \left\|\sum_{\mathbf{i}} \mathbf{C_i}\right\| = \mathbf{L_a} \tag{5.13}$$

where $\mathbf{C_{i\Sigma}}$ and $\mathbf{B_{i\Sigma}}$ are the position vector of the spherical joints $\mathbf{C_i}$ and $\mathbf{B_i}$, with respect to the global system $\sum$.

Equation (5.13) can be expressed as

$$\mathbf{B_i C_i}$$
$$= \sum R_{\sum_i} = \begin{bmatrix} L_a \cos(\alpha_i) \\ 0 \\ L_a \sin(\alpha_i) \end{bmatrix}$$
$$+ \begin{bmatrix} \Delta r \cos(\theta_i) \\ 0 \\ \Delta r \sin(\theta_i) \end{bmatrix} - \begin{bmatrix} x \\ y \\ z \end{bmatrix}, \quad \text{for } i = 1, 2, 3 \tag{5.14}$$

where $\Delta r = R - r$ and $\sum R_{\sum_i}$ is the rotation matrix among $\sum_i$ and $\sum$ systems.

If forearms are considered rigid bodies, constraint equations for the robot are given by

$$\|B_i C_i\|^2 = L_b^2, \quad i = 1, 2, 3 \tag{5.15}$$

which can be expressed as

$$(x - x_i)^2 + (y - y_i)^2 + (z - z_i)^2 = L_b^2$$
$$i = 1, 2, 3 \tag{5.16}$$

with

$$x_i = (\Delta r + L_a \cos(\alpha_i)) \cos(\theta_i)$$
$$y_i = (\Delta r + L_a \cos(\alpha_i)) \sin(\theta_i) \tag{5.17}$$
$$z_i = L_a \sin(\alpha_i)$$

Equation (5.15) (Morales et al., 2014) refers to Figure 5.13 is the equation of sphere with $\mathbf{L_b}$ radius and centered on $\mathbf{B}_i$ which is the geometric locus of $\mathbf{C}_i$. A second constraint is imposed by the circular trajectory of $\mathbf{C_i}$ with center on $\sum_i$ and radius $\mathbf{L_a}$:

$$(x - R)^2 + z^2 = L_a^2 \tag{5.18}$$

The point $\mathbf{C_i}$ is located over the intersection of the sphere (4) and the circle (6) in the plane $y = 0$. From the earlier equations, a quadratic equation on x can be computed to determined α_i:

$$\alpha_i = \sin^{-1} \frac{z}{L_b} \tag{5.19}$$

5.3.4 Inverse Kinematic Analysis

As theoretical background, the inverse kinematics problem is briefly reviewed in this subsection. Since the suggested robotic device is symmetrical, as was mentioned in the kinematic analysis part, only one arm and forearm will be taken into consideration (Figure 5.13). Determining angle values $\alpha_i(i = 1, 2, 3)$ for a desired end effector position $\sum_i (X_{\text{eff}}, Y_{\text{eff}}, Z_{\text{eff}})$ leads to the inverse kinematics issue.

In our scenario, each link has two potential location options, or two viable solutions. Furthermore, the manipulator can only reach one of the two solutions please refer. This finding enables us to deduce that a desired can have up to eight different alternative solutions. use the inverse kinematic model to determine the final effector pose (Morales et al., 2014). The following geometrical constraints must be taken into account when selecting each joint pose to avoid singularities and obtain a workable configuration: if $x - R \geq 0$, then α_i can be determined, and if $x - R < 0$, then $\bar{\alpha}_i$ is computed as $\bar{\alpha}_i = \pi - \alpha_i$.

Direct Kinematic Analysis: As given in the design of each angle $\alpha_i(i = 1, 2, 3)$ of the joints, direct kinematics of the parallel manipulator can help determine end-effector pose as $\sum_i (X_{\text{eff}}, Y_{\text{eft}}, Z_{\text{eff}})$. The center of the moving platform is formed at the lower part of the robot with the intersection of three spheres such that the center lies in C_i and the forearm lengths are L_b as radius. There will be two solutions that describe the two intersections.

There are two critical locations of intersection in the interaction of the three spheres: one with the positive coordinate and the other with a negative coordinate.

Prototype: The prototype was built in order to demonstrate the effectiveness of a new robotic solution in real-world applications. The following list shows several components of the prototype in Figure 5.14.

1. A heart of our control system is a quite inexpensive multiaxis control board, DMC 2143, from Galil Motion Control. It can work either independently or attached to a PC via Ethernet 10/100Base-T or RS232. For driving brakes or relays, it includes optically isolated inputs/outputs, and for connecting analog sensors—analog inputs. It also has 8TTL unallocated inputs and 8 outputs supporting 1 to 4-axis control.

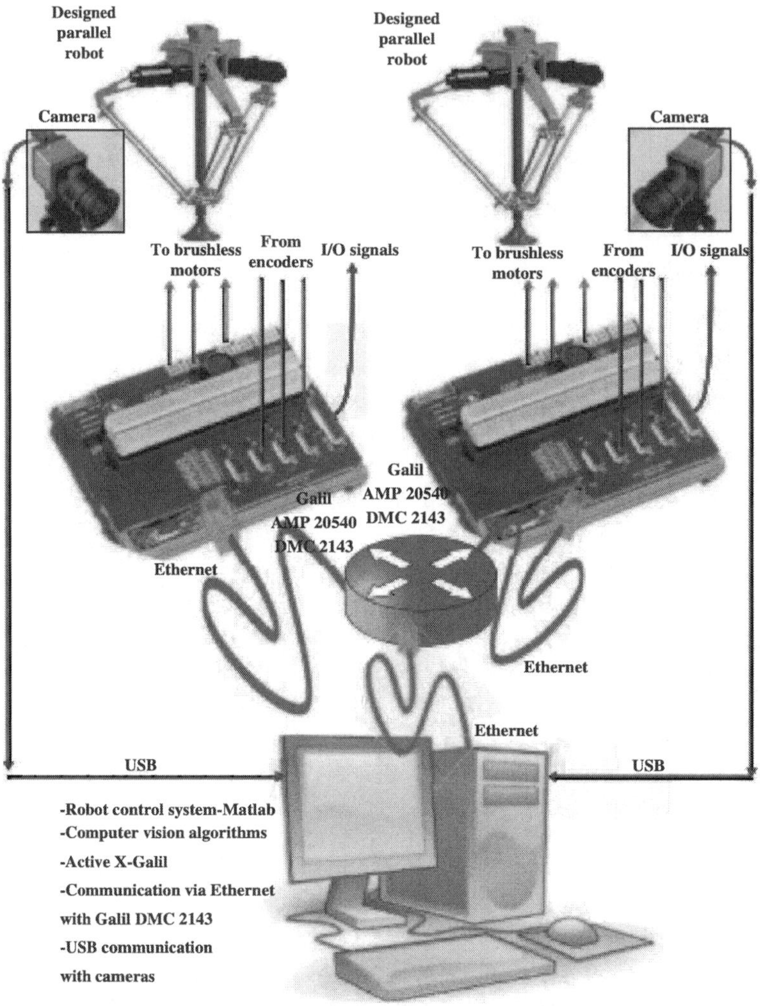

Figure 5.14 Components of the proposed robotic solution (Morales et al., [2014]/SAGE Publications/CC BY 3.0).

2. The robot is equipped with three high-efficiency brushless servo motors BLM N23 that respectively provide a continuous torque of 0.39 Nm. They are equipped with a 1000 line encoder to ensure detailed feedback regarding the position of the motors reaches the Galil controllers. Coupled with an APEX AB042 gearbox from Tecnopower with a ratio of 1:10, it increases torque to about 3.9 N m well more than the 2 N m requirement.

3. AMP 20540 amplifier board, manufactured by Galil Motion Control. Will drive up to 4 brush or brushless motors. The four PWM amplifiers will operate over an 18–80 V DC range and will provide constant current up to 7 Amps, peaking at 10 Amps.
4. Communications software provided by Galil to interface.
5. Matlab will be used to develop software tools for controlling and interfacing the robotic arm system.
6. The prototype uses computer vision software for extended capabilities.
7. An Ethernet cable is used to connect the multiaxis board with the PC.
8. A 24 VDC power source energizes the actuators while a ± 12 VDC and 5 VDC power source supplies the multiaxis board.

The prototype takes a major role in the parallel robot's control system. It computes the inverse kinematics using a Matlab m-function. Such a procedure calculates certain joint variables basing on translation of desired Cartesian coordinates which were made with Matlab Guide. Subsequently, these joint coordinates are transferred to the multiaxis board through an Ethernet connection enabled by an activeX control from Galil. This arrangement allows for accurate placement of each of the motors. Figure 5.14 depicts the functioning of the robot in a simulation laboratory. The work rate can be attained here to be nearly equal to the given optimum of 60 cycles a minute, at 40 cycles per minute. Considering the fact that optimized algorithms for delta acquisition and image processing are implemented in appropriate hardware and software of Matlab, it's assumed that the robot will achieve the goal of 60 cycles per minute without any hindrances.

5.4 End Effectors in Food Industry

5.4.1 Robotic Grippers for Large and Soft Object Manipulation

Grasping is one of the fundamental actions through which cyber-physical systems interact with the physical world, so it can be done. This paper critically reviews the latest very advancements in robotic gripping solutions specifically developed for food-industry applications, with special attention given to handling red meat. It provides a systematic study of grippers available in practice, which can be directly adapted or used as inspiration for robotic meat processing. The paper lists several solutions and gripping principles aimed primarily at low-payload applications, yet the classic shape-locking and force-locking grippers mostly under discussion are those capable of lifting and handling heavier objects. While identifying outstanding and/or novel designs for grippers, or revolutionary new concepts for gripping, this literature review will identify fully developed, commercially available alternatives from patent examination.

The high-technology development marketplace has experienced technological mutations at breakneck speeds. Traditional approaches to the factory automation environment—where a rigid known-shape object is positioned at a fixed location and the robotic gripper performs a manipulation task on the object—have seen fundamental changes. Current robot systems face much more difficult challenges compared to the past. Target objects whose shapes, sizes, and properties are either variable or unknown have to be handled. The materials range from fragile and slippery to highly elastic, while the operations are within environments that are either not so well defined or changing over time. As such, now modern grippers are supposed to work with devices that would offer various desired functionalities, such as 3D imaging systems, additional robotic arms, and even human employees in collaborative robotics settings.

In such scenarios, the role of "intelligent grippers" has appeared in order to fully exploit the advantages of cloud computing, feature advanced sensors allowing for real-time feedback on the application of forces as well as slip detection, and work within an integrated control loop. Specifically, this study explores these advanced grippers in the treatment of sensitive tissues, but with an emphasis on the red meat industry and in particular with automated grabbing and manipulation of large amounts of pork meat or intestines-an operation critical to be accomplished in slaughterhouse automation. The paper attempts to outline optimal design strategies for the gripper systems within the RoBUTCHER project, which aims to design the world's first completely automatic slaughter cell. The most characteristic gripping challenges in this study included securely holding and manipulating limbs and parts of the trachea of the carcasses so that the internal organs could be removed; therefore, the grip must be firm and secure on objects soft, slippery, and roughly cylindrical in shape.

For the latest developments in soft material grippers, the present study utilized the PRISMA methodology to carry out a comprehensive literature review with the PRISMA-flowchart (Llanos et al., 2025). Although this methodology can be applied excellently in screening papers within selected databases, it, however, requires the consideration of overlooking studies that were available but not published in these databases. In this regard, the researcher accessed online databases and search engines, such as Google Scholar, NCBI Pubmed, IEEE Xplore, among others, for the last five years' research output. Certain specific keyword terms were used for the search, with patents excluded by using the following terms: "Meat gripper," "Soft tissue gripping," "Meat processing gripper," "Red meat processing robot," "Robotic pig slaughter," and "Pig slaughter automation."

Despite strong filtering applied in the search process to eliminate concepts that are not relevant to designs, some very potent gripping strategies theoretically incompatible with the particular field of application appeared in our detailed survey (Takacs and Haidegger 2020). These mechanisms often function extremely well for small objects under conditions of laparoscopic surgery but fail to scale to the desired applications, such as grippers using the Coanda or capillary effects. The critical features of a standard robotic gripper have been identified and classified according to their relevance to the food processing and meat-handling industries before discussing related designs and research papers in Table 5.2.

It should be noted that not all the technically critical specifications are quantitatively elaborated in patent literature, and it becomes difficult to understand which of the designs proposed could be used for specific tasks. In this regard, our review is based on the conceptual and principle-based foundations of the presented grippers, with a secondary focus on the physical models of the grippers themselves.

Probably, the most advanced among robotic gripper designs is that of anthropomorphic and similar-to-advanced prosthetic-hand features, as well as hand-like mechanisms of similar shape. They are intended to simulate the function and appearance of natural human hands to the maximum extent possible. Such a design could be useful in either of two ways: either as a total substitute for human hands or even as an end effector equipped with performing sophisticated handling tasks robot arms. The thrust might be different at a core level, but the result as far as finger dynamics and actuation procedures are concerned is virtually the same, which leads to a conclusion that principles behind the prosthetic hand design can be generally transferred to industrial grippers, and vice versa. Still, despite above-mentioned huge efforts to mimic the human hand, complexity and effectiveness are impressively set aside, and to create one seems to be well-nigh impossible.

5.4.2 A Robotic Gripper with Scooping and Binding Capabilities for Managing Diverse Food Items

Food products have different shapes, sizes, textures, and softness. This makes it a tough task for robots to handle food products since they come in so many shapes and sizes. It would be very advantageous if there were an all-purpose gripper that takes on various food products yet proved versatile. We here present a brand-new robotic gripper which can be used for scooping and binding, designed for this type of variability. This gripper generates a pneumatic parallel actuator coupled with two symmetrical scoopingbinding mechanisms. Thin scooping plates and numerous rubber bands were attached to every machine on the two sides which loop around and hold the object being grabbed. First, the food is shaped by the scooping mechanism which is mainly held by the scooping plate. Then, around the food item, rubber strings extend so that they can tighten up for a secure hold. This architecture renders high-speed and effective pick-and-place operations feasible. For considering the performance and working of this gripper, two models were used: a simple beam model and a more intricate finite element model. The

Table 5.2 Types of IR end-effector grippers applied in food handling (Campos et al., 2011; Lien, 2013; Salvietti et al., 2019).

Gripper mechanism	Pinching	Enclosing	Pinning	Pneumatic	Freezing	Levitating	Scooping
Description	Mechanical grip between two or more "fingers," hold established via friction between the "fingers" and the component, release achieved by opening "fingers"	Claw/jaw-like attachments encompass components to achieve partial or full grip, release achieved by opening of apparatus	One or more pins are injected into components, the penetration establishes a surface or deep grip, release achieved by extraction of pins	Air or pressurized gas is used to establish a grip via vacuum, release achieved by removal of pressure	Ice is formed between the gripper and the foodstuff component via instant freezing points, release achieved by instant melting of ice	Based on the Bernoulli concept, grippers lift components via differential air velocities, release by interruption of air flow	Gripper design can be flat or parabolic, allows for pick-up of foodstuff in "sweeping" motion, release by tipping
Advantages	Simple, effective, reliable, nonintrusive		Strong single-sided grip, precise grip, quick, hygienic	Surface grip only, compact gripper, reliable, nonintrusive	Surface single-sided grip, hygienic		Portion control, simple, effective, nonintrusive, quick

Disadvantages	Possible bruising of food components; Food residue may collect in grooves (dependent on design of gripper)		May cause visual quality decline, may tear tissue, and can collect food residue	Loss of grip can occur, may collect food debris	May have unpredictable releasing action	May have unpredictable releasing action, most suitable for flat foodstuffs	Food residue may build up, control of material can be difficult, foods may spill during motion, may deform foods
Foodstuff component characteristics	Rigid, semi-rigid, Non-deformable, deformable, Nonsticky, slippery		Rigid, semi-rigid, Nondeformable, slippery	Rigid, semi-rigid, Nondeformable, deformable, smooth surface, Non-sticky, slippery	Rigid, nonrigid, semi-rigid, Nondeformable, deformable, smooth surface		Rigid, non-rigid, semi-rigid, non-deformable, slippery, Nonsticky
Example application	Pick and place of confectionary items and small decorative foodstuff	Pick and place of fruits and vegetables for sorting, packaging and palletising	Pick and place of meats and poultry, fish and seafood	Pick and place of eggs, packaging and palletising	Pick and place of meats, poultry, fish, seafood or frozen fruits and vegetables	Pick and place of soft, light foods such as pastas and baked goods	Loose foods, Sauces, powders, cooked rice, portioning ready meals

results obtained from the analyses were verified via real-world testing. An analytical model was developed based on the elastic properties of the rubber strings, accounting precisely for the binding force that ensures safe handling, under consideration of the physical dimensions and displacement of the food items during binding. Twenty different foods were then subjected to intense testing with some sporting smooth surfaces or very thin profiles. After each work executed in the order of almost 4 s average operating time, the scooping-binding gripper undoubtedly demonstrated its capability. The results demonstrate potential application of the gripper in the food processing industry in general.

In order to include versatility and accommodate objects of different sizes and forms, the gripper mechanism discussed in the study is very adaptable as it consists of a flexible string in its design. Using a single motor to wind the string while the rigid rods close, this gripper secures things within its tight loop. Traditional binding grippers have problems when handling thin, heavy, or slippery objects. The former has more strings arranged in a parallel manner to provide greater gripping and friction forces (yellow and purple arrows, respectively). This allows wrapping of a 3D object, thus enhancing stabilization of an object during grasp. Additionally, as suggested by red arrows, this device has two thin plates at the bottom of the gripper which slide beneath the object for provide needed support. Because they support the weight of the object and stability by preventing it from sliding, such plates are very essential when dealing with thin, heavy, and slippery objects.

The detailed design for the scooping-binding gripper is shown in Figure 5.15. It employs a single two identical scooping-binding mechanisms with an elastic thread that form a closed loop so it can be used as a lifter to grip the pneumatic parallel actuator. Mechanisms have ultrathin plate, stiff rods, springs, nuts, and several strings running between rods. Springs help in permanent contact and to minimize impact. The rods are perforated to accept the strings; the bolts secure the rods to the shoulder connection and provide spring tension adjustment. One string is wound around the rods in a winding pattern and thin plates are attached, though a slot at one end of the rod figure. The string is spaced apart by an exact measurement of 6 mm. Dimensions of the thin plate are defined to contain the 10 mm wide and 1 mm thick and slant length 3 mm (Figure 5.15d).

The shoulder connecting piece, the strong rods, and the thin plate were printed and assembled with a 3D printer (Prusa I3 MK3, Prusa Research, Prague) using PLA as building material. After production, these parts were mounted together with springs, nuts, and a string, as presented in Figure 5.15a, forming one scooping-binding mechanism. The spring used for this assembly has an outside diameter of 7.5 mm, a resting length of 19 mm, and a stiffness rating of 2.55 N/mm. In the string part, a 3 mm diameter, Shore A55 hardness silicone rubber cable (3-2316-02, AS ONE, Osaka) was inserted through the appropriate holes in the rods. A typical pneumatic parallel actuator (HLC-12ASL1, Kondoh Mfg.Co., Ltd.,

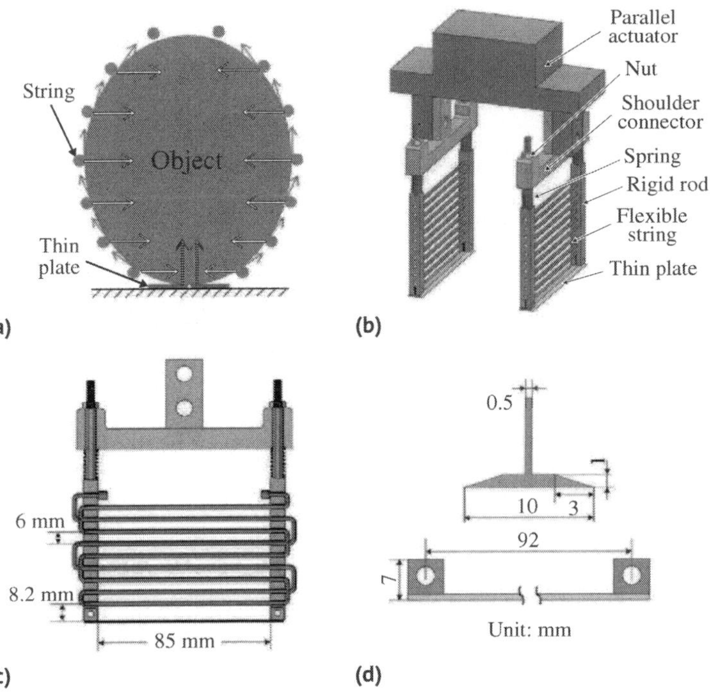

Figure 5.15 A scooping-binding gripper. (a) shows the concept, (b) depicts the CAD design of the gripper, (c) presents a section view of the gripper showing the dimensions and the string winding, and (d) depicts the design and dimensions of the thin plate (Morales et al. (2014)/Sage Publications/CC BY 4.0).

Nagoya) with a stroke of 60 mm powers the open and close movements of the mechanism. Figure 5.15b shows that this mechanism, having a flange base, was attached to the parallel actuator to complete the scooping-binding gripper assembly. The changes in the grasp width are determined by the scoopingbinding mechanisms' configuration on the flange base. As can be seen in Figure 5.15c,d, the connection at the outer edge has a maximum opening of 84 mm with a minimum closing at 24 mm. On the other hand, as presented in Figure 5.16 an inner side connection offers a smaller opening of 60 mm, and the fully closed condition is at 0 mm. The aggregate weight is 536.8 g, and it incorporates a 440 g pneumatic parallel actuator that is included in the assembly.

5.4.3 Beam Model Analysis

We modeled this scooping-binding mechanism as a simple bending problem in a beam. As the contact indentation is vertical to the displacement of x, a vertical contact force $N = kx$ is created by the end of the rigid rod as described above,

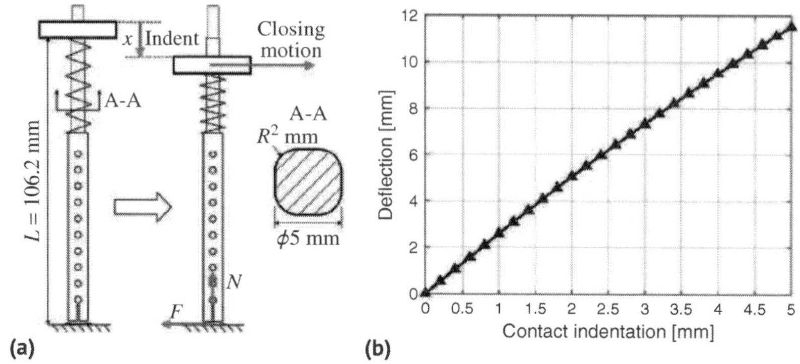

Figure 5.16 A gripper that binds and scoops. The concept is shown in (a), the gripper's CAD design is shown in (b), a section view of the gripper is shown in (c) with the string winding and dimensions, and the thin plate's design and measurements are shown in (d) (Morales et al. (2014)/Sage Publications/CC BY 4.0).

and as shown schematically in Figure 5.16, here k is the spring constant. When the gripper closes, the shoulder connector induces horizontal motion, producing a friction force $F = \mu N$ at the end of the rigid rod, where μ is the friction coefficient. This friction causes the rod to bend. We use simple beam theory to find the bending deflection d of the rod at any indentation x.

$$d = \frac{\mu k(L - x)^3}{3EI} x \tag{5.20}$$

5.4.4 Determining Binding Force Estimation

We first developed analytical models to find the appropriate binding displacement for known objects. For example, when the string is used to bind the object, we have studied three objects differing in shape. The red dashed line is the undeformed string and the line in color is the deformed string. When the gripper closes, the right rod moves to point $P_1(L, d)$, where L is the half length of the undeformed string and d is the displacement associated with the binding. Point $P_2(x, y)$ represents the last contact between the object and the string. For a circular object, the coordinates (x, y) can then be computed by the two following equations:

$$\begin{aligned} (L - x)x + (d - y)(y - R) &= 0 \\ x^2 + (y - R)^2 - R^2 &= 0 \end{aligned} \tag{5.21}$$

The first equation represents the normal overlap of the two vectors of $\overrightarrow{P_1P_2}$ and $\overrightarrow{CP_2}$. The second one represents the distance from point P_2 to the object's center.

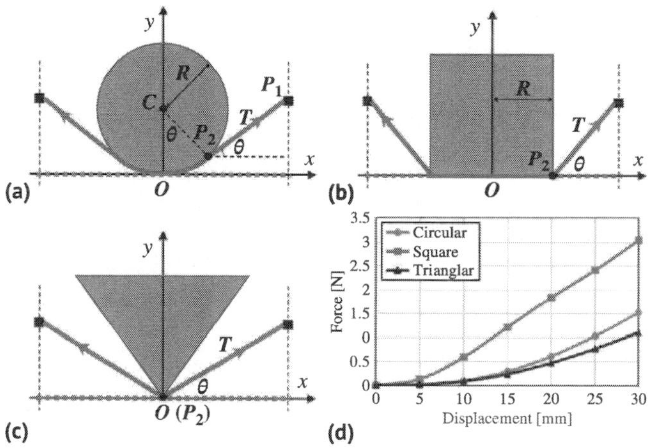

Figure 5.17 Model diagram for calculating the binding force while grasping a (a) circular, (b) square, and (c) triangular object, respectively. (d) depicts the calculated binding forces against different binding displacements (Wang et al. (2021b)/Frontiers/CC BY 4.0).

From Figure 5.17 for the square and triangular objects, respectively, we can determine directly that the coordinates of the contact point are $(R, 0)$ and $(0, 0)$. We can then find θ knowing the coordinates of P_2.

Circular : $\theta_c = \arctan \frac{d-y}{L-x}$, Square: $\theta_s = \arctan \frac{d}{L-x}$, Triangular : $\theta_t = \arctan \frac{d}{L}$. The deformed length (L_c, L_s, L_t) and strain $(\varepsilon_c, \varepsilon_s, \varepsilon_t)$ of half string can be then calculated by

Circular : $L_c = R\theta_c + \frac{d-y}{\sin\theta_c}$, $\varepsilon_c = \frac{L_c-L}{L}$ Square : $L_s = R + \frac{d}{\sin\theta_s}$, $\varepsilon_s = \frac{L_s-L}{L}$ Triangular : $L_t = \frac{d}{\sin\theta_t}$, $\varepsilon_t = \frac{L_t-L}{L}$, where R denotes the radius of the circular object and the half length of the square object. The tension force of the deformed half string can be then obtained by substituting the strain values into equation $T = 7.8929\varepsilon^3 - 8.2711\varepsilon^2 + 5.4335\varepsilon + 0.0514$. We can disregard the impact of friction if we assume that there is no slippage between the object and the string. As a result, we can ultimately determine the binding force produced by the string's expansion as

$$F = 2T\sin\theta \tag{5.22}$$

Based on the above analysis, we conducted simulations by supposing the parameters of $L = 42$ mm, d varying from 5 to 30 mm with an interval of 5 mm, and $R = 30$ mm. The binding displacement d grasping a square object generates much larger binding force comparing to a circular or triangular object.

5.5 Robotic and Vision Systems in Use for Fish Processing

A new technology has been developed which will automate a vision system for checking the cleanliness of fish processing lines. The system works with advances in neural networks in identifying trace amounts of fish blood on sanitized surfaces. We experimented with configurations and settings for different neural networks to determine the best scenario. We created datasets using data augmentation methods to mimic various conditions that are present during the processing of fish. In our experiments, we added a great amount of augmented data to the training dataset and found that it improved the performance of the neural network significantly. This was significantly more than the use of traditional color thresholding or training the neural network with no augmented data. To enhance the reliability and effectiveness of the system further, we conducted industrial-level trials. We recommend the use of the system to process images by using neural networks with augmented data because of its effectiveness. We therefore recommend employing an area scan camera for image acquisition. We captured our dataset using a Sony IMX214 image sensor at resolution 4160 × 2336 pixels. We used a F2.0 28 mm wide angle lens at about 800 mm from the subject in capturing the images. Testing equipment used included a conveyor along with an electric stunner, widely implemented in fish processing lines. These machines have a stainless steel belt and plastic parts, which often stain with blood while being manufactured.

Convolutional Neural Network

CNN is a feedforward neural network architecture that is a combination of convolution, pooling/subsampling and fully connected layers. The output of a convolution is given by

$$\mathbf{S}(i,j) = \sum_{m}\sum_{n} \mathbf{I}(i-m, j-n)\mathbf{K}(m,n) \tag{5.23}$$

where the input from the previous layer is **I** and the convolution kernel is **K**. The stride moves the locations (i, j) at which the result is computed. Once the kernel has processed all the input, the whole feature map is created. Many different kernels within a layer can yield many different output feature maps. One result of layer l of a convolution layer when it receives several feature maps as input is:

$$\mathbf{I}_l(i,j) = \sigma\left(b + \sum_{k}\sum_{m}\sum_{n} \mathbf{I}_{l-1}(i-m, j-n, k)\mathbf{K}_l(m,n,k)\right) \tag{5.24}$$

where b is the bias, σ is the activation function, and k is the index of the feature maps in the input $\mathbf{I}_{l-1}$ from the preceding layer. A nonlinear activation function, such as a Rectified Linear Unit (ReLU), typically follows the convolution, cutting down on computation time without sacrificing accuracy. The ReLU function is:

$$\sigma(x) = \begin{cases} x, & x \geq 0 \\ 0, & x < 0 \end{cases} \tag{5.25}$$

One of the techniques to downsample the feature maps is pooling. In addition, the pooling policy makes the network less sensitive to input translations. One of the most widely used pooling policies is max pooling. Pool operates by sliding a window over the feature maps; in max pooling, the highest value from that window is preserved. The output maps' size is determined by the stride and window sizes. A CNN's fully connected layers are analogous to a Multilayer Perceptron's (MLP) layers. The middle layer l neuron j output is

$$a_j^l = \sigma\left(z_j^l\right) \tag{5.26}$$

where σ is the activation function, and where the argument is the weighted sum

$$z_j^l = b_j^l + \sum_k w_{jk}^l a_k^{l-1} \tag{5.27}$$

of the outputs a_k^{l-1} from neuron k in layer $l-1$. Here, w_{jk}^l are weight factors, and b_j^l is the bias. In the first layer $l = 1$, and the input is $a_j^0 = x_j$, where $x_j, j = 1, \ldots, N$ are the inputs of the neural network. As the layer is fully connected, the weighted sum of the outputs of layer $l-1$ is over all neurons k. Backpropagation is a supervised learning technique that allows neural networks to generalize on training data. Backpropagation is a technique that computes the network prediction error and then minimizes it by computing the gradient through the network backward and adjusting the weights to reduce the error. In terms of the cross entropy, the loss function has been employed.

$$C = -\sum_{j=1}^{n} y_j \log\left(\hat{y}_j\right) \tag{5.28}$$

where $\hat{y}_j = a_j^L = \sigma\left(z_j^L\right)$ is the predicted probability value for class j in the final layer L, and y_j is the true probability. For our classification problems, the activation function in the output layer is the softmax function:

$$\sigma(\mathbf{z}))_j = \frac{\exp\left(z_j\right)}{\sum_{k=1}^{K} \exp\left(z_k\right)} \tag{5.29}$$

The true probability has a one-hot encoding, i.e. $y_j = 1$ for the correct class, while $y_j = 0$ for all other values of j[4]. This error is propagated backward in the network and the weights of the neurons are updated to minimize the error. The flow in a CNN is shown in Figure 5.18.

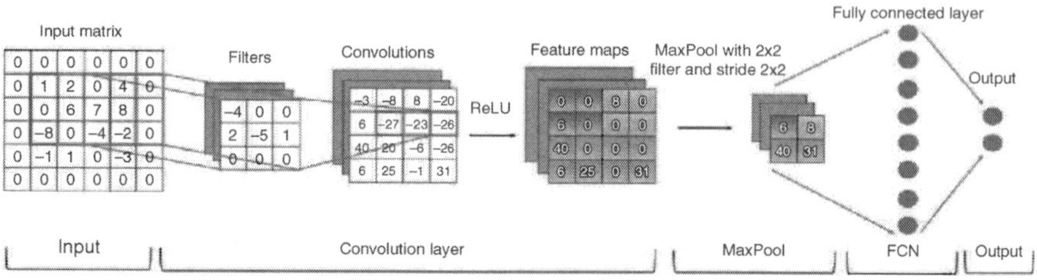

Figure 5.18 Figures for a 6 × 6 × 1 input in a basic CNNconvolutional neural network with three 3 × 3 filters with ReLU activation in the convolution layer, two 2 × 2 filters and stride 2 × 2 in the MaxPool layer, eight neurons in the fully connected layer, and two neurons in the output layer. The fully connected layers are not calculated. Convolutions, feature maps, and extra filters are represented by the light shade planes (Bjørlykhaug and Egeland, 2019 Bjørlykhaug and Egeland, 2019).

Weight Initialization

The capacity of the network to train is significantly affected by how the weights are initialized, especially as the network gets deeper. If the weights are too small, the signal will get smaller as it moves through the layers until it is no longer usable. As the signal descends, excessively large weights will cause its amplitude to skyrocket. To solve this problem, Glorot and Bengio developed the Xavier weight initialization.

$$\mathrm{Var}[w] = \frac{2}{n_{\mathrm{in}} + n_{\mathrm{out}}} \tag{5.30}$$

where n_{in} is the number of connections in to a neuron, and n_{out} is the number of connections out of a neuron. The equation is often simplified to

$$\mathrm{Var}[w] = \frac{1}{n_{\mathrm{in}}} \tag{5.31}$$

The weight initialization, on the other hand, is based on the assumption of linear activation and has been shown to work fairly well for traditional nonlinear activation functions such as the sigmoid function and the hyperbolic tangent. This assumption is no longer valid for more recent activation functions, such as ReLU and PReLU, which require weight initialization to be tailored depending on the activation function. This can further enable convergence of even deeper CNNs. The initialization of the ReLU weight is

$$\mathrm{Var}[w] = \frac{2}{n_{\mathrm{in}}} \tag{5.32}$$

because a rectified linear unit would output zero for half of its inputs when the distribution has a mean of zero.

Dropout

According to Srivastava et al., dropout technique might increase the generalization of a network from training data, thus preventing overfitting. Unlike traditional regularization techniques, which add penalties to weights on neurons, dropout effectively drops a set fraction of neurons in a given layer during the forward pass at random. This usually results in more neurons in affected layers and therefore increases training time.

Enrichment of Training Set

Some projects may face difficulties in acquiring sufficient labeled training data (Hemamalini et al., 2022). Among various approaches to overcome this, one of the most efficient ones is data augmentation. In effect, it increases the size of the training set by creating new artificially created data that enhances the performance of the neural network by achieving better accuracy during the cross-validation and testing phases. Initially, our training data consisted of 43 images in class 1 and 57 images in class 2. As our image classification task is independent of the spatial orientation or rotation of features separating the classes, we simply followed a simple protocol to increase the number of data images. We first rotated all copies of all images thrice, each 90° apart, which effectively multiplied our dataset by a factor of 4. We then applied fourfold cropping with each crop retaining 80% of the corners.

How to Deblur

Camera lenses become foggy due to wet and humid environments present in the later stages of cleaning in the processing plant. This may sometimes cause images to appear misty, significantly compromising their quality. To address this issue, another training set comprises images that are out of focus. The blurring method uses Gaussian filtering

$$I(x,y) = \frac{1}{\sqrt{2\pi\sigma}} \exp\left(-\frac{d^2}{2\sigma^2}\right) \tag{5.33}$$

where $d = \sqrt{(x-x_c)^2 + (y-y_c)^2}$, (x_c, y_c) is the center of the filter and σ is the standard deviation. For our experiments, $\sigma = 2$ is used.

Adjusting Contrast

To enhance the CNN's adaptability to different environments, we crafted a dataset through the application of contrast adjustments. This process involved creating images with both elevated and reduced contrasts. Specifically, images designated for reduced contrast had their contrast levels decreased to 50% of that found in the original images. Conversely, images slated for increased contrast were adjusted so that the lowest and highest 1% of pixel values reached saturation points.

Lighting Variability

Lighting conditions can greatly differ across processing facilities. Furthermore, the presence of mist within a facility following cleaning routines can alter the lighting conditions. To ensure the dataset accounts for these variations in illumination, we manipulated the lighting within our images. This entailed generating images that were both lighter and darker in appearance. This manipulation allows for the creation of new images by

$$I_{\text{out}}(x,y) = I_{\text{in}}(x,y) + \gamma \tag{5.34}$$

where γ is the change in illumination. The effect of changing the illumination is shown.

To enhance the robustness of the neural network and mitigate overfitting issues within the training dataset, further data augmentation methods were applied to the training data. This process is henceforth known as Image Transform (IT). IT involves the manipulation of images through flipping and warping. The various steps in the Image Transform are sequentially applied, each with distinct probabilities.

$$I_{\text{out}} = I_{\text{in}} * \text{Flip}(x_1) * \text{Flip}(x_2) * \text{Warp}(x_3) \tag{5.35}$$

where x_1 is the probability of flipping the image using the first transform, x_2 is the probability of flipping it again, and x_3 is the probability of warping the image. There is an equal probability that the flip transform will not flip it, flip it horizontally, flip it vertically, or flip it both horizontally and vertically. The function for the warp transform is

$$I_{\text{out}}(x,y) = I_{\text{in}}\left(\frac{M_{11}x + M_{12}y + M_{13}}{M_{31}x + M_{32}y + M_{33}}\ \frac{M_{21}x + M_{22}y + M_{23}}{M_{31}x + M_{32}y + M_{33}}\right) \tag{5.36}$$

where M is the transformation matrix

$$\begin{bmatrix} t_i x_i' \\ t_i y_i' \\ t_i \end{bmatrix} = M \cdot \begin{bmatrix} x_i \\ y_i \\ 1 \end{bmatrix} \tag{5.37}$$

where (x_i, y_i) are the quadrangle vertices

$$\begin{bmatrix} 0 & \text{cols} & \text{cols} & 0 \\ 0 & 0 & \text{rows} & \text{rows} \end{bmatrix} \tag{5.38}$$

where cols and rows are the number of columns and rows in the image, respectively, while (x_i', y_i') are the quadrangle vertices

$$\begin{bmatrix} \delta * r_1 & \text{cols } + \delta * r_3 & \text{cols } + \delta * r_5 & \delta * r_7 \\ \delta * r_2 & \delta * r_4 & \text{rows } + \delta * r_6 & \text{rows } + \delta * r_8 \end{bmatrix} \tag{5.39}$$

where $r_i, i = 1, \ldots, 8$, are random numbers that are uniformly distributed in $[-1, 1]$ and δ is the maximum deviation from the original position.

Early stopping one straightforward method to prevent overfitting involves early stopping. This approach entails ending the training process as soon as the accuracy on the validation set begins to decline.

Down-sampling

Before all the data manipulation steps, the images are downsampled to 100×100 by bicubic interpolation to reduce the computational cost.

Using the techniques described (Bjørlykhaug and Egeland, 2019) several datasets are generated. Dataset 1 is the base dataset where rotation and cropping has been applied to increase the size, without increasing the complexity of the dataset. Dataset 2 consists of blurred images obtained by the technique explained in Subsection III-E. Dataset 3 has both images with higher and with lower contrast as explained by Subsection III-F, and thus has twice as many elements compared to dataset 1 and 2. Similarly, dataset 4 also has twice as many elements as dataset 1 and 2, using the technique described in Subsection III-G. Dataset 5 is a combination of dataset 1, 2, 3 and 4, and thus the largest and most complex.

Name	Description	Elements of class 1	Class 2
Dasaset 1	Base augmented dataset	688	912
Dataset 2	Blurred dataset	688	912
Dataset 3	Contrast manipulation	1376	1824
Dataset 4	Illumination	1376	1824
Dataset 5	All	4128	5472

For our performance evaluation, we use accuracy:

$$\text{Acc} = \frac{\text{TP} + \text{TN}}{\text{P} + \text{N}} \tag{5.40}$$

where TP is true positive, TN true negative, P all the images with blood and N all the images without blood.

For the suggested CNN technique, the datasets are separated into 80% training data and 20% cross-validation data. Both the training data and the cross-validation data are used to train and assess the CNN technique. Overfitting is a significant issue in ML. When a ML system learns the peculiarities and noise of the training data rather than broad patterns, this is known as overfitting. When this happens, the network's accuracy on the training data is significantly higher than its accuracy on the cross-validation data. Since it demonstrates how effectively the network can predict on data it has never seen before, the accuracy of the cross-validation data is thus the most crucial performance metric.

5.5.1 Handling, Sorting, and Inspection in Food Industry

Inspection for flaws prior to product shipment or consumer receipt is a crucial part of guaranteeing product quality. Delivering goods of superior quality is more important for satisfying customers than quantity. The consumer is a major factor in determining what makes a product quality. In essence, quality is the culmination of all the characteristics that lead to goods that satisfy consumers. Utilizing machine vision and image processing technology to improve the surface quality of fruits and other food items has been increasingly popular recently. These technologies' capacity to outperform human vision abilities has led to a major improvement in areas that were previously difficult for human inspectors, which has contributed to their rapid acceptance. By utilizing computer vision and image processing, these technologies enable more objective and efficient industrial quality control techniques, which simplify and expedite traditionally laborious and subjective operations. The paper under discussion examines how to inspect and assess food products using ML and image segmentation, with a focus on classifying fruits and spotting rotting indicators. Histogram equalization is used to improve image quality after Gaussian elimination techniques are used to reduce image noise. Next, the K-means clustering technique is applied to segment images. Then, fruit photos are classified using ML algorithms like KNN, SVM, and C4.5, which identify the fruit's quality and whether or not it is damaged.

As seen in Figure 5.19, the +e approach is explained in terms of five main stages. In-depth information is as follows:

- Step 1 entails gathering a varied dataset made up of photos featuring food.
- Gaussian elimination is the image preprocessing technique used in Step 2, which is essential for removing noise and improving image clarity. This phase is essential for improving image quality, which makes categorization more accurate.

- Image enhancement, accomplished through histogram equalization, becomes the main focus of Step 3. This method increases the accuracy of the categorization while simultaneously enhancing the visual quality of the photos.
- Step 4 presents the K-means algorithm, which divides an image into several portions by image segmentation. This segmentation is an essential phase in the process since it helps pinpoint specific image sections that are indicative of sickness.
- Lastly, Step 5 uses a variety of classifiers to segment and preprocess the photos. Once the image segmentation and noise-reduction processes are finished, the dataset is ready for classification. At this point, three different classification techniques are being used: C4.5, SVM, and KNN.

From initial preprocessing to final classification, this organized technique makes it easier to handle images in a systematic manner, guaranteeing high-quality inputs for precise illness identification and classification.

The agriculture industry faces numerous obstacles when using traditional techniques of assessing fruit quality, which rely on manual visual examination and are labor-intensive and prone to inconsistent evaluation. With the ability to significantly alter the visual inspection process, the study of automated defect detection using computer vision and ML technologies has become one of the most promising areas of research. In this work, we provide a novel and effective machine vision system built on state-of-the-art deep learning methodologies and stacking ensemble techniques. This technology seeks to cost-effectively and noninvasively automate the process of evaluating the overall quality and freshness of fruits. In order to determine the best model for fruit quality grading, we assessed and benchmarked the performance of a number of cutting-edge deep learning architectures, such as ResNet, DenseNet, MobileNetV2, NASNet, and EfficientNet. A real-time visual inspection capability is also included in our system. It makes use of a low-cost Raspberry Pi device that has a touchscreen interface and a camera for simple user engagement. This real-time module can accurately classify and segment each fruit in a picture. With the EfficientNet model, the system achieved impressive average accuracies of 99.2% for apples and 98.6% for bananas during training and validation using datasets containing both fruits. Additionally, incorporating stacking ensemble deep learning techniques resulted in a little improvement in detection accuracy, increasing the identification rate by 0.03% for apples and 0.06% for bananas. When our system was applied to the same datasets, it performed better than previous methods. Furthermore, actual fruit samples tested in the real world produced accuracy rates of 93.7% for bananas and 96.7% for apples, confirming the system's efficacy.

5.5.2 Materials and Methods

The diagram in Figure 5.19 shows the workflow of the system proposed. It comprises three primary elements: (i) a classifier that utilizes computer vision and deep learning to identify different fruits, (ii) a Raspberry Pi camera module

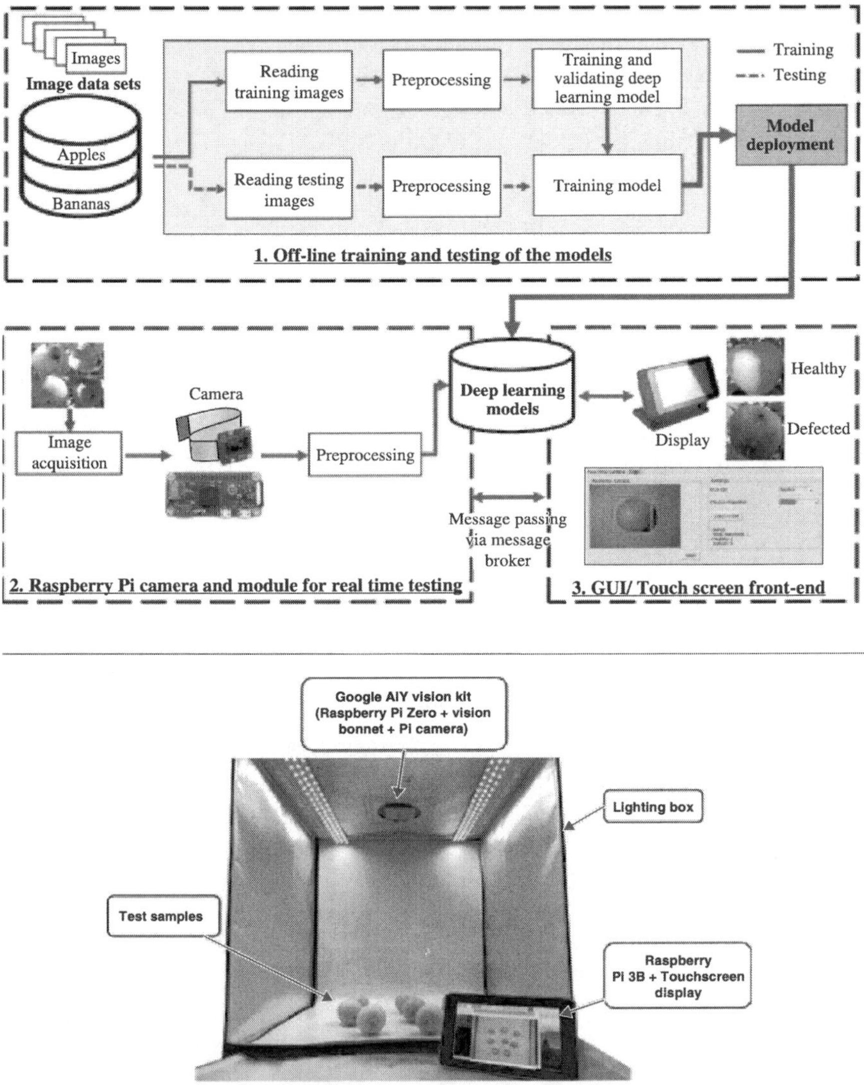

Figure 5.19 Hardware of the proposed system and experimental set up (Ismail and Malik [2022]/Elsevier).

responsible for capturing images in real-time for analysis, and (iii) a graphical user interface on a touchscreen that presents the classification results as they happen. Initially, cutting-edge deep learning methodologies underwent training with fruit image data sets. Afterward, the refined models specific to each technique were preserved and utilized for real-time analysis. The following subsections delve deeper into the system's components.

Hardware Configuration: A prototype built for testing the deep learning framework involves the assembly of two Raspberry Pi units: one is a Google AIY Vision Kit (inclusive of a Raspberry Pi Zero with a vision bonnet for executing neural network operations and a Pi Camera), and the other is equipped with a touchscreen for real-time display of feedback captured by the camera. To ensure optimal lighting conditions for precise real-time analysis, a lightbox was incorporated to produce even lighting across samples. Developing the Classification Model The purpose behind crafting this real-time visual evaluation system for fruit grading was to harness deep learning models trained on existing data sets. Detailed below are the specifics regarding the data sets, how the images were processed before training, and the approach to training/testing the models. Data Collection Two distinct data sets were the focus of training and analysis. The first included 8791 images of apples, split between 4693 images of apples with defects and 3946 of those considered healthy. Sourced from the Internal Feeding Worm (IFW) database as part of the Comprehensive Automation for Specialty Crops (CASC) project, permission for usage was granted by Purdue University. This collection showcases images of four apple types: Fuji, Golden Delicious, Red Delicious, and York, across different ripeness stages for both healthy and defected specimens. Defects in apples, such as dark spots caused by worms, were visually identifiable in the images. For processing needs, photos were resized to 120x120 pixels. Figure 5.19 illustrates the distribution among the varied types of both healthy and defected apples. Given that the type of cultivar had minimal impact on algorithm learning, classification was simplified to a binary system—healthy versus defected. The second data set featured images of Egyptian bananas annotated with ripeness levels. It consists of 300 banana images classified into four ripeness stages: unripe (green banana), yellowish-green, mid-ripe, and overripe. Curated internally using a high-resolution camera, Figure 5.19 depicts the classification distribution, highlighting a notable class imbalance between the yellowish-green and overripe categories. This imbalance was addressed through data augmentation.

Image Preprocessing: To tackle the issue of undesirable noise that often contaminates images—affecting contrast, intensity, and illumination—a Gaussian filter was applied for noise elimination. This filter, a type of linear smoothing filter, uses weights from a normal distribution that are isotropically or circularly symmetrical, providing uniform smoothing across the image. When applied to an image pixel (x, y), the Gaussian kernel functions as a 2D, zero-mean discrete function, as illustrated in Figures 5.19 and 5.20.

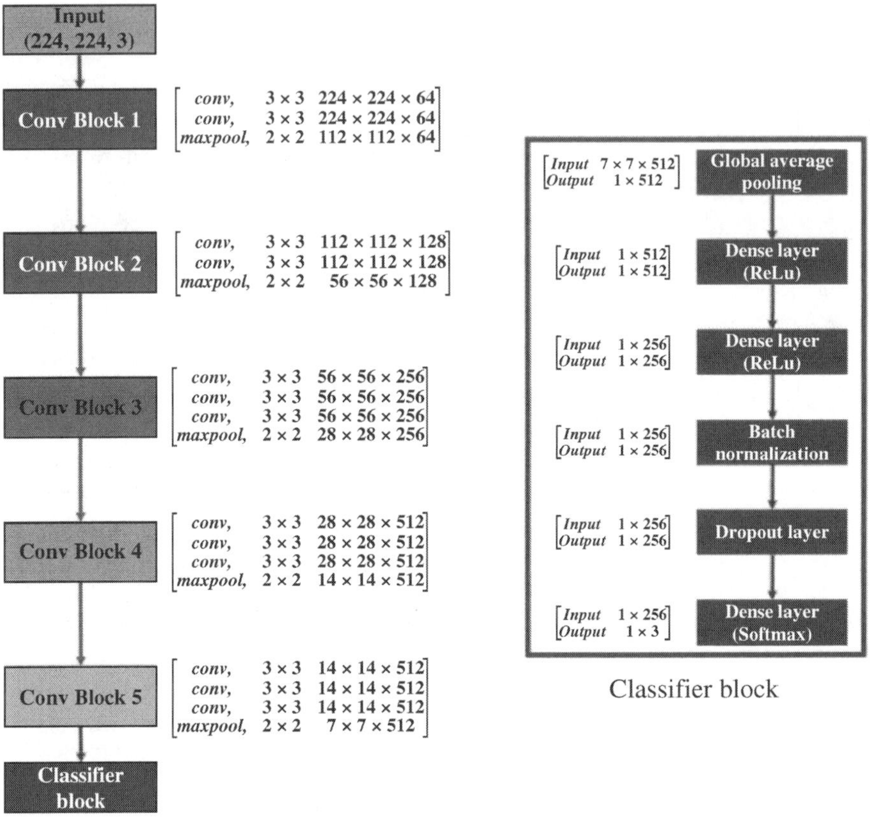

Figure 5.20 The architecture of our deep learning model based on VGG16 to classify eggs (Nasiri et al. [2020]/Elsevier).

System Overview: The diagram for the process undertaken by the proposed system has been presented as shown in Figure 5.9. The three major constituents of the system are: (i) a classifier that uses deep learning and computer vision to classify the different fruits; (ii) a module that is integrated with a Raspberry Pi camera module, captures photographs of the fruits in real-time for analysis; and (iii) a touch-screen-based graphical user interface that plots the classification results in real-time. State-of-the-art deep learning techniques were first fine-tuned on fruit image datasets. Next, the improved models unique to each method were saved and applied to real-time analysis. A more detailed description of the various components of the system can be found in the following subsections.

Hardware Setup: Two Raspberry Pi units are packaged within a prototype to be used to test the deep learning framework. The first is a Google AIY Vision Kit, which includes a Raspberry Pi Zero provided with a vision bonnet designed for neural network operations, along with a Pi Camera; the second is fitted with a touchscreen to visually feed back camera views in real time. A lightbox was used to allow for uniform illumination over samples such that the condition would be optimum for live-time analysis with no inaccuracy.

Classification Model Building: This real-time fruit grading by visual assessment system was designed with the objective of using deep learning models, which have been trained on already existing data sets. In the following subsections, information on the data sets can be found with further elucidation regarding the pre-training image processing and the training and testing methodology on the models.

Information Regarding Data Sets: Two different data sets were taken for training and analysis purposes. The first data set was of apples with images, where 8791 images of apples were used and divided into two categories: 4693 photos of apples with defects, and 3946 photos of apples that appeared to be healthy. Purdue University permitted the use of this material taken from the Internal Feeding Worm database that forms part of the Comprehensive Automation for Specialty Crops. Images of the four apple varieties: Fuji, Golden Delicious, Red Delicious, and York, both healthy and defective, at different stages of ripeness, are included in this collection. It includes data collected over years at Penn State University Fruit Research and Extension Center. In some of the pictures, it was apparent that the apples were not flawless, like those affected by the apple worm with the dark spots showing in some pictures. Images were also resized to 120 x 120 pixels for analysis purposes. A bar chart outlining the various classes for both healthy and defective apples is illustrated in Figure 5.19. Since cultivar type played a minor role in algorithm learning, the classification was reduced to a simple two-class system: defective vs. healthy. Four stages of ripening are recognized based on the 300 photos of bananas: unripe (green banana), yellowish-green, mid-ripe, and overripe. The class distribution is shown in Figure 5.19, taken using a camera with high resolution. Yellowish-green and overripe classes can be seen obviously having a terrible class imbalance. As mentioned in Section 2.3.4, data augmentation was adapted to this issue. Image preprocessing shows some images from the corpus of banana drawings. For the elimination of unwanted noise which often drenches the photographs and changes their contrast, intensity and brightness, a Gaussian filter was applied. Such a linear smoothing filter will affect uniform smoothening throughout the image with circularly or isotropically symmetrical weights from a normal distribution. Gaussian kernel is applied to an image pixel (x, y) as a 2D zero-mean discrete function. The architecture of the deep learning model based on VGG16 to classify eggs is presented in Figure 5.20.

$$G(x,y) = \frac{1}{2\pi\sigma^2} e^{-\frac{(x^2+y^2)}{2\sigma^2}} \tag{5.41}$$

By replacing a pixel in the original image with the weighted average of neighboring pixels, the Gaussian Kernel smooths in such a manner that the weights of the pixels that are farthest from the core pixel decrease monotonically. A cross-correlation function applies the entire smoothing process to the input images $f(x, y)$, and the results can be summed up as follows:

$$f'(x,y) = \sum_{s=-a}^{a} \sum_{t=-b}^{b} G(x,t) \otimes f(x-s, y-t) \tag{5.42}$$

5.5.2.1 Visual Inspection System for Grading of Fruits

A novel approach that applies deep learning and strong computer vision to mechanize the visual inspection of fruits, especially bananas and apples. The main points are as follows:

Problem Statement: Precision and effectiveness in agriculture are hindered by the sector's labor-intensive and unreliable traditional manual fruit grading (Ismail and Malik,2022). There is thus a need for an automatic solution.

To evaluate the freshness and visual appearance of fruits, the authors developed a machine vision system that depends on state-of-the-art deep learning models, such as ResNet, DenseNet, MobileNetV2, NASNet, and EfficientNet. The proposed system employs a low-cost Raspberry Pi module for real-time operation, including equipping this module with a camera and touchscreen display for direct human interaction.

Preprocessing Techniques of Image Smoothing Operational: The images of the fruit were filtered using a minimum Gaussian filter in order to enhance quality and remove noise. The type of smoothing achieved here is through the replacement of the weighted average of every neighboring pixel by this filter (Yan et al., 2017).

Histogram Equalization: Illumination problems are addressed with contrast limited adaptive histogram equalization (CLAHE), which tends to equalize the histograms of areas to enhance contrast in the images (He et al., 2016).

Training and Testing of the Model: Using the model of Efficient Net, this system was able to reach an accuracy rate of as high as 99.2% for apples and 98.6% for bananas from training and testing using the datasets of apples and bananas. Using stacking ensemble techniques significantly enhanced such performance.

Once more, the results in real testing are as follows: for real testing, the system produced real-time grading accuracy of 96.7% in apples and 93.8% in bananas. This clearly reflects its effectiveness in real-world use.

What are some problems associated with grading fruits manually and how might such problems be overcome?

Challenge: Manual grading of fruits is labor-intensive and inconsistent, necessitating automation in the agricultural sector.

Proposed Solution

System Overview: A machine vision system utilizing deep learning techniques for real-time fruit inspection.

Key Technologies

Deep Learning Models: ResNet, DenseNet, MobileNetV2, NASNet, EfficientNet.

Hardware: Low-cost Raspberry Pi with a camera and touchscreen display for user interaction.

Image Preprocessing Techniques: Smoothing Operations

Gaussian Filter: Reduces noise by averaging pixel values with their neighbors, enhancing image quality.

Histogram Equalization

CLAHE: Improves contrast by equalizing histograms in localized regions, addressing illumination issues.

Model Training and Evaluation Datasets: Trained on apples and bananas.

Performance Metrics: Accuracy: 99.2% for apples, 98.6% for bananas using EfficientNet.

Improvement: Stacking ensemble methods enhanced recognition rates by 0.03% for apples and 0.06% for bananas.

Results of Real-time Testing

Accuracy: 96.7% for apples and 93.8% for bananas during actual sample testing, demonstrating system efficacy.

Conclusion Impact: The system offers a nondestructive, cost-effective solution for fruit grading, showcasing the potential of automated visual inspection in agriculture.

Visual Elements (for Infographic) Icons: Fruits (apples and bananas), Deep learning models (brain or neural network icon), Raspberry Pi (small computer icon), Graphs showing accuracy rates.

Color Scheme: Green and yellow to represent freshness and agriculture. This infographic encapsulates the key aspects of the research paper, providing a clear and engaging overview of the automated visual inspection system for grading fruits.

5.5.2.2 Overview of the Automatic Sorting System for Unwashed Eggs

Sorting unwashed eggs using deep learning techniques, specifically through the application of a CNN. Here are the key points from the research:

Objective: The primary goal is to enhance the quality and safety of eggs by developing an automatic sorting system that classifies eggs based on their condition, including intact, bloody, and broken (cracked or with holes) eggs, even when they are dirty.

Methodology:
Image Processing: The study employed a negative Laplacian of Gaussian (LoG) operator to enhance the visibility of cracks in egg images. Additionally, a machine vision system is utilized to assess both internal and external quality attributes of the eggs, such as yolk density, pH, area index, and weight (Nasiri et al., 2020).

Data Augmentation: To improve the training dataset, various image processing techniques are applied, including rotation, zoom, and flipping, which help in generating additional training images without incurring extra labeling costs.

CNN Architecture: On top of the pretrained VGG-16 model, a global average pooling layer is stacked. Then dense layers, a batch normalization layer, and a dropout layer are added to it. The used pretrained model has been tuned to enhance its classification effectiveness. For the assessment of the CNN model, the fivefold cross-validation technique is adopted, which proves to have excellent outcomes. The bar graph reflects the performance by comparing the area under curve, sensitivity, specificity, accuracy, and precision in the order of 96.55, 95.59, 94.92, 97.39, and 96.16% respectively. The total accuracy attained with this procedure is 94.84%. As can be seen in Figure 5.20, the second and third columns in the convolutional block notation provide information about the kernel size, as well as the output parameters of the layer, namely height × width × number of channels, respectively.

5.5.2.3 Predictive Maintenance

Industry 4.0, or the so-called Smart Factory, is characterized by advanced computing and AI technologies integrated into networks. This new generation enables the management of Big Data coming from various sources such as sensors and intelligent devices, which are communicating sources in the environment, thereby enhancing decision-making and operational effectiveness.

The application of AI to supply chain optimization aims at streamlining processes, cutting costs, and shortening delivery times.

ML algorithms enable the choice of the correct crops based on factors, thereby maximizing agricultural productivity.

AI streamlines routes and delivery schedules in logistics and food delivery to ensure faster and just-in-time delivery of foods.

Predictive maintenance on food processing machines through AI reduces machine downtime and costs.

It provides a summary overview of ML techniques—a precursor for enhancing classification—resulting in the use of decision trees, for example, during classification. It implies creating a model that would predict specific outcomes based on incoming data from which it would be easier to categorize and analyze information about food.

This case study elevates how artificial intelligence and ML transformed the food processing and manufacturing industry. Such technologies help challenge some of the hardest problems facing food waste: inefficiency in the supply chain and food safety.

The system introduces cost-effective measures through the mechanization of processes associated with supply chains. Low prices and a short period to deliver are basics that can ensure one serves the demands of consumers well without wasting anything.

Therefore, farmers are enabled by ML techniques to adequately choose crops depending on environmental factors and trends in the market, hence increasing and making agriculture productive and sustainable.

Supply Chain Optimization
AI algorithms analyze data to streamline supply chain processes, reducing costs and improving delivery times. This is essential for meeting consumer demands and minimizing waste.

Crop Selection
ML techniques assist farmers in selecting the most suitable crops based on environmental conditions and market trends, enhancing agricultural productivity and sustainability.

Logistics and Food Delivery
AI optimizes logistics by determining the most efficient routes for food delivery, ensuring timely distribution and reducing transportation costs, especially in the perishable food sectors.

Predictive Maintenance
AI technologies enable predictive maintenance of food processing machinery, allowing for timely repairs and reducing downtime, which is vital for maintaining operational efficiency in manufacturing plants.

ML Techniques
Decision trees used for classification tasks: Decision trees help in making data-driven decisions by predicting outcomes based on historical data. This simplifies complex data analysis, making it accessible for industry stakeholders.

Density-based Clustering: Techniques like DBSCAN categorize data points based on density, which is useful for identifying patterns in food processing data.

Predictive Maintenance and Intelligent Sensors in Smart Factory
Focus on Smart Factories: The paper specifically targets smart factories, excluding studies related to smart cities, homes, agriculture, and logistics, ensuring a concentrated analysis of manufacturing and production processes (Pech et al., 2021).

Methodology: It employs a systematic review methodology, utilizing burst analysis, co-occurrence analysis of keywords, and cluster analysis to synthesize the literature. The PICO logic grid was adapted to frame the research questions.

Trends and Challenges: The paper identifies contemporary trends in predictive maintenance and highlights future research challenges. It emphasizes the growing importance of predictive maintenance in relation to Industry 4.0 technologies.

AI-enabled Predictive Maintenance Framework for Autonomous Mobile Cleaning Robots
AI-powered Predictive Maintenance Framework for Autonomous Mobile Cleaning Robots

Objective: Develop an AI-based predictive maintenance system that identifies operational safety concerns and performance degradation in mobile cleaning robots through vibration signals.

Methodology: Utilizes a four-layer 1D CNN framework trained on a dataset of vibration signals collected from the autonomous steam-mopping robot "Snail."

Data Collection Vibration Classes: Five classes of vibrations were identified: Normal operational vibration, Hazardous terrain-induced vibration, Collision-induced vibration, Loose assembly induced vibration, Structure imbalanced vibration.

Data Sources: Collected using an IMU sensor under various conditions, including different surface types (tile, concrete, carpet, etc.) and operational speeds (0.02–0.4 m/s). A total of 2500 samples for each class were recorded, with 80% used for training and 20% for validation.

Case Studies Lobby Environment
Setup: Glass sidewall and carpet floor.

Findings: Collision-induced vibrations were recorded due to incorrect mapping of the glass wall, leading to navigation issues.

SUTD Food Court

Setup: Smooth concrete flooring with dining furniture.

Findings: Collision-induced vibrations occurred during peak hours due to accidental collisions with humans and furniture.

SUTD ROAR Lab

Setup: Dynamic objects and vinyl flooring.

Findings: Regular drive without abnormal vibrations for three weeks; however, loose assembly patterns were observed in the fourth week due to wobbled wheels.

Results Performance Evaluation: The predictive maintenance framework accurately predicted performance degradation and operational safety issues by analyzing vibration signal patterns (Pech et al., 2021).

Predictive Maintenance Map: A map was generated by fusing vibration signal classes with the cartographer SLAM algorithm-generated 2D environment map, enhancing navigation and operational safety.

Prediction with Test Dataset

The class-classification ability of the trained model was evaluated with a separate set of test datasets. In this assessment, 500 test samples for each class were made, which were not used in the process of training and cross-validation. The performance of the model is measured in terms of some statistical metrics, such as accuracy, precision, recall, and F1 score defined in Eq. (5.43). In this context, we shall use the notation TP, FP, TN, FN to denote the number of true positives, false positives, true negatives, and false negatives, respectively, in keeping with the usual usage for the standard confusion matrix.

$$\begin{aligned}
\text{Accuracy} &= \frac{TP + TN}{TP + FP + TN + FN} \\
\text{Precision} &= \frac{TP}{TP + FP} \\
\text{Recall} &= \frac{TP}{TP + FN} \\
\text{F1 Score} &= \frac{2 \times \text{ Precision } \times \text{ Recall}}{\text{Precision } + \text{ Recall}}
\end{aligned} \tag{5.43}$$

The findings reveal that the model achieves an 89% accuracy rate in identifying normal operational vibrations. For vibrations induced by hazardous terrains and collisions, the model's accuracy stands at 95 and 92%, respectively. Additionally, vibrations resulting from loose assemblies and structural imbalances are accurately identified with a 94 and 91% accuracy rate. The average accuracy across all five vibration categories is reported to be 92.2%. The analysis indicates that the trained model is highly effective in distinguishing between the five types of vibrations encountered by a mobile cleaning robot operating on various surfaces. Therefore, this model is deemed highly suitable for real-time application in mobile

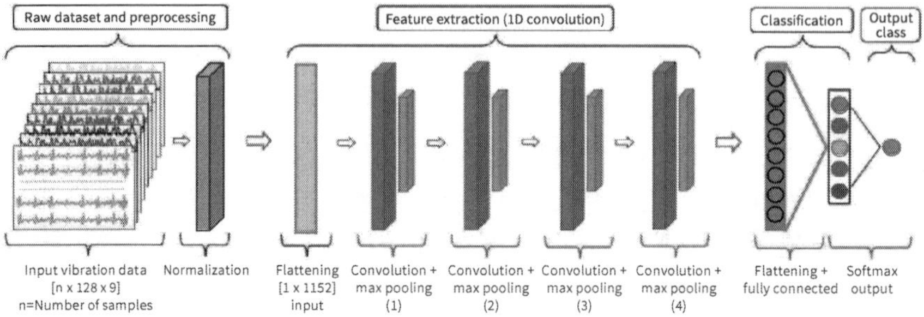

Figure 5.21 1D CNN architecture (Pookkuttath et al., (2021)/MDPI/CC BY 4.0).

cleaning robots to predict performance degradation and identify hazardous operational areas. Figure 5.21 represents the 1D CNN architecture.

5.5.3 AI and IoT Applications in Food Processing Case Studies

5.5.3.1 The Role of IoT, Big Data, and AI in Agriculture and Food Industry

Utilizes sensors to capture real-time data regarding soil moisture, temperature, and crop health. Examples include autonomous machinery such as the Deepfield robotics weed controller, which uses GPS and cameras for precision weeding.

Big Data Analytics: It analyzes large amounts of data coming from various sources such as social media to understand consumer behavior and makes further improvements in the field of food safety. The technique of UV-visible-near infrared spectroscopy (UV-vis-NIRS) is used for the assessment of food quality.

5.5.3.2 Applications of AI

Using ML techniques, ML systems work with past data and come up with predictions for future events. The result is enhanced farming operations where the decision making is framed as a better prediction.

AI can automatize processes such as yield prediction and crop monitoring. Utilize blockchain technology to ensure end-to-end tracking with vast reduction in food waste as well as increased safety. Blockchain was tested by Walmart and IBM whereby traceability time was reduced from 7 days to 2.2 s.

5.5.3.3 Impact of Social Media

Customer Engagement: Nutrition campaigns use social media channels to raise awareness of healthy eating. Social media data can be used to focus messages on the specific target audience and, as such, make health interventions very effective.

Gathering Data: Information that social media analytics could supply about consumer preferences and behavior facilitates the development and marketing of products.

In agricultural IoT applications, various sensors are important for collecting real-time data from the fields to facilitate farmers' decisions. A few of the best sensors utilized in this situation are as follows:

Most Effective Sensors for Data Collectors

1. Soil Moisture Sensors
   Function: This sensor checks the moisture level of the soil, which allows farmers to make decisions on the appropriate irrigation time for the crop.
   Effect: Repeated watering will be avoided, and hence water resources will be saved, while sustainable agriculture will be promoted.

2. Temperature Sensors
   Function: Measures the temperatures in the air as well as in the soil, where high temperatures will have a direct influence on crop development.
   Impact: It gives advance notice of frost events or heat stress that may be experienced, and this way, the farmer can prepare sufficient ways to conserve their crops.

3. Hygrometric Sensors
   Purpose: The hygrometric sensor calculates the humidity of the air, the one factor affecting plant health and the incidence of disease presence.
   Effect: Enables maintaining microclimate conditions for optimal crop development.

4. Environmental Sensors
   Purpose: Sum up data of all environmental features, such as light intensity, wind speed, and rainfall volume.
   It gives the farmer knowledge of general agricultural conditions and directs him to precision agriculture approaches.

5. Hyperspectral Imaging Sensors
   Purpose: To allow for high-resolution spectral crop acquisitions for plant health and nutritional condition determination.
   The impact could directly allow for timely identification of diseases and nutrient deficiencies through correction through expedient measures.

6. Integrated Field Sensors
   Function: Sensor data relays send information back to a central system for processing.

Impact: It advocates for the analysis of farming conditions from an integrated perspective, and then data-backed decisions are made to optimize the operation process.

5.5.3.4 Summary of Internet of Nonthermal Food Processing Technologies (IoNTP)

The article "Internet of Nonthermal Food Processing Technologies (IoNTP): Food Industry 4.0 and Sustainability" revolves around the discussion of implementing cutting-edge technology in food processing with respect to efficiency and sustainability.

It discusses the development of Industry 4.0 as a means whereby nonthermal processing technologies can be integrated into food production, by implanting smart factory and IoT elements into the management of food processing for optimization.

Nonthermal Technologies: Some of the mentioned nonthermal technologies include high-power ultrasound, pulsed electric fields, and UV-LED. They are considered because they can maintain food quality without excessively consuming energy.

Additive Technologies: The paper emphasizes the role of additive technologies, particularly 3D printing, in creating personalized and nutrient-rich food products. It discusses the processes involved in 3D printing, including the design and slicing of models, and the potential for these technologies to produce ready-to-eat products or those requiring further processing.

SWOT Analysis: A SWOT analysis is presented, identifying strengths such as energy savings and optimized environmental performance, weaknesses like high machine costs, opportunities for eco-friendly product development, and threats related to the limitations of current technologies.

Sustainability Focus: The importance of sustainability in food processing, aligning with the United Nations' sustainable development goals (SDGs). It advocates for life-cycle assessments to evaluate the environmental, social, and economic impacts of food processing technologies.

Future Directions: The review calls for further research into integrating smart sensors, artificial intelligence, and big data with nonthermal technologies to enhance the efficiency and sustainability of food production processes.

5.5.4 Case Study: Real-time Sorting of Broiler Chicken Meat

5.5.4.1 Overview Title: Real-time Sorting of Chicken Meat Prepared from Broiler Chickens Using a Robotic Arm

Focus: Validating the combination of convolutional neural networks (CNN) and local interpretable model-agnostic explanations (LIME) for the detection of freshness in poultry processing.

Objective: Automate the sorting process for an improvement in the safety of food from contamination risks.

Methodology System Architecture: Uses the Inception V3 CNN model with the application of LIME toward explainable artificial intelligence.

High precision for the identification of chicken meat as fresh or spoiled to be obtained.

Data Preprocessing

Images resized to 224 × 224 × 3 pixels for uniformity.

Training and Testing

The model was trained on a number of images of chicken meat.

Accuracy in sorting was attained with a robotic arm testing 1000 samples of fresh ones and 300 samples of rotting meat was achieved 96.3%.

Key Findings

Metrics of Performance

Accuracy: fresh meat 94.19% and rotting meat 97.24%

Sorting precision is also increased compared to traditional methods of sorting.

Interpretability:

LIME provides explainable model predictions, therefore becoming a source of scrutiny of automated processes.

Conclusion Application 1: Food Safety

Technology reduces human interaction with potential pollutants, thus making food products safer.

Operational Efficiency:

Technology helped optimize the sorting process by upgrading the benchmark set for the automated systems within food production.

Problematic Areas

The technology still struggles with its inability to decrease variability in texture and illumination, which will inherently hamper its performance.

Future Studies

Real-time learning capabilities and adaptability should be tested across various poultry products to increase accuracy levels.

Final Appraisal

This case study has indicated a new application of AI and robots in poultry processing and has realized tremendous improvements in food safety and operational

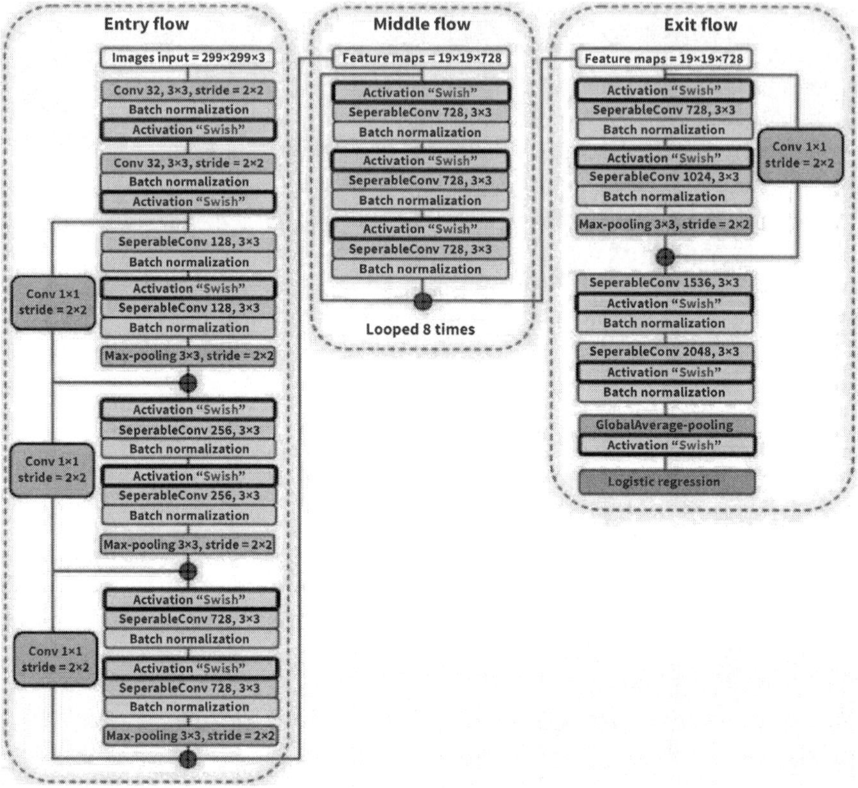

Figure 5.22 The CNN architecture (Hasan et al. [2024]/Elsevier/CC BY 4.0).

efficiency as reported by Hasan et al. (2024). The integration of CNN and LIME improves (the CNN architecture represented in Figure 5.22) sorting precision but is also guaranteed to ensure transparency in decision-making for advancement in technology in automated food processing.

5.5.5 Case Study: Fish-robot-based IoT Platform

Overview of the FishBot Objective: The FishBot should be a mobile collector in the IoUT, supporting smart ocean applications. It swims and glides through water with efficiency through behaviors mimicked from a fish and buoyancy-driven gliding.

Design Features

Multilink Structure: This is a BCF configuration with three degrees of freedom, which gives flexibility to the swimming and gliding pattern of the FishBot.

Pectoral Fins: These fins facilitate gliding attitude control and enhance maneuvering capability within the vertical plane.

Control and Modeling

Control Framework: A behavioral control framework is designed to achieve various fish-like swimming behaviors and gliding motions. These include: State Feedback Algorithm: Utilizing LQR control method for precise regulation.

Locomotion Controller: It is implemented using central pattern generators (CPG) in order to control swimming and gliding motions.

Modeling: The FishBot is modeled for three swimming modes: The Cruise Swimming Model, The Pure-Pitching Swimming Model, and 3D Swimming Model.

Experimental Validation Gliding Motion Experiments: These experiments were conducted in a 50 m × 20 m swimming pool to test gliding performance. With the help of buoyancy control and mass placement, the FishBot demonstrates smooth gliding.

Swimming Behavior Tests: Extensive experiments on hybrid motions of propulsion validate the two types of behavior gliding as well as fish-like swimming of the FishBot.

Dynamic Modeling: The future work will be the development of enhancements in the dynamic model for 3D gliding motion, and more complex environments, such as oceans, will test the performance of the LQR controller using state tracking. The FishBot is a big step toward underwater robotics. It exhibits innovative design, such as sophisticated mechanisms of control, and enables smart ocean applications. The ability to adapt swimming and gliding modes under different scenarios makes it a useful tool for the IoUT (Wang et al.,2021a).

5.5.6 3D Swimming Model for the FishBot

In studying the dynamics of the FishBot, we assume no slippage in the motion. We introduce a difference in lift experienced by the left and right side gliders by setting the angles on the left and right pectoral fins to be different. The difference-though we neglect the effect of induced drag-introduces a tilt to the FishBot, which generates a sideways force through the side component of the lift, and this promotes sideways movement. The vertical component of the lift is responsible for moving the FishBot vertically upward. Changes in the pectoral fins and the tail fin are defined with deviations: deviations δ_1 and δ_2 describe the deviation of the left and the right pectoral fins, and δ_3 describes the tail fin deviation. Deviations provide a basis for establishing equations of motion for the FishBot.

$$
\begin{aligned}
\dot{V} &= \frac{1}{m}(P\cos\alpha - D) \\
\dot{\psi}_V &= \frac{P\sin\delta_3\cos\gamma - F\sin\gamma - P\sin\alpha\sin\gamma}{\text{mV}} \\
\dot{\omega}_x &= \frac{F_{1l}\cdot\frac{1}{4}a - F_{1r}\cdot\frac{1}{4}a}{I_x} - \frac{M_{\omega x}}{I_x} \\
\dot{\omega}_z &= \frac{M_z}{I_z} \\
\dot{\gamma} &= \omega_x \\
\dot{\theta} &= \omega_z \\
\dot{\vartheta} &= \frac{F\cos\gamma + P\sin\alpha\cos\gamma - P\sin\delta_3\sin\gamma}{\text{mV}} \\
\alpha &= \theta - \vartheta
\end{aligned}
\tag{5.44}
$$

where the drag force is D and the propulsion force is P. The angle of attack of the robotic fish body is α, its angle of velocity in the horizontal plane is ψ_V, its roll angular velocity is ω_x, its pitching angular velocity is ω_z, its roll angle is γ, its pitching angle is θ, its vertical plane angle of velocity is ϑ, and its left pectoral fin lift is

$$F_{1l} = S_2\pi\rho V^2\sin(\delta_1 + \alpha) \tag{5.45}$$

and the lift of the right pectoral fin is

$$F_{1r} = S_2\pi\rho V^2\sin(\delta_2 + \alpha) \tag{5.46}$$

and the combined force of lift is

$$F = F_{1l} + F_{1r} \tag{5.47}$$

and the roll damping torque is

$$M_{\omega_x} \approx 2\int_0^{0.5a} 2\pi\cdot\frac{\omega_x r}{V}\cdot\frac{1}{2}\rho V^2\cdot dr\cdot c\cdot r = \frac{1}{6}\pi\rho V\omega_x S_2 a^2 \tag{5.48}$$

where c is the mean aerodynamic chord, and the pitching moment is

$$M_z \approx \frac{(F_{1l} + F_{1r})L_1\cos\gamma\cos\alpha}{I_z} - \frac{2}{3I_z}\pi\omega_z\rho V\frac{S_2^3}{a^2} \tag{5.49}$$

and the combined force of lift is

$$F = F_{1l} + F_{1r} \tag{5.50}$$

and the roll damping torque is

$$M_{\omega_x} \approx 2 \int_{0}^{0.5a} 2\pi \cdot \frac{\omega_x r}{V} \cdot \frac{1}{2}\rho V^2 \cdot dr \cdot c \cdot r = \frac{1}{6}\pi\rho V\omega_x S_2 a^2 \tag{5.51}$$

where c is the mean aerodynamic chord, and the pitching moment is

$$M_z \approx \frac{(F_{1l} + F_{1r}) L_1 \cos\gamma \cos\alpha}{I_z} - \frac{2}{3I_z}\pi\omega_z\rho V\frac{S_2^3}{a^2} \tag{5.52}$$

where P represents the force of propulsion and D denotes the force of drag. The angle at which the robotic fish's body attacks is represented as α, while ψ_V stands for the velocity angle in the horizontal plane. The roll angular velocity is given by ω_x, and ω_z signifies the angular velocity of pitching. The angle of roll is denoted as γ, and the angle of pitching is represented by θ. ϑ indicates the velocity angle in the vertical plane, and the lift generated by the left pectoral fin.

5.6 Review Questions

Q1 What are the recent advancements in robotics and automation technology that are being adopted by the meat processing industry? Which developments show increased efficiency and safety?

Q2 How is computer vision aiding in quality control procedures while fish are being processed? What algorithms have been used with much success in such an application?

Q3 How does a robotic manipulator improve poultry product handling and sorting, and what specific problem are they solving?

Q4 What is the role of an end-effector in food processing automation and in the meat and poultry industries?

Q5 Describe how predictive maintenance can be integrated into a robotic food production system along with improving operating efficiency?

Q6 Describe the successful application of AI and IoT in food processing, specifically in the meat and fish industries?

Q7 What are the major barriers that will face the massive adoption of robotics in food processing, and what do the barriers have to do to face them?

Bibliography

I Adhikari, B Koirala, and MR Panda. Decoding plant diseases: unleashing the power of resnet-50 for superior disease detection. In: *2024 International Conference on Advancements in Smart, Secure and Intelligent Computing (ASSIC)*, 1–6, 2024. IEEE.

BA Aly, T Low, D Long, C Baillie, and P Brett. Robotics and sensing technologies in red meat processing: A review. *Trends in Food Science & Technology*, 137:142–155, 2023.

E Bjørlykhaug and O Egeland. Vision system for quality assessment of robotic cleaning of fish processing plants using cnn. *IEEE Access*, 7:71675–71685, 2019.

CB Campos, MM Arteche, C Ortiz, and ÁV Fernández. Technologies for robot grippers in pick and place operations for fresh fruits and vegetables. *Spanish Journal of Agricultural Research*, 9(4):1130–1141, 2011.

C Chen, H Seo CH Jun, and Y Zhao. Pavement crack detection and classification based on fusion feature of LBP and PCA with SVM. *International Journal of Pavement Engineering*, 23(9):3274–3283, 2022.

MT Das and LC Dülger. Mathematical modelling, simulation and experimental verification of a SCARA robot. *Simulation Modelling Practice and Theory*, 13(3):257–271, 2005.

TP Haidegger, P Galambos, JK Tar, LA Kovács, M Kozlovszky, Z Zrubka, G Eigner, DA Drexler, A Szakál, V Reicher, et al. Strategies and outcomes of building a successful university research and innovation ecosystem. *Acta Polytechnica Hungarica*, 21(10):13–35, 2024.

SS Harakannanavar, JM Rudagi, VI Puranikmath, A Siddiqua, and R Pramodhini. Plant leaf disease detection using computer vision and machine learning algorithms. *Global Transitions Proceedings*, 3(1):305–310, 2022.

M Hasan, N Vasker, and MSH Khan. Real-time sorting of broiler chicken meat with robotic arm: XAI-enhanced deep learning and lime framework for freshness detection. *Journal of Agriculture and Food Research*, 18:101372, 2024.

K He, X Zhang, S Ren, and J Sun. Deep residual learning for image recognition. In: *Proceedings of the IEEE Conference on Computer Vision and Pattern Recognition*, 770–778, 2016.

V Hemamalini, S Rajarajeswari, S Nachiyappan, M Sambath, T Devi, BK Singh, and A Raghuvanshi. Food quality inspection and grading using efficient image segmentation and machine learning-based system. *Journal of Food Quality*, 2022(1):5262294, 2022.

MA Hossain, MI Sabik, M M Rahman, SN Sakiba, AKMM Islam, S Swakkhar, S Islam, and A Ahmed. An effective leukemia prediction technique using supervised machine learning classification algorithm. In: *Proceedings of International Conference on Trends in Computational and Cognitive Engineering: Proceedings of TCCE 2020*. 219–229, 2021. Springer.

N Ismail and OA Malik. Real-time visual inspection system for grading fruits using computer vision and deep learning techniques. *Information Processing in Agriculture*, 9(1):24–37, 2022.

Y Jang and Y-S Seo. Machine vision based fish cutting point prediction for target weight. *Computers, Materials and Continua*, 75(1):2247–2263, 2023.

A Kadam, AP Dwivedi, A Shekhar, DH Sree, and A Prajapat. Noval approach using image detection to identify diseases in fruit plants (apple). In: *2024 IEEE 9th International Conference for Convergence in Technology (I2CT)*. 1–8, 2024. IEEE.

Y Liu, H Pu, and D-W Sun. Efficient extraction of deep image features using convolutional neural network (cnn) for applications in detecting and analysing complex food matrices. *Trends in Food Science & Technology*, 113:193–204, 2021.

G Manduca, L Padovani, E Carosio, G Graziani, C Stefanini, and D Romano. Development of an autonomous fish-inspired robotic platform for aquaculture inspection and management. In: *2023 IEEE International Workshop on Metrology for Agriculture and Forestry (MetroAgriFor)*. 188–193, 2023. IEEE.

R Morales, FJ Badesa, N Garcia-Aracil, JM Sabater, and L Zollo. Soft robotic manipulation of onions and artichokes in the food industry. *Advances in Mechanical Engineering*m, 6:345291, 2014.

A Nasiri, M Omid, and A Taheri-Garavand. An automatic sorting system for unwashed eggs using deep learning. *Journal of Food Engineering*, 283:110036, 2020.

I Nyalala, C Okinda, L Nyalala, N Makange, Q Chao, L Chao, K Yousaf, and K Chen. Tomato volume and mass estimation using computer vision and machine learning algorithms: cherry tomato model. *Journal of Food Engineering*, 263:288–298, 2019.

I Nyalala, C Okinda, N Makange, T Korohou, Q Chao, L Nyalala, Z Jiayu, Z Yi, K Yousaf, L Chao, et al. On-line weight estimation of broiler carcass and cuts by a computer vision system. *Poultry Science*, 100(12):101474, 2021.

K Pame, SK Laskar, S Borah, K Kakoty, and S Choudhary. Machine-vision based quality evaluation of meat and meat products-a review. *Journal of Livestock Science*, 14, 2023.

M Pech, J Vrchota, and J Bednář. Predictive maintenance and intelligent sensors in smart factory. *Sensors*, 21(4):1470, 2021.

S Pookkuttath, MR Elara, V Sivanantham, and B Ramalingam. Ai-enabled predictive maintenance framework for autonomous mobile cleaning robots. *Sensors*, 22(1):13, 2021.

D Romanov, O Korostynska, OI Lekang, and A Mason. Towards human-robot collaboration in meat processing: Challenges and possibilities. *Journal of Food Engineering*, 331:111117, 2022.

A Taner, YB Öztekin, and H Duran. Performance analysis of deep learning cnn models for variety classification in hazelnut. *Sustainability*, 13(12):6527, 2021.

C Wang, J Lu, X Ding, C Jiang, J Yang, and J Shen. Design, modeling, control, and experiments for a fish-robot-based IoT platform to enable smart ocean. *IEEE Internet of Things Journal*, 8(11):9317–9329, 2021a.

Z Wang, H Furuta, S Hirai, and S Kawamura. A scooping-binding robotic gripper for handling various food products. *Frontiers in Robotics and AI*, 8:640805, 2021b.

M Wu and H Jiang. Research on food safety prediction method based on k-means clustering algorithm. In: *MATEC Web of Conferences*, vol. 355, 03018, 2022. EDP Sciences. doi: 10.1051/matecconf/202235503018.

B Yan, X-H Wu, B Ye, and Y-W Zhang. Three-level supply chain coordination of fresh agricultural products in the internet of things. *Industrial Management & Data Systems*, 117(9):1842–1865, 2017.

L Yang, Y Liu, H Yu, X Fang L Song, D Li, and Y Chen, Computer vision models in intelligent aquaculture with emphasis on fish detection and behavior analysis: a review. *Archives of Computational Methods in Engineering*, 28:2785–2816, 2021.

6

Principles of Fuzzy PID and Its Algorithm

After reading this chapter, you should be able to:

- Insight into Fuzzy PID Applications: Overall understanding of the way the fuzzy PID control system is used for different processes in the food industry, like fresh food atmosphere packaging, heat exchangers' temperature control, and milk pasteurization with an increase in accuracy and versatility than conventional PID.
- SCADA Systems Overview: Understanding SCADA, its history, standards, applications, and importance in automating and monitoring processes within the food industry, future trends that can lead to its evolution.
- Laboratory studies, as well as real-time case studies on SCADA and fuzzy PID implementation: This would really help in bridging theory with practical applications and all the associated challenges in food process control engineering.

6.1 Fresh Food Atmosphere Packaging Gas Distribution System Utilizing Fuzzy PID Control

The fresh food atmospheric packing uses an advanced fuzzy proportional integral derivative (PID) control algorithm that's applied to the engineered gas distribution system. The system seeks to improve the performance of gas distribution monitoring through comparative simulations which will compare the conventional methods of PID control.

Food Engineering Automation with Robotics and AI, First Edition. Abir Chakravorty.

6.1.1 System Structure

Mathematical Model: It was developed with the step response method to complement the gas distribution process.

Control Mechanism: Aims at the second stage of the process, like preparing pure gas, flow control, and data management.

The system provides for its remote control, using the 4G network in a mobility device, thereby enhancing convenience and efficiency.

6.1.2 Performance Evaluation

Experimental Setup: Carried out a series of tests in order to check the stability of the system and its ability to distribute gases of different concentration indices and output flows.

The fuzzy PID controller had a time of stability at about 38 s whereas the traditional PID controllers had at 85 s. A concentration inaccuracy value of $\pm 0.25\%$ was also attained, and the gas distribution velocity was improved by about 50%.

The method proposed here demonstrates applicability in many preservation scenarios in foods, which would enhance the accuracy, quality, and safety of fresh food during storage and supply chain processes.

Optimization of the self-governing gas distribution control mechanism and refinement of the fuzzy rules in the controller architecture remain for future work.

6.1.3 Key Take-aways

Fuzzy PID control improves the flexibility and effectiveness of distribution systems for perishable foodstuffs.

The capability of a system to monitor and control gas distribution remotely represents an advancement in food preservation technology.

More research needs to be conducted in order to make the control mechanism more effective and more diverse, making it applicable for other types of fresh food.

6.1.4 Case Study: Fresh Food Gas Distribution Systems Improvement

The study goes successfully as the fuzzy PID control proves to be an effective controller for system performance improvement and dependability.

6.1.4.1 Fuzzy-PID-based Atmosphere Packaging Gas Distribution System

Analysis of the Control Model for Parameter Identification: The first step is to determine the control model parameters. For this, a step signal is applied, with defined conditions; for example, 60% of CO_2 and an initial data point of 0.2%. The system curve is established with the value 57.2% along with a delay time of 7 s. The S-shaped characteristic belonging to a class of first order pure hysteresis systems is indicated by the response curve of the gas distribution system. Based on this analysis, the mathematical model will be chosen as a step response method.

Mathematical Model Construction: The mathematical model of the gas distribution system is built making use of the step response method, as well as applying equation using identified parameters.

$$\frac{0.94 \cdot e^{-7s}}{25S + 1} \tag{6.1}$$

Various methods for obtaining the mathematical model are discussed, including the step response method, approximation methods, and others, highlighting the flexibility in modeling approaches.

A Process of Gas Distribution Control Analysis: This paper focuses on the process of preparing and controlling gas distribution, in fresh preparation, the control of gas flow, and the design of a data management and transmission system. Indeed, this step is where an optimal mechanism for a control model, including the incorporation of a fuzzy PID control mechanism, will be added. Optimization of the control strategy is performed in order to improve the performance of the gas distribution monitoring system. In order to prove this, comparative simulation experiments are carried out using fuzzy PID control and traditional PID control algorithms; it is confirmed that the fuzzy PID approach is quite efficient.

Simulation and Performance Evaluation: An experimental simulation test bed has been developed to check up the fuzzy PID controller and compare it with a conventional PID controller. Experimentation results show that the fuzzy PID control method offers high speed and accuracy in gas distribution by reducing the concentration error to ±0.25% with a stability time of 38 s as compared with the conventional method at time of 85 s.

Implementation of Remote Monitoring: The system shall enjoy remote monitoring of gas distribution processes through communication networks using mobile phones. Such a feature increases reliability and ease of operations, contributing to overall improvements in efficiency in the gas distribution system.

Gas Distribution Flow Control Model: In a fresh food gasification system, the aim should be to have control objectives as well as requirements: dynamic process for gas distribution with mass flow controllers adjusting the raw gas stream in such a way that the desired concentration ratio is obtained promptly. Dynamic gas distribution methods can continuously output a fixed concentration of gas. In this method, different gases could mix appropriately based on a given flow ratio to improve the distribution of gas better than its use without needing large volumes for storage tanks.

The distribution of raw gases in the oxygen-free preservation process consists of only two raw gases, including CO_2 and N_2; thus, the formula for gas mixture concentration calculation is given by Equation (1) [1]:

$$\alpha = \frac{Q_1}{Q_1 + Q_2}\alpha_1 + \frac{Q_2}{Q_1 + Q_2}\alpha_2$$

The gas concentration after mixing is denoted by α, the first raw gas concentration by α_1, the second by α, the first raw gas flow of material by Q_1, and the second by Q_2.

The gas distribution starts at the same time for both gases, as demonstrated in the equations are as follows:

$$t_2 = t_1 = t$$

$$Q_1 = v_1 \cdot t_1$$

$$Q_2 = v_2 \cdot t_2$$

$$v_1 = \frac{\alpha - \alpha_1}{\alpha_1 - \alpha_2}v$$

$$v_2 = \frac{\alpha_1 - \alpha}{\alpha_1 - \alpha_2}v$$

where v_1 is the flow rate of CO_2 and v_2 is the flow rate of N_2. The gas flow rate can be directly controlled to achieve the required gas concentration since it is clear from the above equations that the two gases have almost the same distribution times. This study simplifies the gas distribution process by allowing only the rate of CO_2 to be regulated with the rate of N_2 fixed at some value.

6.1.5 Principle of Fuzzy PID

Three links make up standard PID, as is well known: proportional, integral, and differential. Furthermore, the error between the feedback value and the system's set value serves as the foundation for all system regulation. The formula is as follows:

$$u(t) = Kp \cdot e(t) + \frac{K_P}{T_i}\int e(t)dt + K_P \cdot T_d\frac{de(t)}{dt}$$

where the coefficients of the proportional link and the integration link are denoted by K_p, $K_i = \frac{K_P}{T_i}$, and $K_d = K_p \cdot T_d$, respectively. When the system is in error, K_p has the effect of changing the error in the direction of reduction. When K_p is too small, the system approaches the step value more slowly and takes longer to stabilize. Despite being simpler to control, high K_p might result in excessive overshoot, which reduces system stability. All these can be brought within limits only

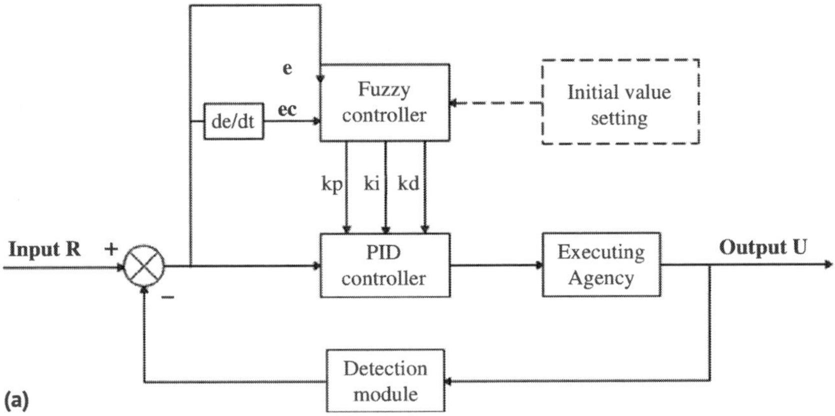

(a)

ΔK_p fuzzy rule table.

EC	E	NM	NS	ZO	PS	PM	PB
NB						PS	ZO
NM	PB		PM		PS	ZO	
NS	PM	NM	PM	PS	ZO	NS	NM
ZO		PS	PS	ZO	NS	NM	
PS			ZO	NS			
PM		ZO	NS	NM		NM	NB
PB	ZO	NS				NB	

(b)

Figure 6.1 (a) Schematic of fuzzy PID control architecture (b) $\delta\ k_p$ fuzzy rule table (Zhang et al., 2023).

by appropriate selection of K_p also reducing overshoot and the steady-state error with an increased speed of regulation (Zhang et al., 2023).

By adjusting the coefficient, the integral link coefficient Ki eliminates the steady-state inaccuracy of the system. If K_i is too high, the output of the control system is unstable and creates oscillations; even if a K_i is too small, it takes more time to eliminate the error, thereby increasing the regulation time of the system. K_d, the differential link coefficient, has three functions: it speeds up regulation, decreases overshoot, and increases system stability.

The differential link coefficient K_d has the following functions: it speeds up-regulation, decreases overshoot, and increases system stability. When it comes to controlling closed-loop negative feedback systems, the fuzzy PID control algorithm outperforms the traditional PID in terms of accuracy and efficiency. This algorithm compares the preset value at real-time feedback loop detection to determine the system deviation *e* and the rate of change of deviation *ec*. The proportional link coefficient K_p, the integral link coefficient K_i, and the differential link coefficient K_i are updated in real-time using fuzzy logic control methods (fuzzification, fuzzy inference, and defuzzification) based on the deviation e and the deviation rate of change ec as inputs. The online self-tuning of the PID control parameters is the outcome of such a scheme. The stable PID control parameters are also used to determine the fuzzy PID controller's initial settings. The fundamental theoretical domain for the concentration error e of the gas distribution system chosen for this work is $[-60\%, 60\%]$, which is based on the gas preservation criteria of the target aquatic product. $\mathrm{A} = \{-6, -5, -4, -3, -2, -1, 0, 1, 2, 3, 4, 5, 6\}$ is the domain of E due to the fuzzy design. Consequently, the concentration error e of the gas distribution system has a quantization factor of 0.1. Negative big NB, negative medium NM, negative small NS, medium ZO, positive small PS, positive medium PM, and positive large PB are the variable languages of E that correspond to its theoretical domain. In the same way, the nominal theoretical range for the change in error e in the gas distribution system is $[-6\%, 6\%]$. The quantization error of e is 1 if the theoretical range of ec is $B = \{-6, -5, -4, -3, -2, -1, 0, 1, 2, 3, 4, 5, 6\}$. The value language of NB, NM, NS, ZO, PS, PM, and PB also expresses the change in E.

6.1.6 Fuzzy Linguistic Variables and Fuzzy Design

If the CO_2 concentration of an aquatic product gas distribution system is set to y, the detection value is g, the deviation is e, then, $e = y–g$, the error rate is $ec = \frac{y-g}{t}$, and the mass flow controller's flow output is u.

6.2 A Fuzzy-PID Controller for Temperature Control in a Shell and Tube Heat Exchanger Simulation

Shell and tube heat exchangers are widely used for heat transfer in many industrial applications. The basic heat exchanger has basically two main subunits: the mechanical unit and the control unit. As such, it is very important to model these units correctly so that the shell and tube heat exchanger may work accordingly. The design of the mechanical unit is application-specific, whereas the design of the control unit is independent of the application and is based on properties of input and output fluids, such as temperature and flow rate. In this chapter, the improvement of the performance of a heat exchanger through an optimization of its control unit is discussed. A new fuzzy-PID control method, employing multiplication operations, is proposed and developed to significantly reduce the settling time and overshoot of the desired temperature. The performance of the new approach was tested and compared with the traditional PI control method employing several tuning algorithms. The PID controller is vastly used across all industries because of its simple modeling, the flexibility of tuning constants, as well as its controllability. The typically tuned constants for such controllers are those obtained by the Ziegler–Nichols method that guarantee nominal values of response and adaptation time. However, the technique has numerous drawbacks; constant optimization is one of them, and apart from that, it is good for linear systems, but less effective as the order of nonlinearity increases.

6.2.1 Mathematical Model

It uses a temperature sensor to draw feedback from input fluid leaving the heat exchanger. The operating range of sensor's temperature is from 500 to 1500 °C. This transducer further converts the signal of temperature into the signal of current by the range 4–20 mA. They mention the gain for feedback and the time constant:

$$K_{\text{feedback}} = \frac{(20-4)\,\text{mA}}{(150-50)\,°\text{C}},\ T_{\text{feedback}} = 10\,\text{sec} \tag{6.2}$$

The transfer function of feedback is given as follows:

$$G_{\text{feedback}} = \frac{1}{10s+1} * 0.16 \tag{6.3}$$

The input fluid transfer function, which is taken as a disturbance model, is given by,

$$T_{\text{inputfluid}} = 30\ \text{sec},\ G_{\text{inputfluid}} = \frac{1}{30s+1} * \text{Varying input fluid gain}$$

The transfer function of fluid flowing into the shell is given by,

$$K_{\text{shellfluid}} = \frac{50°\text{C}}{\left(\frac{\text{kg}}{\text{s}}\right)},\ T_{\text{shellfluid}} = 30\,\text{sec},\ G_{\text{shellfluid}} = \frac{1}{30\,\text{s}+1} * 50$$

It is calculated by converting the measured sensor value into a current signal, which is then converted back into a pressure signal.

$$K_{\text{valve}} = \frac{(15-3)\text{psi}}{(20-4)\text{mA}}$$

The valve capacity gain is given by the below equation, and the transfer function of the valve which is used to increase the flow rate of fluid coming into the shell is given by:

$$K_{\text{valve capacity}} = \frac{1.6\ \text{kg/s}}{(15-6)\text{psi}},\ T_{\text{valve}} = 3\,\text{sec},\ G_{\text{valve}} = \frac{1}{3s+1} * 0.75 * 0.13$$

This part uses a fuzzy-PID controller to manage the temperature that exits the tube outlet and a feed-forward controller to reduce the error caused by the input fluid disturbance entering through the tube inlet. The general block diagram is shown in Figure 6.2 for easier comprehension.

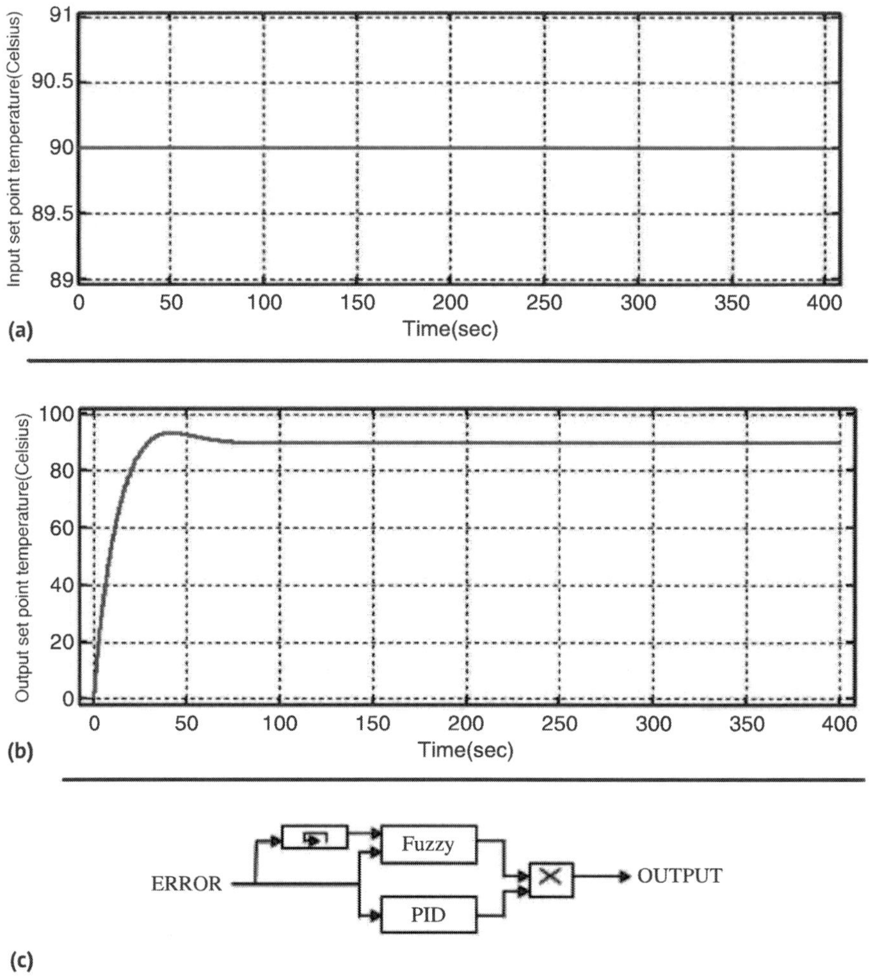

(d)

Fuzzy set rules for the proposed model

IF-THEN rules		
Δe	n	p
E		
N	n	p
Ze	p	ze
P	ze	p

Figure 6.2 (a) Input set point temperature, (b) output set point temperature, (c) proposed controller block diagram, (d) fuzzy set rules of the proposed model (Reddy and Balají, [2021]/Institute of Advanced Engineering and Science/CC BY-SA 4.0).

6.2.2 Fuzzy PID Controller Simulation Analysis

6.2.2.1 PID Controllers

For industrial control systems, the PID controller is frequently chosen because it is straightforward to develop and implement. Gaining values can be adjusted, despite the PID controller's seemingly straightforward architecture. Three different signal types are usually used with PID controllers in the majority of applications: proportional, integral, and derivative. A correction signal for the plant model is generated by adding all these signals. Every signal has derivative gain (K_d), integral gain (K_i), and proportional gain (K_p). Error is the input of the PID controller, while the error corrective value is the output of the PID. The following is the mathematical equation:

$$PID_{\text{output}} = K_p * \text{ error } + K_i * \int \text{error } * dt + K_d * \frac{d_{\text{error}}}{dt}$$

Some of the tuning methods experimented for tuning the PID gains are there. In that, the reason behind using the Ziegler–Nichols tuning in this paper is the faster and adaptable gain process with a less complex structure. The method of Ziegler–Nicholas consists of two types of tuning methodology: This was achieved by finding the dead time and time constant. Critical gain and critical time period are to be found.

First Method

In this method in Figure 6.2a, the dead time (D) is calculated based on the time period used for the initial response of the set point. The time constant is found by dropping the tangent line on the response curve up to set point and subtracting the time where the tangent line will meet up with set point, and dead time. K_{process} is the gain of the process, that is again determined by the process of finding distance to meet up the set point. This method of tuning is basically an open loop tuning method. The equations of the gain of the first method can be presented as follows:

$$K_p = \frac{1.2 * T}{K_{\text{process}} * D}, K_i = 2 * D, K_d = 0.5 * D$$

Second Method

The critical gain (K_{critical}) and critical time period (P_{critical}) are determined using two ways in the manner depicted in Figure 6.2. To determine the point at which the system becomes more unstable, the first way involves setting the proportional gain to zero and setting the integral and derivative gains to zero. K_{critical} is the proportional gain with the system that becomes unstable as the number of oscillations increases. The time interval between those oscillations is used to determine P_{critical}.

$$K_p = 0.6 * K_{\text{critical}}, K_i = 0.5 * P_{\text{critical}}, K_d = 0.125 * P_{\text{critical}}$$

The second approach determines the K_{critical} and P_{critical} using the Routh criteria, also referred to as the R-H criterion. The first approach is used in this study to implement the second methodology, with the K_p, K_i, and K_d values being 13.25, 0.36, and 56.25, respectively.

Feed Forward Controller

The input fluid is regarded as a disturbance, and the feed forward controller further minimizes the inaccuracy it causes. This controller will function as an open loop rather than a closed one, which will lessen the complexity and strain on the controller. The transfer function for the feed forward controller is provided by:

$$G_{\text{feedback}} = \frac{-18.461s^2 - 6.769s - 0.205}{30s^2 + 31s + 1}$$

The above transfer function is obtained by the process transfer function and input fluid transfer function,

$$G_{\text{process}} = \frac{4.875}{90s^2 + 33s + 1}, G_{\text{feedback}} = \frac{G_{\text{inputfluid}}}{\left(G_{\text{process}}(s)\right)^{-1} * (\lambda s + 1)}$$

λ ranges from 0 to 1, which is used as a parameter for filtering the transfer function to be accurate.

Fuzzy Controller

Many businesses use fuzzy controller systems in control applications to identify the exact points of fit for the problem. The fuzzy logic (FLC) system, which blends expert and human knowledge systems, was used by fuzzy. The way the expert system works is very different from how a normal Boolean statement works. For example, the Boolean expression shows the solution as either tasty or not tasty, whereas the FLC provides the additional response that is also somewhat tasty and slightly not tasty. These findings allow us to pinpoint the precise solutions to the problems. The invoking fuzzy system converts crisp input values into fuzzy values based on the membership functions used for this process. The shell and tube heat exchanger model-based rules are put into practice to produce the fuzzy set output values. The fuzzy set output values are converted into clear outputs using the defuzzification approach. The regulations of the proposed system are illustrated in Figure 6.2d.

Fuzzy-PID

The PID controller is one of the most conventionally used controllers in industrial control systems; basically, it works due to its simple nature and how easy it is to use. It is rudimentary, yet it averages the proportional, integral, and derivative signals, adds them together, and puts out an error correction signal for the plant model.

On the basis of the operation that is logical, there are two control output scenarios through this approach where the correct output for a plant model is decided based on a Boolean operation. The choice between which output to use is done by cross-referencing the error signals between two controllers. Specifically, if the error signal is less than some pre-set critical value, then the input(i) is false and can be valued at 0. If it is greater, then the input(i) must be a true value that equals 1. The equation for the former is given as

$$\text{Fuzzy} - \text{PID}_{\text{out}} = (\text{PID}_{\text{out}} \wedge \bar{i}) \vee (\text{Fuzzy}_{\text{out}} \wedge i)$$

In this instance, the PID output demonstrates great performance on the dynamic state, while the fuzzy output demonstrates good performance on the steady state. The operation is the same in the second instance as well; however, the equation is different and is provided by:

$$\text{Fuzzy} - \text{PID}_{\text{out}} = (\text{PID}_{\text{out}} \wedge i) \vee (\text{Fuzzy}_{\text{out}} \wedge \bar{i})$$

In this case, the fuzzy output and PID output are contradicted when compared to the first case, i.e. fuzzy shows high performance in dynamic state and the PID output shows high performance in steady state.

Based on the Arithmetic Operation

This approach uses an arithmetic operation to obtain the output that is supplied to the plant model. There are two cases in it as well. In this instance, the fuzzy and PID outputs are added together using the addition procedure. The overflow's saturation limit is set when the addition operation is performed. The formula is:

$$\text{Fuzzy} - \text{PID}_{\text{out}} = \text{PID}_{\text{out}} + \text{Fuzzy}_{\text{out}}$$

This fuzzy-PID combination combines the advantages and disadvantages of each controller for the plant model. In this case, multiplication was involved in the multiplication procedure as the outputs of the fuzzy and PID were multiplied. The overflow saturation action of the amplifier is performed by this multiplication operation. After the fuzzy enhances the output of the PID, the amplified signal is sent to the plant model.

The equation is given by:

$$\text{Fuzzy} - \text{PID}_{\text{out}} = \text{PID}_{\text{out}} * \text{Fuzzy}_{\text{out}}$$

Because the overshoot and settling times are smaller compared to addition operation-based fuzzy-PID, here the multiplication operation-based fuzzy-PID is implemented in this section. The block diagram of the proposed controller is as shown in Figure 6.2c. A shell and tube heat exchanger using a transfer function model has undergone simulation testing. The set point temperature is maintained at 900 °C, the simulation run time is set at 400 s, the input fluid gain is set at 20, and the $K_p, K_i,$ and K_d values are set at 13.25, 0.36, 56.25, respectively. MATLAB/SIMULINK has been used to obtain the results for conventional verification. Figure 6.2a,b is the set point temperatures for input and output. Figure 6.3a,b shows the output of the fuzzy-PID controller and fuzzy-PID+feed forward controller, respectively. The standard technique has been compared with the settling time and peak overshoot shown in Figure 6.3c,d. The traditional method takes 81.6 s to settle. With the proposed controller, the set point temperature settles in 74 s. The suggested peak overshoot of the set point temperature is 3.54%, while the typical peak overshoot is 5.4%.

The peak overshoot percentage is calculated as:

$$\%\ \text{peak overshoot} = \frac{93.2 - 90}{90} * 100 = 3.54\%$$

6.2.3 PID Control of Milk Pasteurization

The subject of milk pasteurization involves a highly complex, power-wasting system in which temperature control is critical. In the milk pasteurization plant, a milk dairy operation seeks accurate temperature control in order to avoid the consumption of large amounts of electricity and fuel. However, plant disturbances often conflict with the objective of achieving pasteurization temperatures that are close to being exact. The paper thus presents an application of an ES-based PID controller that was designed to stabilize temperature dynamics for both milk and water, hence ensuring the milk attains the desired temperature in a reasonable time with very low energy expenditure. This was to design, test, and tune an ES-based PID controller to decrease the amount of liquefied petroleum gas (LPG) used for the boiler's burner, which generates steam in the process of heating pasteurized products. This was represented through modeling, based on foundational principles for every component within the system of pasteurization for the process in terms of mass and energy balance. Using MATLAB R2021a, the parameters of ES-PID were adjusted to provide a systematic comparison between the systems with and without the controller. The simulation validation of the obtained results was used to prove that the controller provides better performance when its merits were measured by rise time, settling time, peak time, and percentage overshoot. This indicates that the ES-PID controller holds vast potential for the improvement of efficiency in the dairy industry pasteurization process. It is significantly better than what may be obtained using traditional PID controllers.

6.2.3.1 Model Designed and Objective Function

Equation 6.4 illustrates the model of a milk pasteurization plant that has been constructed with the goal of minimizing the yearly energy expenditure associated with LPG and power costs for the manufacture of pasteurized milk. The cost of

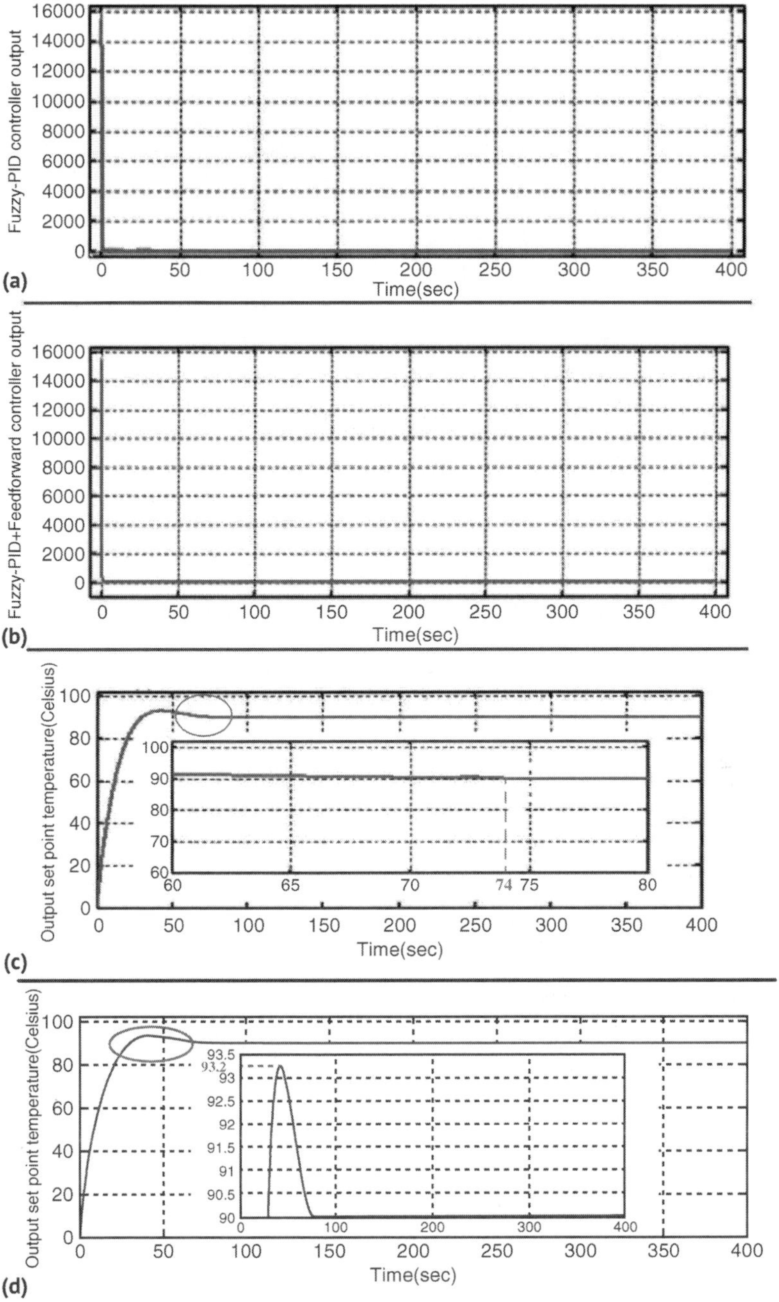

Figure 6.3 Fuzzy PID feed forward controller output model (Reddy and Balají, [2021]/Institute of Advanced Engineering and Science/CC BY-SA 4.0).

fuel in the boiler is shown in the first objective function, and the cost of electricity, which includes the ripple plate's overall electrical consumption, is shown in the second objective function. The operational units of the process model were applied to Eq. 6.4. The process models are represented by the optimization constraints, which include the need for pasteurized temperature for both the bottom and upper bounds of the LPG feeding rate at the boiler's burner (Adegbola et al., 2022).

6.2.3.2 Mathematical Modeling Equations of the Operational Units

This section outlines the mathematical equations used to model the dynamic temperature changes in milk and water across various stages of the pasteurization process. Based on the system shown in Figure 6.4, each stage is mathematically represented by ordinary differential equations (ODEs), grounded in the principles of mass and energy conservation. These equations aim to accurately reflect the operational behavior of each stage within the milk pasteurization facility. The development of these equations for each operational stage is predicated on several key assumptions: (i) There is a perfectly mixed condition of the liquid at each stage of the process. (ii) The physical and chemical properties of the liquid, such as its density and heat capacity, remain constant. (iii) Any changes in volume on either side of critical components such as the plate heat exchanger, water tank's heating coil, boiler, ripple plate, and cooling tower are considered negligible. (iv) Due to the effective insulation of the boiler's hot water pipes and the ripple plate's iced water pipes, temperature variations along the pipe lengths can be disregarded.

$$f_{\min} = n_w \sum_{t=1}^{36,000} \left(C_{\mathrm{LPG}} m_f + C_{\mathrm{elect}} \frac{(CL + W_e)}{3.6 \times 10^3} \right) \tag{6.4}$$

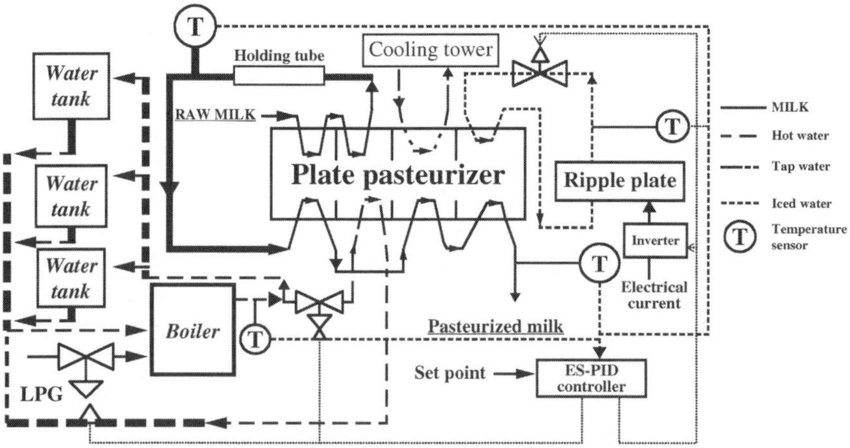

Figure 6.4 Process model of milk pasteurization plant-based ES-PID controller (Adegbola et al. [2022]/IAEME Publication).

Pasteurized temperature °C $= 72 \leq T_{m,7} \leq 76$; Cooling temperature °C $= 4 \leq T_{m,7} \leq 6$; Boiler fueling rate kg/s $= 0 \leq m_f \leq 0.00180$.

The multivariable dynamic pasteurization plant depicted in Figure 6.1 describes the process model of milk pasteurization used in the dairy industries, where $f_{\min}$ = Minimum annual energy expense (baht/year), n_w = Rate of pasteurized milk production (kg/day), t = Time of pasteurized milk production (s), C_LPG = Cost of LPG (kg/s), m_f = Boiler fueling rate (kg/s), C_{elect} = Electricity cost (kWH), W_e = Total electrical power for the equipment (J/S), and CL = Refrigeration load of ripple plate (J). The plate pasteurizer sequence identifies the general sections within a pasteurization unit such as regeneration, heating, precooling, and cooling section. To ensure proper heating and cooling of the milk, hot and cold utilities were provided in the form of a boiler and cooling tower in figure (Adegbola et al., 2022), respectively, while a ripple plate cooling system is used for the final cooling.

The mass and energy balance for each unit operations and utilities is expressed in Eqs. (8.72 and 8.73).

(a) Water Tank

$$\frac{dT_{w,o}}{dt} = \frac{U_{HC}A_{HC}\Delta T_{HC}}{\rho_w C_{p,w} V_{WT}} - \frac{U_{WT}A_{WT}}{\rho_w C_{p,w} V_{WT}}(T_{w,o} - T_a) \tag{6.5}$$

$$\frac{dT_{hw,o}}{dt} = \frac{F_{hw}}{V_{HC}}(T_{hw,i} - T_{hw,o}) - \frac{U_{HC}A_{HC}\Delta T_{HC}}{\rho_{hw} C_{p,hw} V_{HC}} \tag{6.6}$$

6.2.3.3 Graphical Representation of Energy Minimization

According to the optimization result in Figure 6.5, the upper-bound pasteurized temperature of 76 °C corresponded with the optimal LPG consumption rate of 1.25×10^{-3} kg/s to save 92×10^{-4} metric tons CO_2 equivalent of global warming

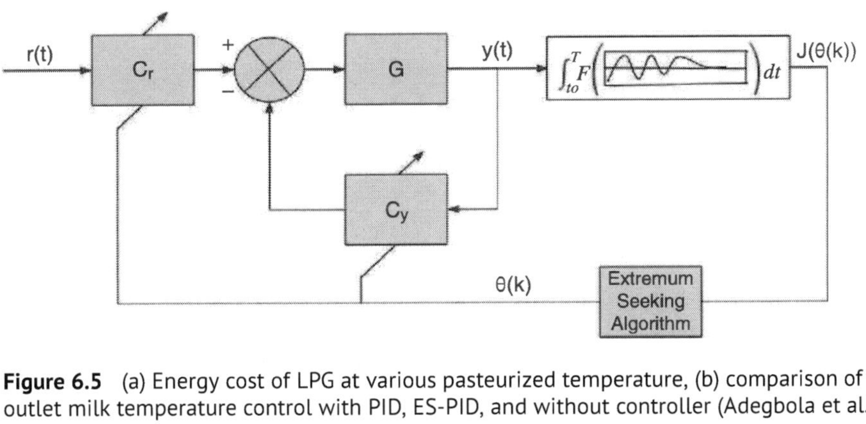

Figure 6.5 (a) Energy cost of LPG at various pasteurized temperature, (b) comparison of outlet milk temperature control with PID, ES-PID, and without controller (Adegbola et al. [2022]/IAEME Publication).

potential (GWP) annually and 4.23×10^5 Baht in energy costs. Similarly, the optimal LPG consumption rate of 1.16×10^{-3} kg/s to save 4.13×10^5 Baht/year energy cost and 89×10^{-4} metric tons CO_2 equivalent of GWP annually matched the lower-bound pasteurized temperature of 72 °C. This minimal pasteurized temperature resulted in the lowest annual energy cost when the pasteurized temperature was 72 °C. Since greenhouse gases (GHGs) are gases that absorb and release radiant energy in the thermal infrared spectrum, GWP is defined as the heat absorbed by any GHG in the atmosphere. Table 6.1 represents the symbols, parameters, and their associated values for this chapter.

Table 6.1 Parameter table.

Symbols	Parameters	Values
d_f	Ripple plate impeller diameter (m)	0.4
A_f	The ripple plate's flow area (m^2)	0.002
A_{RP}	The ripple plate's area (m^2)	2.5
A_B	Area of boiler (m^2)	2.3
A_T	Area of storage tank (m^2)	2.2
A_{HC}	Water tank and heating coil transfer area (m^2)	0.507
V_{HC}	The water tank and heating coil volumes (m^3)	4.9
V_B	Volume of boiler (m^3)	1.20
V_{RP}	Volume of ripple plate (m^3)	0.50
V_{CT}	Volume of cooling tower (m^3)	0.05
d_p	Holding tube diameter (m)	3.83×10^{-2}
L_p	Length of holding tube (m)	12
n_w	The water tank and heating coil volumes (kg/day)	1×10^4
$T_{m,i}$	Milk's temperature at the inlet (°C)	72
F_m	Milk's volumetric flow rate (m^3/s)	4×10^{-4}
$F_{m,i}$	Milk's volumetric flow rate at the inflow (m^3/s)	5×10^{-4}
$F_{m,o}$	Volumetric flow rate of milk at outlet (m^3/s)	4×10^{-4}
F_{hw}	Volumetric flow rate of hot water (m^3/s)	1.60×10^{-3}
F_{tw}	Volumetric flow rate of tap water (m^3/s)	4.80×10^{-3}
F_{iw}	Volumetric flow rate of ice water (m^3/s)	1.92×10^{-3}

$F_{hw,return}$	Hot water volumetric flow back to the boiler (m^3/s)	3.20×10^{-3}
T_w	Temperature of water make-up (°C)	27
T_a	Ambient air temperature (°C)	30
E	Evaporation rate for cooling tower (m^3/s)	1.7×10^{-5}
h_A	Heat transfer coefficient at water surface (W/K)	2000
λ_v	Heat vaporization of water (kJ/kg)	2410
L_D	At the cooling tower, mechanical drift is lost (%)	2
n_s	Rotating speed (rpm)	5000
m_f	Boiler fueling rate (kg/s)	1.80×10^{-3}
LHV	Low heating value of fuel combustion (KJ/kg)	49,888
CL	Refrigeration rate at ripple plate (J/s)	5.20×10^{-4}
ρ_m	Density of milk (kg/m^3)	1015.96
ρ_{tw}	Density of tap water (kg/m^3)	997
ρ_{iw}	Density of ice water (kg/m^3)	917
ρ_{hw}	Density of hot water (kg/m^3)	977.3
$C_{p,m}$	Specific heat capacity of milk (J/kg·°C)	3.855
$C_{p,tw}$	Heat capacity of the tap water (J/kg·°C)	2.356
$C_{p,hw}$	Heat capacity of the hot water (J/kg·°C)	108.376
$C_{p,iw}$	Heat capacity of ice water (J/kg·°C)	2.108
ΔT_{PHE}	Difference in heat exchanger temperature (°C)	15.6
$T_{m,i}$	Temperature of milk in the holding tube (°C)	76
U_{HC}	Heat transfer coefficient of heating coil and water tank $(W/m^2 \cdot K)$	490
h_{fg}	At the ripple plate, the latent heat of vaporization (kJ/kg)	217

6.3 An Introduction to Automated Process Control in the Food Sector Using Supervisory Control and Data Acquisition and Associated Systems

Supervisory control and data acquisition (SCADA) systems, however, come in handy when there is a need to automate processes within the food industry. This mainly offers SCADA systems being supervisory control and data acquisition or supervisory control and data acquisition systems with much development from their humble beginnings, since they now deploy the use of advanced technologies, for instance, the Internet of things (IoT) to improve monitoring and control capabilities. Below are key aspects of SCADA systems relevant to the food industry.

6.3.1 Overview and History of SCADA

SCADA systems are designed to automate industrial processes, collecting real-time data and control. Its importance lies in different industrial fields, such as food processing, giving guarantees about the quality while respecting safety standards.

6.3.2 SCADA Standards and Applications

Standardized SCADA systems, such as DNP3, enable secure communication and interoperability. Applications of SCADA systems in the food processing industry include temperature and humidity monitoring among other critical parameters that ensure the safety of the product in the processing, cooking and packaging phases.

Recently, research has demonstrated the actual implementation of SCADA in food processing by using IoT, where control and monitoring can be done from a distance and subsequently reduce the cost of operation. AI, including machine learning, can enhance predictive maintenance and decision-making strategies in SCADA. Cybersecurity remains a big issue as researchers continue their efforts on how to protect SCADA systems against threats. Cybersecurity risks with SCADA systems are a public health and national security issue. All in all, the development of SCADA systems in modern food processing has nothing to do with new technological developments alone but also depends on security conditions.

6.4 Introduction to SCADA

SCADA systems enable the monitoring and control of industrial processes. Software and hardware can be integrated together to gather and analyze real-time data, thereby increasing the productivity of operations within manufacturing and utility industries. SCADA employs HMIs that ease interaction with devices such as sensors and motors in the management of any process.

6.4.1 Main Elements of SCADA

1. Data Gathering: SCADA devices gather data from sensors and measuring equipment, thereby enabling real-time monitoring of a process.

2. Control Actions: SCADA provides for remote industrial control. This helps to improve responses and can make it easier to increase flexibility in operation.
3. Communication Procedures: SCADA employs a lot of communication techniques. The main ones involve wire communications as well as wireless connections that facilitate reliable transfer of data.

6.4.2 Applications of SCADA

1. Industry Applications: SCADA is used in most of the industries that include power generation, water treatment, and manufacturing to optimize production processes.
2. Microbrewing: The latest research findings show how SCADA is used to enhance the quality of craft beer through monitoring temperature and fermentation time.

Though SCADA technology has improved efficiency significantly, there are a number of cybersecurity concerns and system complexity issues, so it is very much a hot-spot for research and development, even in these fields.

6.5 SCADA and Its History

SCADA systems have evolved significantly since their origin in the 1960s. From a conceptual formulation for industrial process control, the term "SCADA" was first referred to in the mid-1970s to refer to the processes of automatic control as well as data acquisition. Over the ages, SCADA systems have evolved from stand-alone networks to sophisticated interconnected systems that improve productivity but pose security risks (Anton et al., 2017, 2019).

6.5.1 Historical Development 1960s–1970s: First Industrial Process Control Introduction; Initial "SCADA" Terminology Initiated

1980s: SCADA first implemented in electric power systems, especially in Indonesia (Febrian et al., 2023).

2000s: Growing connectivity to the public Internet was resulting in increasing cyber threats. Therefore, security measures have to be improved.

6.5.2 SCADA Components and Functionality

The SCADA systems are comprised of four major components that include the remote terminal units, the communication systems, central stations, and human-machine interfaces. They also allow the effective observation and regulation of operations, thereby greatly enhancing operational dependability and response times in the industry. Although SCADA systems have dramatically improved the performance of numerous operations, the general increasing trend toward complexity and connectivity raises concerns over cybersecurity and hence creates a need for continuous evolution in protectory measures.

6.6 SCADA Standards and Applications

SCADA systems are core to industrial operations, particularly in utilities like power supply and chemical processing industries. Standardized communication protocols like IEC 61850 enhance SCADA capabilities by enabling more efficient data handling and improved visual representation.

Important Standards of SCADAIEC 61850: This is an international standard which assists the interoperation of devices in a substation so it works efficiently with protection and control actions.

OPC (OLE for Process Control): It provides a universal interface for SCADA systems that can integrate anything and everything smoothly.

Applications of SCADA Substation Management: SCADA systems use augmented reality for real-time data visualization to further amplify maintenance and operational efficiency.

Industrial Process Control: SCADA systems monitor and control those processes using data retrieved from intelligent electronic devices.

SCADA systems have developed hugely, but on the other hand, there remain some issues associated with their reliability and sensor faults as well as public safety concerns.

6.7 SCADA in Food Processing

SCADA systems play a very essential role in food processing, improving efficiency and quality control and enhancing automation. Real-time monitoring and control of all kinds of processes makes their results high-quality products with maximum efficiency.

Automation and Efficiency: SCADA systems make filling and packing very easy by eliminating human error and achieving higher levels of hygiene. The SCADA system was intended to only pack multigrain rice with controlled grain composition and a precise packaging process. This paper discusses an automatic food processor equipped with a SCADA system that was originally designed for process control within cooking processes. SCADA can be very versatile in various applications in the food industry.

Quality Control: SCADA systems play a critical role in quality control, especially among small food producers. It is proposed here as a low-cost SCADA system for monitoring food quality and process tracking as well as data management, which keeps the cost very low for access by smaller enterprises. Moreover, the applications of SCADA in olive oil manufacturing are significant for the improvement of the final product quality, particularly key parameters: tocopherols and phenolic compounds.

Distributed Control Systems: Implementing distributed control systems with an SCADA architecture has been proven to optimize the food industry, particularly in the process of wheat grinding and bread manufacturing, by enhancing overall system performance and communication at various levels of control.

While SCADA systems offer a number of advantages in food processing, the high setup costs and the requirement for technical sophistication pose barriers to their adoption, particularly among smaller producers. However, the potential for increased efficiency and quality control remains the compelling reason to bring SCADA systems into the food industry.

6.8 Laboratory Study: Implementation of SCADA

SCADA constitutes the most important technology used to enhance the reliability and efficiency of several industrial processes. Recent studies have shown that SCADA has applications in various fields, including power, water supply, and wastewater treatment, among others, showing its versatility and importance in modern automation.

SCADA in Power Systems: It demonstrated the success in the integration of the IEC 61850 protocol into the SCADA systems by demonstrating it within the IEEE 14-bus system that integrates IEDs for effective communication, which would ensure not to compromise fault detection. The management of decentralized renewable power sources requires a SCADA system. Optimization for performance and integration of renewable technologies entails monitoring coupled with control in real time.

SCADA in Wastewater Treatment: The PLCs have been widely used within SCADA systems to ensure that the wastewater treatment process is well automated. In this way, precision control and monitoring of biochemical processes significantly improve operational efficiency.

Design and Implementation: Good designs of SCADA require proper planning for the careful integration of hardware and software components for the establishment of increased industrial productivity and creativity. However, deploying SCADA systems is highly difficult and mainly due to their deployment, as very complex simulation tools have to be used to test functionalities before deployment, as established in the case of a microgrid system.

6.9 Future Trends in SCADA

Advances in SCADA systems in the future will majorly be seen in security, communication, and data utilization. As various industries adopt automation through SCADA for their operations, the need for advancement in security and innovations in data-driven technologies becomes increasingly widely emphasized.

Security Improvement: Advanced IDS integration forms a critical aspect to some of the vulnerability solution issues with SCADA networks, especially when connected to diverse external networks. Future security protocols will be based on encryption techniques specialized for SCADA systems to provide integrity and confidentiality of data.

Communication Advancements: Satellite communication is rising as an option for remote SCADA access, thus bringing enhanced monitoring capabilities across dispersed locations. It is already transforming SCADA operations, and further advancements are sure to increase the efficiency and reliability.

Data-driven Approaches: Machine learning and data analytics in SCADA data promise to revolutionize the monitoring of conditions and predictive maintenance, especially in wind energy.

There is a growing interest in the exploitation of high-resolution data but it creates computational challenges that need to be overcome to be fully utilized.

Although these trends look bright for the future of SCADA systems, data confidentiality and the standardization of protocols are some of the evergreen issues that should be carried out in research and development regarding these systems.

6.9.1 Case Study

Problem Statement: One of the more basic methods to share the types of liquids into the desired proportions between bottles or jugs is manually. Because human factors are present, it hampers the speed at which production will take place, and sometimes leads to a final product of similar quality. This study describes a simulated prototype of a SCADA and PLC-based bottle filling process used in the beverage industry to fill different types of containers with liquids of different compositions. The various stages of the simulated operation include labeling different types of containers, capping, filling liquids, sorting containers, and counting units per batch. To grade the metallic base and nonmetallic bottles, an inductive proximity sensor is used. While the nonmetallic bottles are filled with only the first two types of liquid composition, the metallic base bottles are filled with all three different liquid compositions in equal volume during the filling stage. Bottles for later phases have separate labels. Bottles are detected by both kinds of photoelectric retroreflective sensors during the filling, capping, and labeling processes.

6.9.1.1 Automated Beverage Industry Operation-filling and Capping

In nearly all sectors of the beverage industry, filling various liquids is considered to be a routine process. Beverage companies are mostly engineered to serve the filling of drinks with different material compositions into various forms of containers. The whole process of filling and the handling of these containers is categorized into several unique processes:

Sortation Phase: During this phase, the containers or bottles traveling on the conveyor belt would be segregated into two categories, namely, metallic-based containers and nonmetallic containers. This would be done using an inductive proximity IME08 sensor, which senses only metallic objects, leaving no nonmetallic material. During this phase, the containers or bottles traveling on the conveyor belt would be segregated into two categories, namely, metallic-based containers and nonmetallic containers. This would be done using an inductive proximity IME08 sensor, which senses only metallic objects, leaving no nonmetallic material.

Liquid Filling Stage: Fill the containers with different liquid compositions. Metallic base containers are filled with three different types of liquids and each type makes up 33% volume; nonmetallic base containers are filled with two different types of liquids and each type takes up 33% volume. A photoelectric retroreflective sensor with valves is used in this stage.

Capping Stage: Once filling is complete, the capping unit is activated to put and close caps on the filled container.

All those stages are so designed that each drink goes through it in such a way that the type of material with which each kind of can or bottle is made, is taken care of properly and filled appropriately. Here, again photoelectric retro-reflective sensor will be used.

Labeling Stage: After the capping of containers, the labeling work unit comes. To distinguish containers, we will be using labels of different colors for the specific container type. Here, also, photoelectric retro-reflective sensor will be used for detecting containers.

Counting Stage: Finally, the counting process of metallic and nonmetallic containers at the end of each and every case.

Such communication from SCADA systems with PLCs is done using a PC, using either wired or wireless connections. For operation, PLCs require a stable power supply of 24V DC and is usually given by SMPS.

The PLC is the backbone of this operation. It takes input data or signals from devices that include sensors, processes that data, executes logical operations with inputs, and comes up with the appropriate responses for its output terminals to devices connected to them. Such output devices may be solenoid valves to fill tanks, LEDs for capping and labeling units, and even variable frequency drives (VFD), which control the speed of motors that drive conveyor belts. The motion of the conveyor system is regulated by a DC motor to a high degree of precision. The precision of the DC motor is assured by regulating it with a drive.

Two types of sensors detect the containers on the conveyor: Inductive proximity sensors (IPR) and photoelectric retro-reflective sensors. Two sensor types are connected to a device known as a sensor splitter, which monitors the status of all sensors. These sensors feed their signals to the PLC through its input ports. Similarly, the PLC links the output devices to it through its output ports. Different kinds of liquids are filled in containers through valves connected to the filling tanks. Also, labeling and capping units are regulated by signals from the reflective sensors conditioned to send output signals toward the PLC.

6.9.1.2 Operation

The process flow is illustrated through the following flow charts: This particular flow chart reveals how the conveyor belt motor operates. The precision in the conveyor belt's movement is crucial to the system's operation. The movement of the conveyor belt is primarily determined by the input and output signals from sensors (including IR and reflective sensors) and the Drive, respectively. The inductive proximity sensor (IPR S1) doesn't directly affect the conveyor belt's stop-start mechanism. However, when any of the five photoelectric retro-reflective sensors (S2, S3, S4, S5, S6) detect an object, such as a bottle, metallic, or nonmetallic item, the conveyor belt motor halts. The belt resumes movement after a predetermined pause, which allows for processes such as bottle filling, capping, and labeling to be completed.

The second flow chart indicates the sorting procedure of containers and filling them with different contents. As the containers move on the conveyor belt, IPR S1 concludes whether the container is metallic or not. A metallic container sets the status of inductive PR Sensor S1 to green; conversely, a nonmetallic one does not change its status. After this detection, the containers are moved forward and hence detected by a photoelectric retro-reflective type sensor, S2. This detection causes the solenoid valve V1 to open; hence, filling from tank1 to the container until it nearly reaches about one-third or 33% of its fullness. Then, the timer of tank1 turns off, the conveyor returns to movement and goes until another photoelectric retro-reflective type sensor, S3 detects the container. This operation also opens the solenoid valve V2. This is how tank2 begins filling the container until it is two-thirds or 66% full.

About two-thirds of the container was filled when the timer for tank2 switched off, restarting the conveyor system. This cycle repeats until the next photoelectric retro-reflective sensor (S4) detects an incoming container. With a light signal from S4, solenoid valve (V3) is opened. This makes tank3 to start pouring liquid into the arriving container. Its fill-up continued, almost filling up 99% with the capacity of the container reached. However, once the IPR S1 fails to identify a container, the solenoid valve V3 will not activate the tank3, and hence that percentage of liquid will not be released into the undetected nonmetallic container. As a result of this, in that situation, the nonmetallic container will fill up to only 66%, or two-thirds, during the entire cycle.

The flow chart above outlines the procedure for capping and labeling as shown below: This commences when the photoelectric retro-reflective sensor, S5, reads a container. At this stage, the LED L1 of the capper lights green, signifying that capping is to be started. The cap is automatically fitted onto the containers as the conveyor belt momentarily stops to allow fitting. Once the capping is completed, the conveyor belt motor is back running. Similarly, when the photoelectric retro-reflective sensor (S6) senses a container, the LED L2 of the labeling unit turns green to indicate that the labeling cycle has started.

Labeling: In this process, labels are affixed to the containers with the conveyor motion stopped. When the labeling is complete, the motor is again turned back to life.

1. Inductive Proximity (IPR) Sensor (S1): It bypasses other bottles and detects metallic base bottles.
2. (S2): The conveyor motor stops, the photoelectric retro-reflective sensor detects both containers, the (V1) valve opens, and the containers became one-third full.
3. (S3): The photo-electric retro-reflective sensor detects both containers, the conveyor motor stops, the (V2) valve opens, and the containers become two-thirds full.
4. (S4): Photoelectric retro-reflective sensor senses both containers, conveyor motor stops, but the (V2) valve only opens when S1 is high in a metallic base container, allowing it to fill completely. In case of a nonmetallic container, S1 is low, (V2) valve doesn't open and the container remains two-thirds full.
5. (S5): The photoelectric retro-reflective sensor detects both containers. The conveyor motor stops. (L1) LED becomes green and the capping unit caps both the containers.
6. (S6): After the conveyor motor has stopped and when the photoelectric retro-reflective sensor has detected both containers, the (L2) LED becomes green. And the labeling unit places a cap on both containers, however different in colors.

6.10 Review Questions

Q1 What is fuzzy logic?

Q2 Describe fuzzy logic control system?

Q3 Application of fuzzy logic system for the pizza production processing optimization? Mention the steps involved and the processes?

Q4 A fuzzy conjunction operators, $t(\mathrm{x}, \mathrm{y})$, and a fuzzy disjunction operator, $s(x, y)$, form a pair if they satisfy:

$$t(x, y) = 1 - s(1 - x, 1 - y)$$

If $t(x, y) = \frac{xy}{(x+y-xy)}$ then $\mathbf{s}(\mathbf{x}, \mathbf{y})$ is given by?

Q5 Match List I with List II.

LIST I		LIST II	
A.	Expert system	I.	Decision tree
B.	Fuzzy system	II.	Scramble
C.	Operator in genetic algorithm	III.	Inference engine
D.	Supervised technique	IV.	Mycin

Bibliography

O Adegbola, E Ojo, D Aborisade, and P Adewuyi. An optimization approach for energy minimization and temperature control of milk pasteurization plant. *Technology (IJEET)*, 139:1–14, 2022.

R Febrian, D Aribowo, MA Satria, and R Badruzzaman. Jaringan telekomunikasi pada scada menggunakan cisco packet tracker. *Populer: Jurnal Penelitian Mahasiswa*, 22:188–196, 2023.

CS Reddy, and K Balají. A fuzzy-PID controller in shell and tube heat exchanger simulation modeled for temperature control. *Indonesian Journal of Electrical Engineering and Computer Science*, 213:1364–1371, 2021.

7

Robots, Ladder Elements, Logic Gates, 3D Food Printing, SCADA, Automation Pyramid in Food Processing

After reading this chapter, you should be able to:

- Understanding of High-tech Food Processing Technology: 3D food printing and various types of robotics as applied in automation and sensor technologies.
- Basics of Automation and Control Knowledge: A foundation from automation concepts like supervisory control and data acquisition (SCADA), programmable logic controllers (PLCs), and the automation pyramid, core programming principles within the food processing field.
- Know-how on Sustainability and Societal Impact: Awareness of sustainability through 3D-printing technologies, food customization, improved social connections through food.

Robotics plays an essential role in the modern world. It is incorporated into all sectors, and the way things are forever run is changing. The food industry is not very aggressive in adopting technologies. However, with the increased demand for food and the growing and aging of the global population, food suppliers have been urged to find ways to cater to the high demand while maintaining and even improving food quality. From farms to processing houses, then packaging to delivery, robotics is slowly finding its way into every form of the food industry. Not just that, but restaurant kitchens too have started to be places of automation.

7.1 3D Printing Technology: The New Era for Food Customization and Elaboration

3D printing or AM and RP is that fast-moving digital technology in which both academics and professionals of the industry and the rest of the world have become increasingly interested. It has already been widely used in medicine, gastronomy,

Food Engineering Automation with Robotics and AI, First Edition. Abir Chakravorty.

engineering, manufacturing, art, and education. It is able to create a product layer by layer from a 3D computer design which might be developed through CAD software or by downloading models from online platforms like Thingiverse, Shapeways, Ponoko, or Sculpteo. Once this design is broken down into layers, the printer systematically assembles them according to the prescribed cross-sectional pattern as elucidated. In food processing, rapid prototyping technology is anticipated to change food processing as items of a particular taste, affordability, convenience,

Table 7.1 Comparison between food 3D printing and non-food 3D printing (Dankar et al. [2018]/Elsevier).

	Food 3D printing	3D printing	References
Differences	Substrate: Food material with high variability in physicochemical properties both before and after the printing process. Flowable and able to hold its structure post-processing (chocolate, cookie dough, cheese, hummus, hydrocolloid). Preprocessing: Fine-tuning of food's recipe; Ingredient mixing with different hydrocolloid. Printing process: Mostly extrusion-based process. Higher retraction for reducing pressure at the nozzle tip as the viscosity of food is less. Post-processing: Some products might be applied to temperature variations (cooking, oven drying, freeze-drying). Appeal, texture, aroma, and taste are critical for evaluation.	Thermoplastic materials with invariable properties: plastic, resin, polymers. • Preprocessing: Usually no preprocessing step is required. • Printing process: Primarily photo-polymerization and selective laser sintering using photo-sensitive (laser and UV light) materials. In extrusion modeling, lower retraction is required because of the high viscosities of plastic materials. • Post processing: Polishing, chemical finishing. • Sensory attributes: Appeal and texture are critical for evaluation.	Yang et al. (2018)
Applications	Industrial sectors, military services, aerospace missions, customized snacks for children, people facing mastication problems.	Bio-printing: the creation of transplanted tissues, bones, and skin grafts.	Severini et al. (2018)
Similarities	Creation of digital 3D models-cutting into distinct layers. Utilize a certain quantity of basic materials. Production runs are shorter. A high degree of autonomy.	López Galdeano (2014)	

and value would be served according to requirement. Moreover, 3D printing has the potential to ease some of the resource and expertise barriers that hinder innovators from bringing their own ideas to life, thus creating a "Democratization of Innovation." Table 7.1 shows the comparison of food 3D printing with other substrates. 3D printing is technology that has focused on the creation of a 3D object from a computer-aided design. Recently, the idea of 4D printing came along as a development of 3D printing, wherein the printed material changes over time. Through this technology applied to food, complicated geometrical structures can be made in high volumes with economic and environmental benefits. It may be possible to tailor food products based on any specific requirements for nutritional needs, caloric intake, shapes, textures, colors, or flavors. Currently, there exist four main techniques under research concerning the 3D food printing technology: extrusion, inkjet, binder jetting, and selective sintering. While most studies focus on improving the resolution quality of the printed food, it severely lacks proper research into technologies that would make mass production feasible. Various methods of 3D printing are shown in Figure 7.1.

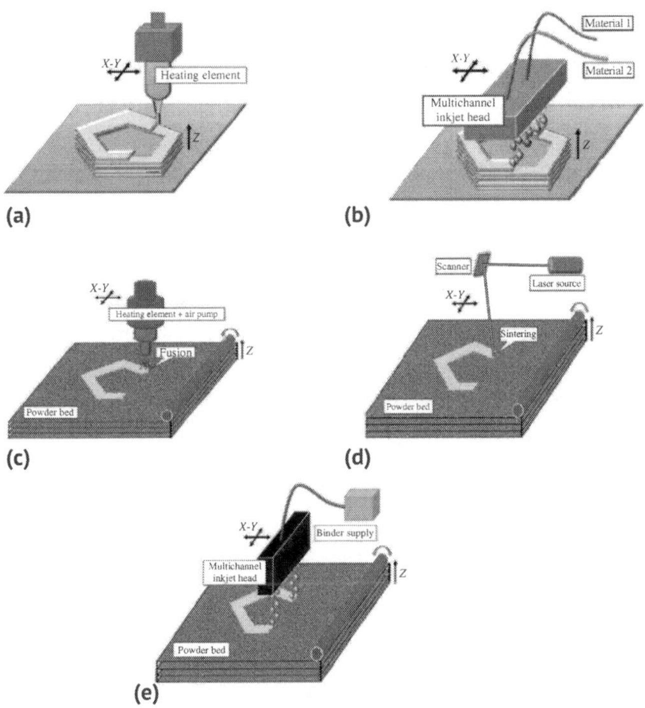

Figure 7.1 A diagrammatic overview of the various methods used in 3D food printing includes: (a) extrusion-based direct printing, (b) inkjet printing, (c) selective sintering with hot air, (d) laser, and (e) along with binder jetting (Sun et al. [2015]/AccScience Publishing/CC BY 4.0).

7.1.1 Why Print Food? Are There Any Benefits?

Development of specialty food products for the diverse consumer population. Personalized food is only the association of self-knowledge with food preferences. 3D printing has the wide possibility of fully personalized food that exactly meets the needs, taste, and dietary pattern of people from different ages, genders, occupations, and healthy lifestyles by adjusting the composition, density, or structure to the preferences and needs of the user (Rodgers, 2016; Sher & Tutó, 2015). Table 7.2 represents the 3D printing technologies used.

About 25% of people aged 50 or more suffer from mastication and swallowing, thus getting the nutrition they require through pureed food, much of which is unattractive and unpalatable. To come up with soft, nutritional, and innovative textured food for the elderly that meets legislative limits for any country is a major challenge that faces the food industry (Aguilera and Park, 2016). Thus, this initiative led TNO to institute the "Performance" project-printed customized pureed food intended for seniors that could bring ease to older citizens with oral impairment problems involving chewing and deglutition difficulties. TNO also proposed the concept of printing customized meals for seniors, athletes, and sportsmen that require recovery products after workout or for expectant mothers with different nutrient component levels like proteins and fats and reduction or complete elimination of unwanted ingredients and adding healthy ones, such as vitamins, fibers, and phytochemicals like anthocyanins, carotenoids and betanidins.

7.1.2 Novel Food Structuring Using a Broad Range of Alternative Food Ingredients

Some products can be created only by using 3D-printing technology. These kinds cannot be produced with any other manufacturing method. The design of food to nicer visual shapes has evolved in multiways in recent times (Periard et al., 2007). 3D printing provides an unlimited platform for creativity, where new shapes, textures, and flavors can be produced both by home users and professional designers. Furthermore, 3D food printers open a new galaxy of mixed foodstuffs and flavor combinations. Companies like Bocusini have created their own marketplace to upload recipes and ingredients, especially designed for 3D food printing. For instance, Bocusini's blog published a precisely crafted potato puree in the shape of an octopus (Molitch-Hou, 2015). The TNO organization also did research concerning new food structures produced from nonconventional foodstuffs. Such unconventional base materials as proteins from algae, insects, fungi, or lupine seeds may give rise to extremely tasty and structured food. These raw materials are healthy and friendly to the environment.

Table 7.2 Comparison of 3D-printing technologies for food fabrication.

Feature	Hot-melt extrusion	Sintering-based technology	Inkjet powder printing	Inkjet-based printing
Materials used	Food-grade polymers like chocolate	Powders with low melting points, such as sugar, fat	Powdered substances, including sugars, starch, corn flour, and flavoring, combined with a liquid binder	Low-viscosity substances such as pastes and purees
Viscosity range	$10^3 \sim 10^5$ cP	Not applicable	1–10 cP (binder)	$5 \times 10^2 \sim 5 \times 10^3$ cP
Printing mechanism	• Motorized stage • Heating system • Extrusion setup	• Motorized stage • Powder bed • Sintering source (laser or hot air)	• Motorized stage • Powder bed • Inkjet printhead for binder application	• Motorized stage • Inkjet printhead • Thermal control unit
Printing resolution	Nozzle diameter: 0.5–1.5 mm	Powder granules: 100 μm	• Nozzle diameter $\leq$ 50 μm • Powder particles $\leq$ 100 μm	Nozzle diameter $\leq$ 50 μm
End products	Customized chocolate designs	Edible decorative objects and toffee-based structures	Full-color sugar cubes	Personalized cookies and precisely shaped food pastes
Advantages	• Affordable • Quick production	• High-quality prints • Intricate designs	• Greater material flexibility • Improved print resolution • Full-color capability • Intricate designs	Superior print resolution

(Continued)

Table 7.2 (Continued)

Feature	Hot-melt extrusion	Sintering-based technology	Inkjet powder printing	Inkjet-based printing
Disadvantages	Lower print quality	• Costly equipment • High power consumption • Material limitations	• Slow production speed • Expensive machinery	• Slow production • High-cost printheads • Expensive equipment • Limited material options
Example Machines	Choc Creator	Food Jetting Printer	Chefjet	Foodjet
Manufacturers	Choc Edge	TNO	3D Systems	De Grood Innovations

7.2 Environmentally Friendly and Sustainable Technology

The benefits of additive manufacturing include a small ecological footprint, lower usage of raw materials and energy. 3D food printing can also contribute to the prevention of food waste-on-demand production, which will reduce water and energy consumption. According to Galdeano, Europe throws away 77 million tons every year, 70% of which is generated from manufacturing and household waste. It is assumed that 3D printers would be useful this case since they are used both in manufacturing and household. Assuming if 3D printer technology accounts for 25% of total European food consumption, it may save 19 million tons of food destroyed, which accounts for 25% of food waste coming from traditional processing modes.

7.2.1 Promoting Higher Social Binding Through Food Messaging

Research has indicated that 3D food printing can actually enhance social connections. According to Huang, Liu, Mokasdar, and Hou (2013), when a message is embedded in food, it evolves with openness. The special ingredients added to social events can be developed by printing customized cakes, chocolates, and cookies with special messages or distributed as free products carrying firm logos. The availability of 3D printing to consumers makes any creative food presentation dispensable, since any creative food presentation requires specialized skills. When the sensory experience of food affects a person's mood and emotions, food creates a more intense connection among people than paper or digital devices, traditional means of communication.

Wei, Ma, and Zhao (2014) explored the influence of edible messaging by 3D printing on subjects from diverse ages (20–38 years), genders, and cultural backgrounds. The study found that "most participants believed that messages represented by food carry more weight and are perceived as a channel of meaningful communication than text displays on screens or on paper." As a respondent relates, food messages are memorable because they remember both the words and the sender. This exposes the vast potential that 3D food printing might bring about in social communication through offering new but efficient means of creating visually striking and emotionally moving messages across a wider range of edibles, from fruits and vegetables, and this would greatly improve the quality of interpersonal exchanges.

7.2.2 3D Food Printing Technology

Inkjet Printing

The Netherlands company that has developed the inkjet printing ("Food jet") technology onto which it prints directly on food. This comes in the form of a pneumatic membrane nozzle jet which deposits on a moving surface extremely small droplets of edible ink, which eventually combine to make up the digital image. This can be a decorative design, a surface coating, or a filling. This method is best suited for low-viscosity materials; hence, more appropriate for food designs on flat products, and not on complex ones. There is an aspect of temperature in this process which changes the properties of the ink as well, so it can be majorly used for confectionery and decorative purposes.

Binder Jetting

Binder jetting is an additive manufacturing technique that applies a liquid binder to stick together successive layers of powder that have been spread evenly onto a building platform. Forming each layer of the object requires spreading, most often using a counter-rotating roller. Subsequently, a liquid binder is precisely deposited onto the powder bed by an inkjet print-head, hence creating the 2D pattern of that layer. It has some benefits, like fast production and low cost of materials. However, it also generates a messy surface texture for the objects and, above all, the equipment used is quite pricey.

Selective Sintering

There are two main classes of selective sintering, depending upon the source utilized: selective laser sintering (SLS) and selective hot air sintering and melting (SHASAM). SLS and SHASAM, though different with respect to the source utilized, are identical in principle, as they both rely on a sintering source to melt and fuse powder particles together to form a solid layer. For each deposited layer, a roller spreads a new layer of powder over the existing one, and the object is built in layers. Then, the object is recovered from underneath the powder bed. An important advantage that this technique has is its ability to generate complex items quite quickly in a relatively short space of time without requiring post-curing after printing. The fabrication process is complicated by requiring many variables.

7.3 Autonomous Mobile Robots

Robotics is the study, design, creation, control, and use of robots. The field is very fast-moving, cutting across science, engineering, industry, manufacturing, transportation, and everyday life. The robots are versatile household helpers who can perform any duty from cleaning to food service, door management, and even assisting human beings with efficiency. In the health and medical field, robots function to cater to patients, assist with operations, manage the equipment in health institutions, and guide and move patients around. In industry, with manufacturing, robots are applied in the automation of processes in dangerous environments where one can be endangered as a human being by hazardous materials or heavy weights. Mobile robotics is on an upward trend as well because scientists are seeking to endow autonomous agents with the capacity to detect patterns, learn through experience, realize their location and chart their surroundings, avoid obstacles, and execute assigned tasks.

GAIT locomotion is the mimicking technology of natural ways of movement observed in bipeds, quadrupeds, and hexapods (Niloy et al., 2021). It majorly applies to a particular leg configuration following a spring-loaded pantograph mechanism with many segments. This allows robots to have high mobility and ensures stability both dynamically and statically to climb up such obstacles as steps or stairs. The complexity of limb coordination in the robot is equal to the number of legs the robot has, and this also determines its GAIT pattern. For n legs in a robot, there are $(2n - 1)!$ possible movement sequences. Figure 7.2 represents various ground mobile robot categories and a comparison of locomotion systems:

1. The one-legged robot, also known as a "hopper," has one leg and hops to move. This type of robot is very handy in missions on the smaller celestial bodies like asteroids and comets, where multilegged or wheeled robots cannot be fully effective.
2. The most commonly viewed locomotion design in robots is actually the two-legged robot, also known as the biped. These human-like robots will walk and

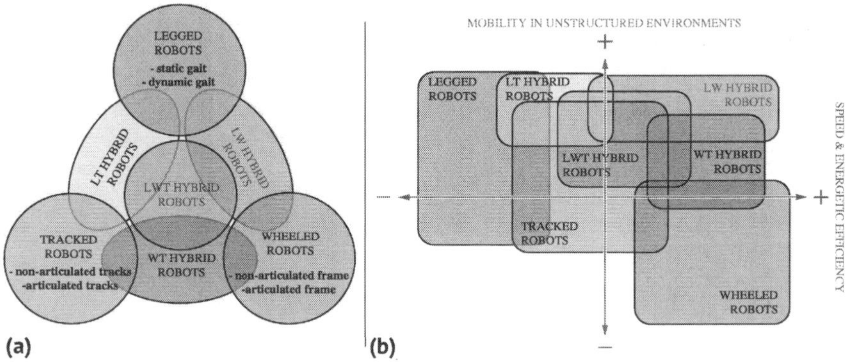

Figure 7.2 (a) Basic control schemes of mobile robots. (b) Various navigation, locomotion schemes (Saadati et al., (2024)/IEEE).

run and even jump, run, or dance. It can go upstairs and cross over varying terrains, but it is difficult for these dynamic, stable robots. Some benefits of biped locomotion over other forms have been established, especially balance.

3. There are few tripods or three-legged robots, and no biological model exists to illustrate them. They keep themselves on an unstable base by moving their center of gravity out of the body to achieve equilibrium. Compared to bipeds, tripods offer a better mass balance in uneven terrains.
4. Quadruped robots have four legs and represent the best dynamic stability of all robotic designs, as they are best at balancing. These animal-like robots are sometimes used as pets. There also exist hexapods that have six legs. A large number of control architectures, based on living organisms, have been applied to quadrupeds, to increase the ability to load heavy weights and traverse harsh terrain.

The geometrical parameters of the Delta robot are represented in Figure 7.3.

Delta Parallel Robot Kinematics: It is made up of two primary processes: inverse kinematics (IK) and forward kinematics (FK). IK determines the necessary joint configurations in order to reach a target point by the end-effector, while FK determines the end-effector position through known joint configurations. The three revolute actuating joints connecting the upper arms to the base of the robot are the three major actuators of the robot. The solution to IK comes from the kinematic closed chain description.

$$\overline{OA_i} + \overline{A_iB_i} + \overline{B_iC_i} + \overline{C_iE} + \overline{EO} = 0 \tag{7.1}$$

Delta Parallel Robot Trajectory Planning: In a delta parallel robot (DPR), trajectory planning is defined as the choice of the proper sequence of desired end-effector positions at any instant. It deals with the kinematics and dynamics of the robot in such a way that it achieves smooth and efficient motion within its operational space. Assuming that the kinematics of the robot are correctly modeled, the calculations in trajectory planning may be transferred from workspace to

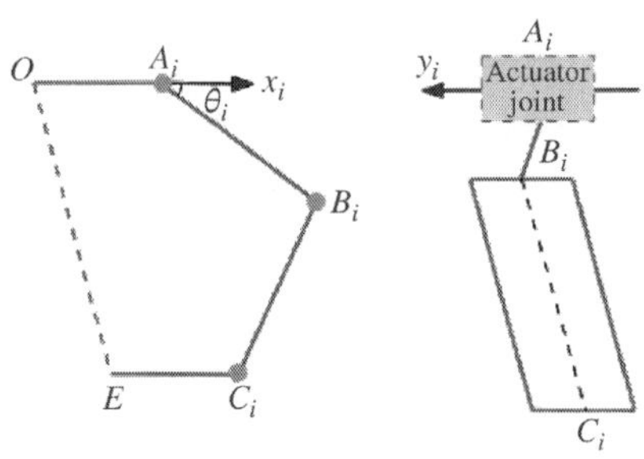

Figure 7.3 Delta parallel robot geometrical parameters.

joint-space. This means we could interpolate using the 4-5-6-7 interpolating polynomial between the initial and final states of the joint-space configuration presented by Θ_I and Θ_F, respectively.

$$\Theta(t) = \Theta_I + (\Theta_F - \Theta_I)\, p\,(t_{\text{norm}}) \tag{7.2}$$

where t_{norm} denotes normalized time, and the interpolating polynomial p can be expressed as:

$$p(t) = -20t^7 + 70t^6 - 84t^5 + 35t^4 \tag{7.3}$$

Alternatively, if there are specific points designated as targets along the path (in addition to the initial and final points), the use of cubic splines becomes a viable option. The cubic spline method efficiently interpolates a trajectory through a series of given points. When dealing with $n + 1$ points, n polynomials with distinct parameters are employed to interpolate the entire path. The mathematical definition is as follows:

$$p_i(t) = a_{i0} + a_{i1}\,(t - t_i) + a_{i2}(t - t_i)^2 + a_{i3}(t - t_i)^3$$

And the overall path will be defined as:

$$p_{\text{path}}\,(t) = p_i(t), \quad t \in [t_i, t_{i+1}] \quad \text{and} \quad i = 0, \ldots, n-1$$

where t represents the overall time, i represents the number of polynomials or the number of the current target point, and t_i is the time when the end-effector will hit the target point number i.

7.3.1 Humanoid Robots

Humanoid robots are those that have an overall or partial human shape, endowed with limbs, a torso, and head, or even only the upper or lower parts of the body. One characteristic of the full-size humanoid robot is that it can be upright-standing and walk on two legs. Most mobile robots rely on wheels, tracks, or other means of locomotion since attempts at simulating the human way of walking have proven historically challenging. Equally significant is the technical challenge of mastering the capability to fall back up and other natural human movements. The configuration of the MIT humanoid robot (Chignoli et al., 2021) and the Honda P2 model (Hirai et al.,1998) is shown in Figure 7.4, including the overall system configuration of P2 and the posture recovery model.

The MIT Humanoid Robot: The MIT humanoid robot mainly does acrobatic behaviors such as flipping and jumping; hence, it perfectly defines state-of-the-art hardware design as well as control systems. This case study intends to take place in robot design, motion planning, and control systems that facilitate dynamic movements.

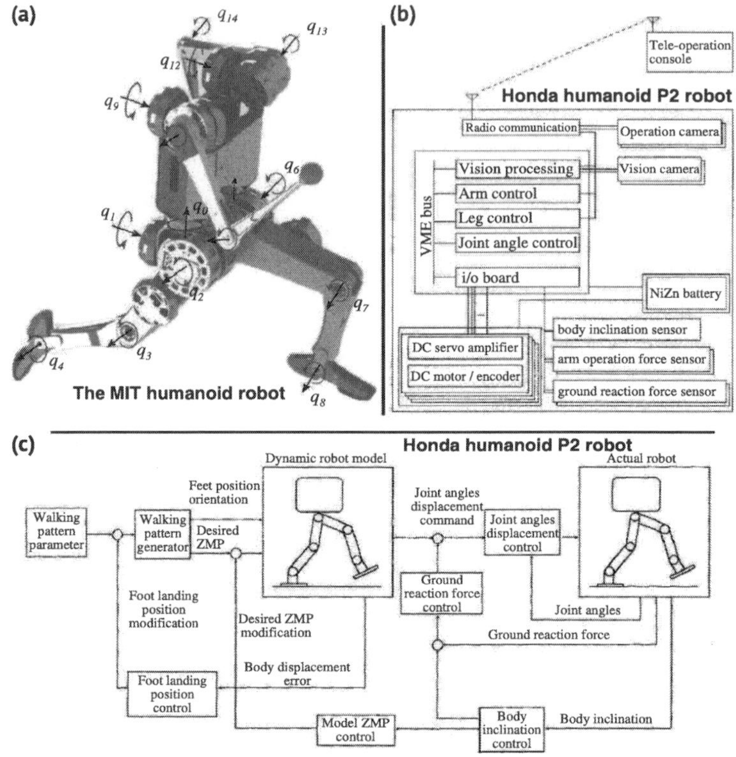

Figure 7.4 The design setup of the (a) MIT humanoid robot and (b) the Honda P2 humanoid model, including the (c) comprehensive system setup of the P2 and its posture recovery framework.

Robot Design

Dimensions and Weight: The length of the robot stands at nearly 0.7 m, and the weight is approximately 21 kg with most of its mass going toward the torso, shoulders, and hips at 75.6%.

Actuator System: The robot is fitted with a customized actuator setup with belt gearing of the knee, ankle, and elbow joints to attain higher torque for dynamic motions.

Power Supply: It is powered by a high-performance 60 V, 3 Ah battery pack that ensures it will have enough energy to power acrobatic activities.

Motion Planning: Kinodynamics Planning: The robot has an actuator-aware kinodynamic motion planner that respects the torque, velocity, and power limits while giving precise control over dynamic motion.

Whole Body Impulse Control: This control architecture integrates impulse control with model-predictive control, ensuring that the robot has full control of long-term and short-term dynamics.

Control Mechanisms: Dynamic Behavior Demonstration: The robot can perform challenging stunts such as backflips and spinning jumps, which demonstrate that its design and control mechanisms are quite effective.

Error Recovery: WBIC can recover from models of errors and small perturbations so that it can maintain stability in acrobatic performance.

Performance Tracking: The robot's behavior at all points during maneuvers like the front flip shows stability and control throughout motion.

Robustness: All dynamic motions track robustly and stay within actuator limits, thus confirming that the control framework is reliable.

As one of the most advanced humanoid robots developed to date, combining innovative design with sophisticated motion planning and control strategies, the MIT humanoid robot is a case study for advancing future developments in dynamic robotic behaviors; it will be ready to lead to yet more complex and capable humanoid robots in its wake.

7.3.2 Educational Robot

7.3.2.1 Case Study: GrowBot Educational Robotic System

Overview of GrowBot Purpose: The GrowBot is designed as an educational robotic system to facilitate hands-on experimentation with environmental conditions for food plant growth. It's a tabletop-sized greenhouse equipped with various sensors and actuators to control plant growth conditions. The GrowBot educational robot is presented in Figure 7.5.

Components Sensors: Humidity, CO_2, temperature, water level, and RGB camera. Actuators: Full spectrum lights, IR lights, UV lights, nutrient pump, water pump, air pump, air change pump, and fan.

Design Methodology

Comprehensive System Design: System design is produced through the initial discovery, followed by designing and then verification of domain issues which are critical for the incorporation of the system in education. The approach thus taken is quite unlike most engineering approaches, which inherently tend to leave out the higher-level details of deployment.

Integrated Development Domains: Hardware development, control software, interaction with the user, application development, and domain adoption.

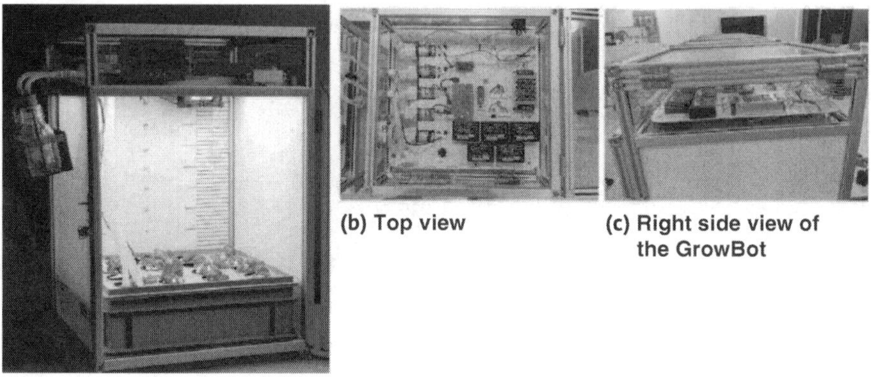

(a) The GrowBot with an open chamber (b) Top view (c) Right side view of the GrowBot

Figure 7.5 The GrowBot with an open door to the chamber, which contains the water tray at the bottom, and small basil plants in the cups in the tray. Water and fertilizer bottles are mounted on the side (Lund et al., (2022) / MDPI / CC BY 4.0).

It is emphasized that all these domains have an interconnection need within this cyclical process of design.

7.3.2.2 Application and Domain Adoption Educational Cases

There are several educational cases offered through GrowBot's website, including: Grow green and become an urban farmer, "Grow green and serve vegetables," "Grow green and eat green proteins," and "Grow green with limited resources." Each case contains 15–20 educational tasks associated with the theme of the case, thus providing structured teaching.

Teaching Resources: It offers a total teacher's guide and 16 different resources to help teachers align the GrowBot activities with curriculum objectives. This is a plug-and-play solution that simplifies the teaching process.

Process of Implementation and Designing of Teaching Intervention: For each of the cases, the teaching intervention spans several weeks and encompasses 18 smaller educational tasks. In the first week, task framing, programming, and the experiment with GrowBots and vegetable-related fieldwork up to the presentations and evaluation take place.

Teachers' Support: As part of a support package to help teachers feel confident about achieving curriculum targets, detailed materials are offered to support the successful integration of the technology into the learning environment.

The GrowBot system is one exciting example of an effective teaching tool that creatively ties high-tech learning and haptic learning in order to interest schools with the concept of food growth and environmental science. It is highly designed and supported for mass uptake in a learning environment.

7.3.3 The Generic Robot

Robotic Ice Cream Portioning

Research Focus: Robotic food portioning systems for ice cream, overcoming product variability and hygiene requirements.

Key Findings: Prototype system commissioned based on programmable logic controller (PLC) control and lab view for data logging. The robot was considered a complete project with respect to a weight-in trials set of 80, 100, and 120 g ±5%.

Commercial Application: Systems were designed. International patents have been acquired and sales have been continued for Europe. Technical Insights:

- Utilized intelligent sensing and feedback systems to control operations in real time.
- Designed an in-vehicle control system for continuity in portion dispensation depicted in Figure 7.6 (Friedrich and Lim, 2001). Robotic ice cream portioning robot and vending operation.

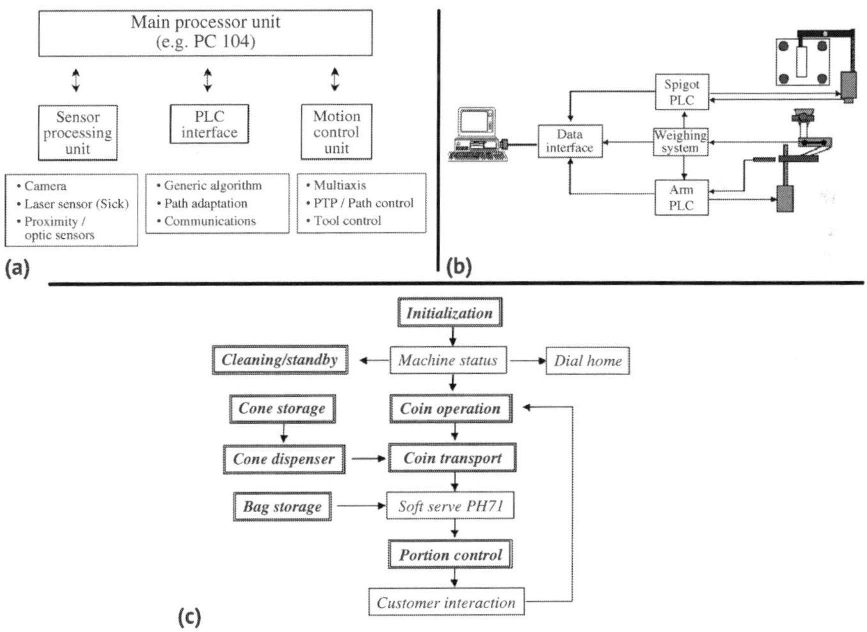

Figure 7.6 (a) Control architecture for variable product processing, (b) robotic ice cream portioning, and (c) vending operation (Friedrich and Lim [2001]/ARAA).

7.3.4 Automation Pyramid

The layers in an industrial automated production plant are usually represented as an automation pyramid. The different functions and the distinct IT infrastructures that they utilize characterize each of these layers. Usually, the automation pyramid has six layers. ISA-95 enterprise/control system integration standard categorizes manufacturing activities into five levels. In general, automation and control systems are usually found at Levels 1 and 2, MOM systems function at Level 3, and ERP systems manage Level 4 activities. The ISA-95 levels are shown in Figure 7.7.

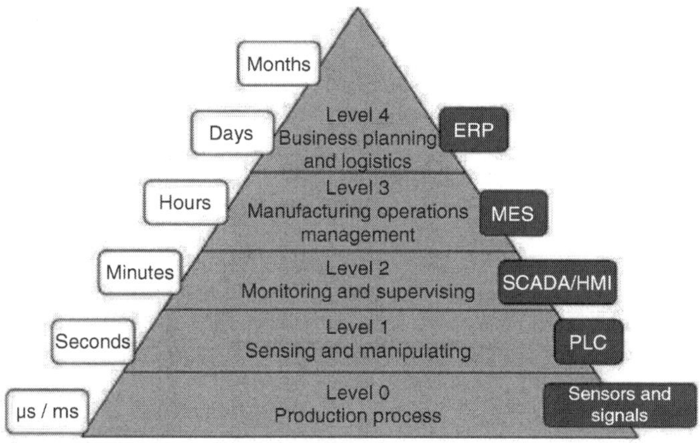

(a)

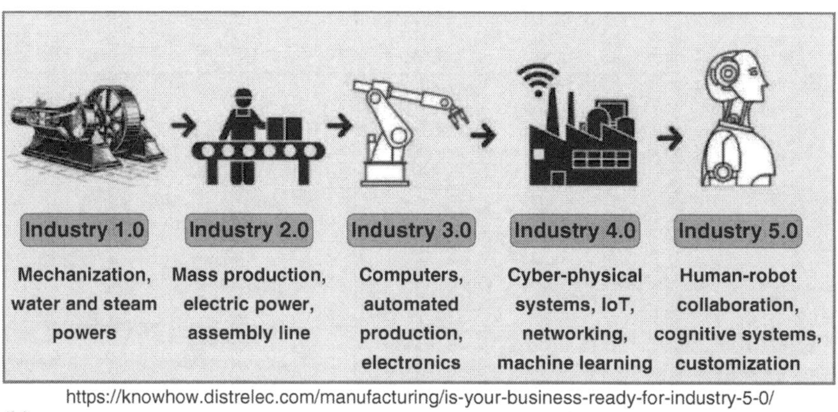

(b)

Figure 7.7 (a) Automation pyramid according to ISA 95 model. (b) The five Industrial revolutions (Åkerman [2018]/Magnus Åkerman).

- Level 0: This is the process level where sensors and actuators are used to collect production data and execute commands. Parallel wiring is usually adopted at this level.
- Level 1: This is the field level that forms an interface to link the production process to the control level. Typical systems include fieldbus and input/output modules.
- Level 2: This is the control level that holds an IT structure designed for controlling the production process via an ICS, typically SPS.
- Level 3: This is the management level of a process that should act as a monitoring, operational, and recipe management tool via a process management system, for example, supervisory control and data acquisition (SCADA).
- Level 4: It is the level that is the management of operations; MES, MIS, or LIMS should be interfaces for extensive planning of production, quality management, data collection, materials management, and KPI documentation.
- Level 5 (it is the highest level): It refers to the company level used for general production planning and order management through the ERP system.

The structure of the automation pyramid can vary. Levels 0 and 5 can be one and separate levels do not exist or the last two levels are one. In the context of Industry 4.0 and the industrial Internet of Things, the automation pyramid is under debate. Experts have been arguing that flexibility and interconnectivity as intended by Industry 4.0 would require flattening of the pyramid.

7.3.5 Difference Between RTU, PLC, and SCADA, HMI Case Studies

SCADA system is simply a system that consists of a number of remote terminal units collecting field data and connecting back to a master station via a communications system; however, the master station shows the acquired data and allows the operator to perform remote control tasks.

Proper and timely adata, typically in real time, enables optimization of plant and process operations. Such optimization increases the efficiency, reliability, and safety of operations, while the cost of operations is reduced compared to earlier nonautomated systems.

There appears to be confusion about the definition of SCADA systems and process control systems. SCADA systems are often associated with remote or distant operation. Other factors that have been attributed to failures of SCADA systems include the improper integration of components, unnecessary complexity, unreliable hardware, and unproven software. Because changes take place in both the performance of all the SCADA systems—the remote terminal units (RTUs), communication links, and master stations—and the field devices—that is, the transducers and control devices—many operators take all of this into consideration when making judgments about the effectiveness of an SCADA system. However, in this manual, field devices fall outside the SCADA scope, so their discussion won't continue further. Here is a usual SCADA system.

In a more advanced SCADA system, there are five fundamental levels or hierarchies:

1. Field-level control and instrumentation devices
2. Remote terminal units (RTUs) and marshaling terminals
3. Communication infrastructure
4. Central master station(s)
5. Computer system of the commercial data processing department

Communication System: The communication system acts as a pathway of communication between the master station and remote sites. Radio, telephone lines, microwave, or satellite technology can be used in this system. Certain protocols and error detection methods are used to make the data transfer more efficient and optimal.

SCADA technology has been available since the early sixties, and is now challenged by two alternative approaches—distributed control systems (DCS) and PLC.

Advantages of SCADA system: A SCADA system requires adequate justification concerning the initial cost incurred when it is mounted. Some of the general justifications for installing a SCADA system include: (i) System optimization that results in saving costs due to a more effective operation of the plant or process. (ii) Increased productivity of personnel. (iii) A safe system due to improved information and better control. (iv) Protection of plant equipment. (v) Safety of environment by avoiding system collapses. (vi) Saving of energy as the plants will be optimally utilized. (vii) Better and faster availability of data which may lead to quicker and more accurate client invoicing. (viii) Safety and gas metering regulations by government in relation to payment of royalties, taxes, etc. Figure 7.8 represents diagram of a typical SCADA system.

Distributed Control System

A DCS is comprised of several units based on microprocessors that perform data acquisition and control functions near the devices being controlled or the instruments from where data are gathered. DCS systems have evolved to be systems that deliver highly sophisticated analog (e.g. loop) control capability. It also employs closely integrated operator interfaces, or man-machine interfaces, to make system configuration easy for operators to use.

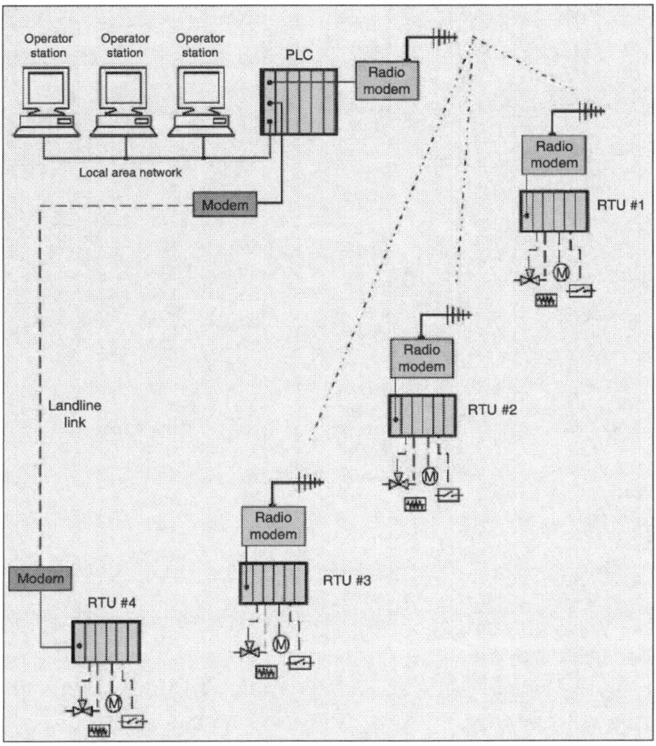

Figure 7.8 Diagram of a typical SCADA system (https://electrical-engineering-portal.com/scada-dcs-plc-rtu-smart-instrument)/Electrical Engineering Portal).

Programmable Logic Controller

Since the late 1970s, PLCs have taken the place of hardwired relays with a mixture of ladder logic software and solid state electronic input and output modules. They are often employed in the realization of a SCADA RTU as they provide a standardized hardware solution which is highly economical. Figure 7.9 represents the diagram of (a) DCS, (b) PLC, and (c) RTU.

Remote Terminal Units

An RTU, or remote telemetry unit, is an independent data acquisition and control unit. This system is basically microprocessor-based and is utilized for monitoring and controlling equipment, located at a distant location, from a central station. The main function of an RTU is to receive data from the process equipment from a remote location and then transmit it back to a central station after processing.

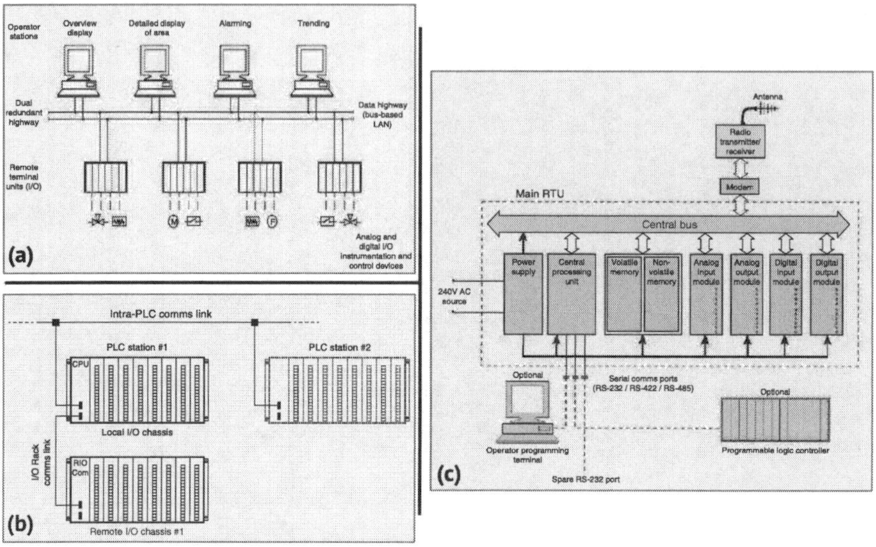

Figure 7.9 Diagram of (a) DCS, (b) PLC, and (c) RTU.

It often accepts the configuration and control software from a master station. In addition, it can be configured locally using an RTU programming unit. Although typically communicating with a master station, it is also capable of peer-to-peer communication with other RTUs. In addition, an RTU can act as a relay station to forward data to another RTU which may not be accessible directly from the central station.

7.3.5.1 Case Study: VFD, PLC, and SCADA in Food Manufacturing

Overview Objective: To implement a controlling and monitoring system for fluid temperature, pressure, and level in the food manufacturing industry using Variable Frequency Drive (VFD), PLC, and SCADA systems (Shaikat et al., 2019). Figures 7.13 and 7.14 represent the SCADA network for milk pasteurization plant.

Key Components

VFD: It controls the speed of the pump-motor, thereby controlling the water flow rate to the required value with accuracy, which is an essential parameter for achieving desired levels and pressure as illustrated in Figure 7.10.

PLC: It works like a central control as it continues processing signals from several sensors, then executes the necessary control instructions.

SCADA: SCADA enables real-time monitoring and graphical representations of the system and this brings to view the operations.

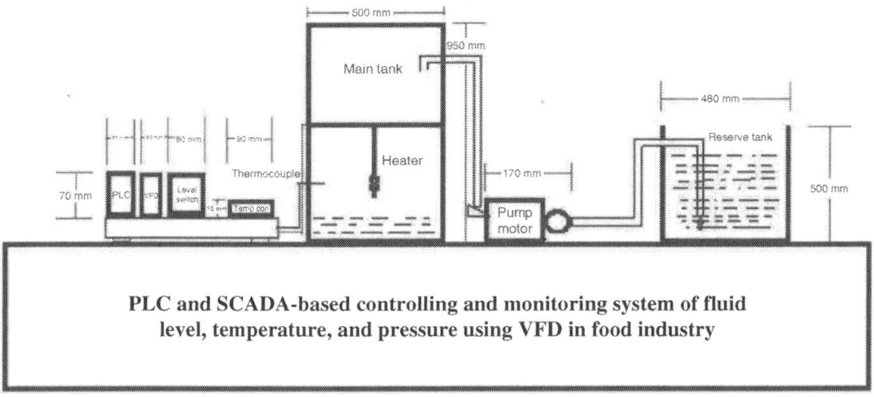

Figure 7.10 PLC and SCADA based controlling of flow, temperature, and pressure in food industry using VLC (Shaikat et al. [2019]/IEOM Society International).

Functions of System

Temperature Control: The temperature is monitored through sensors and the liquid is maintained at a set temperature through PID control, and thus maintaining quality in the food processing process.

Pressure and Level Control: The liquid level in the tank is detected by the switches. Pressure computation is done by the system through fluid density and height. Pump speed can be varied with the help of VFD to regulate the level and pressure, thereby governing optimum levels and pressures in the process.

Process Flow

Initialization: Input desired values for pump-motor and temperature to be achieved.

System On: Click the start button.

Operation: The pump-motor fills up the main tank while the heater is on standby until this required level is adequately achieved. When it is adequate, the heater is turned on to attain the set temperature.

Monitoring: WinCC RT advanced highlights real-time data, including temperature change along with the system status indication. An emergency stop facility is also available if the process needs to be stopped in any situation.

The system will keep consistent temperatures and pressures; therefore, the food production processes will become efficient. The energy saved by using VFDs relates to running the motors at lower speed based on the demand instead of continuously running at full capacity. Therefore, combining these technologies in food manufacturing sets up control over fluid dynamics, which subsequently leads to the quality and operational effectiveness of the system.

7.3.6 Ladder Elements, Logic Gates in PLC Programming

A PLC is a computer-specific device that lacks a monitor and keyboard. It is primarily used within the industrial automation domain, such as manufacturing, automotive, food and beverage, and chemical processing to automate systems. This small computer accepts input data and delivers operating instructions as output. It is programmable. The two primary languages for programming the PLC are "Ladder Logic" and "C". Ladder logic is the most used programming language for PLCs.

Ladder Logic: Ladder logic is a graphical programming language represented through a diagram that looks like a circuit diagram of relay logic. The program makes up two vertical lines known as "rails" and horizontal lines as "rungs," making it look like more of a ladder. This graphical format is called a ladder logic diagram (LLD). Figure 7.11 represents (a) ladder logic diagram, (b) switch and their output, (c) AND gate ladder logic diagram, and (d) OR gate ladder logic diagram.

LLDs are one of the graphical programming languages that work through real-time input. It comprises two vertical lines known as rails. The left rail is the power supply for the circuit, and power passes through to each rung. Every rung includes switches and an output coil. Switches can act like an OR, AND, and NOT; therefore, all kinds of programming logics can be made in PLC. An LLD is performed

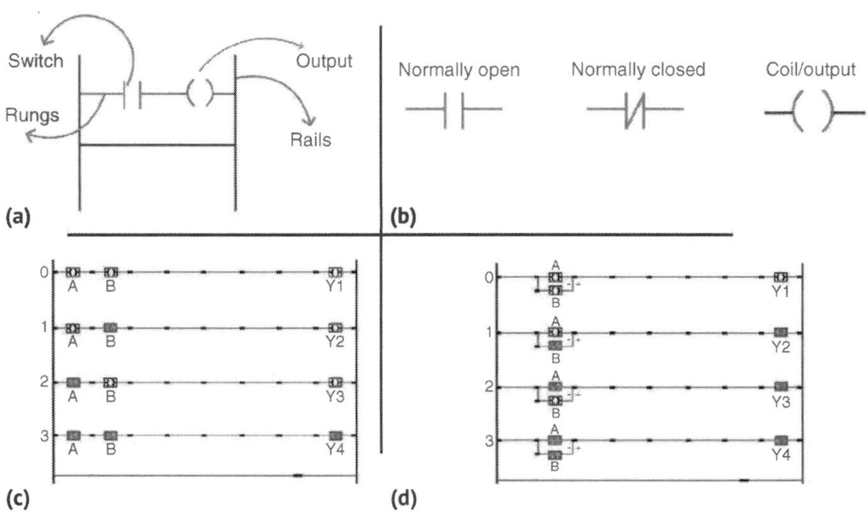

Figure 7.11 (a) Ladder logic diagram, (b) Switch and their output, (c) AND gate ladder logic diagram, (d) OR gate ladder logic diagram.

from left to right and from top to bottom. When all the conditions of a rung are satisfied, then the associated output coil is energized, which in turn energizes the real-world system, such as turning ON/OFF a motor. The PLC and control system scan this process repeatedly to automate the operating system.

Rails and Rungs: Rails are vertical lines while rungs are horizontal lines.

The NO/NC switch indicates whether the flow of current or contact in an electronics circuit is open or closed and is thus used for PLC ladder logic programming: NO (Normally Open) and NC (Normally Closed). These are referred to as whether the switch is open or closed.

NO: A NO contact or switch is open by default unless an action is applied to it, at which time the contact closes the circuit and allows current to flow. In an LLD, a contact that is normally open is illustrated by an open break across that particular contact, and when activated, it represents a solid line.

NC: This switch or contact, NC, closes when not actuated. The actuation of the switch opens the circuit, which would stop current flow. In an LLD, it is portrayed as a solid line; on actuation, it becomes an open gap.

Output/Coil: Outputs, such as motors, valves, indicators, and lights, are depicted with a vertical line and labeled with the name of the output device.

Description and Working of Ladder Logic: Two rails and a rung in between. The left rail has power supplied to it and is connected to an input switch on one end, and the right rail has an output coil on the other end. A wire is connected from switch to coil. Place input switches, which can be either NO or NC based on the required logic, on the left-hand rail of the panel. The quantity and location of the switches will depend on the type of logic to be created. In this problem, a truth table is generated to create this logic and switches located accordingly: For "0" input, typically NO switches are not used, but NC switches may be used, in which case they must be wired made open or false. For "1" input, NC switches are employed, but similarly, NO switches may also be used. An output coil must be connected to the right side rails.

Examples of Some Ladder Logic Programming

Some examples of Ladder Logic Programming are:

Creating AND gate LLD, please refer Table 7.3 and Figure 7.11c. AND logic: Y=A.B (here A and B are the inputs, and Y is output)

In Figure 7.11, the NO switches represent input 0, indicating that no current is passing, while the NC switches represent input 1, indicating that current is passing through the circuit. Each line in the diagram represents a different combination from the truth table. For example, the first line, where A=B=0, indicates that no current is passing. Since these switches are in series (according to AND logic), no current will reach the output module. The other rungs in the diagram are created

Table 7.3 The AND gate in LLD.

Input (A)	Inuput (B)	Output (Y = A.B)
0	0	0
0	1	0
1	0	0
1	1	1

Table 7.4 The OR gate logic in LLD.

Input (A)	Input (B)	Output ($Y = A + B$)
0	0	0
0	1	1
1	0	1
1	1	1

from the truth table to show each switch in series and to check if it matches the table or not.

OR Gate Ladder Logic Diagram

OR Gate logic : Y = (A+B), please refer to Table 7.4 and Figure 7.11.

In an OR gate, a parallel circuit is shown, so two switches are placed in parallel. The rest of the parts work the same as in the previous example. NC switches (green) show current flow, which is represented by 1, and an NC switch (white) shows no current flow, which is represented by 0. Now, go to the truth table to check the output.

Application of LOGIC Gates NOR Gate: The NOR gate is essentially the opposite version of the OR gate. This means that if we were to freeze the output of the OR gate, it would essentially become a NOR gate. The NOR gate is a modified version of the NOT-OR gate; it can carry out NOR operations by taking in two binary inputs to derive a binary signal. It is also known as a universal gate, since it can produce any digital circuit independently.

Applications of NOR Gate:

There are several applications of NOR gates (represented in Figure 7.12) in multiple industries. These are as follows:

1. It is referred to as a universal gate because it can form any other logic gates.
2. Just like the NOT gate, the NOR gate is almost identical to the inverter.
3. It is used for the designing of combinational logic circuits.
4. NOR gates can be utilized for establishing various sequential logic circuits.

7.3.7 Fiber Optic Sensors and Refractometry

Fiber Optics and Types

Fiber optics, also known as optical fiber, is a technology that transmits data as light pulses along a glass or plastic fiber. The optical fiber itself is a thin, cylindrical fiber made of glass or transparent dielectric material. It is commonly used for optical communication as waveguides made of transparent dielectrics. In this article, we will delve into the details of fiber optics.

What Is Fiber Optics?

Fiber optics is the science and manner in which information is transmitted in the form of light pulses along a glass or plastic strand or fiber. These cables are applied to long-distance and high-performance data networking. Fiber-optic cables transmit information over longer distances and higher bandwidths (data rate) as compared to electrical cables. Fiber optics are, therefore, an integral part of the modern telecommunications, internet, and computer networks.

Major Components of Fiber Optics

Core: It is a central tube of very small diameter made of optically transparent dielectric medium, carrying light from the transmitter to the receiver. The diameter of the core can range from about 5–100 μm.

Cladding: It is the external optical material outside of the core. The refractive index of cladding is lower than that of the core. The cladding confines the light to remain inside the core by a process known as total internal reflection.

Buffer Coating: It is the plastic layer that covers the fiber to prevent it from damage due to environmental conditions. The buffer coating is usually silicon rubber. After application of the coating, the diameter of the fiber will be 250–300 μm.

Types of Fiber Optics

There are many types of fiber optic based on several categories as mentioned below:

1. Based on the Number of Modes

   Single-mode fiber: In single-mode fiber, only one kind of light ray can travel through the fiber. This type of fiber has a very tiny diameter for the core—about 5 μm—and has a very high cladding diameter—about 70 μm. The core and cladding have a very small difference between them in refractive index. There is no dispersion and therefore no degradation of the signal while it travels through the fiber. Light is passed through it using a laser diode.

   Graded index optical fiber: Here, the refractive index of the core is not uniform and in turn decreases gradually from the center to the boundary with cladding. The refractive index of cladding is uniform. The light rays travel through it in the form of skew rays or helical rays and do not intersect with the axis of the fiber at any time, depicted in Figure 7.12b.

Input OR gate

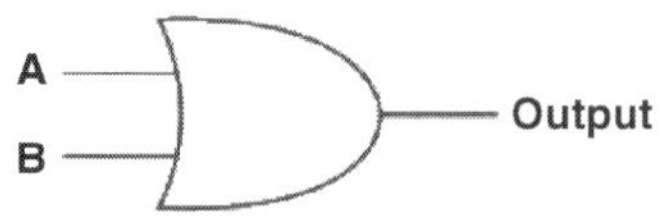

Truth table

Input A	Input B	Output
0	0	0
0	1	1
1	0	1
1	1	1

(a)

Input NOR gate

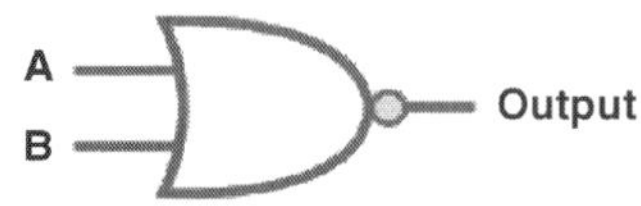

Truth table

Input A	Input B	0 = (A + B)'
0	0	1
0	1	0
1	0	0
1	1	0

(b)

Figure 7.12 (a) Input OR gate (b) Input NOR gate.

7.3.8 Sensors in IoT

Broadly, sensors are used in the architecture of IoT devices. Sensors are used to detect and measure physical parameters and then convert them into signals suitable for processing, such as electric, mechanical, or optical. These signals are converted into a human-readable form indicating changes in characteristics like resistance, capacitance, or impedance.

A *transducer* is an item that transfers energy from one physical form to another, essentially transforming one type of energy into another. It can also be an actuator for various systems.

The sensor has two primary characteristics: static and dynamic.

Static characteristics are those that describe how the output of the sensor will change, given that there has been a change in input which, following a transient condition, has arrived at a steady-state condition.

Accuracy is defined as the capability of measuring devices in determining a value close to the actual quantity being measured. The absolute and relative errors are used to measure accuracy, which explains the correctness of the output against the higher prior system.

Range is the maximum and minimum values of a physical quantity that can be sensed by the sensor and reflected in its response. Resolution is the smallest input change that a sensor can detect, which is an important specification in the selection of sensors.

Precision is the ability of a measuring instrument to provide the same measurement under the same conditions when repeatedly measuring the same quantity.

The *sensitivity* is defined as the ratio of incremental changes in the output of the system to incremental changes in input parameters. Linearity defines how much the curve deviates from a straight line, and this is why it can be determined from the calibration curve, too. Drift can be defined as the difference between the measurements issued by the sensor from a given reading when it is kept at that value over a long time. Repeatability is a difference in measurement in a series of several observations carried out under the same conditions; it is required for reliable measurement.

Properties of the Systems

Zero-order system: The output has no delayed response to the input signal. It involves no energy-storing elements. Applications include potentiometer measures, linear and rotary displacements.

First-order system: The output response to an input signal takes some time before reaching its steady value. It includes a component with energy storage and dissipation.

Second-order system: This system provides an output response of the sensor in which the sensor's output response oscillates before achieving a steady state.

Sensor Classification: Passive and Active, Analog and Digital, Scalar and Vector.

Passive Sensor: Cannot sense the input independently. Examples include accelerometers, soil moisture, water level, and temperature sensors.

Active Sensor: Is able to sense the input independently. Examples include radar, sounder, and laser altimeter sensors.

Analog Sensor: The response or output of the sensor is a continuous function of its input parameter. Examples include temperature sensors, LDR, analog pressure sensors, and analog hall effect sensors.

Digital Sensor: The response is of binary nature. It has been made to overcome some of the disadvantages of analog sensors, so it has additional electronics for a bit conversion. Some of the examples are passive infrared (PIR) sensors and digital temperature sensors.

Scalar Sensor: The sensor measures the input parameter solely based on its magnitude. In this case, the output for the sensor is dependent on the magnitude of some input parameter, independent of the direction of the input parameters involved. Examples include temperature, gas, strain, color, and smoke sensors.

Vector Sensor: The sensor response is a function of the magnitude, direction, and orientation of the input parameter. A few examples are accelerometers, gyroscopes, magnetic field sensors, and motion detectors.

Types of Sensors

Electrical Sensor: Electrical proximity sensors can be either contact-type or noncontact-type.

Simple contact sensors work with the sensor and the components making the circuit electric. Noncontact electrical proximity sensors make use of either induction for metal detection or capacitance for nonmetal detection on the electrical principles.

Light Sensor: Light sensors are also called photo sensors and are one of the important sensors. A light-dependent resistor (LDR) is a simple light sensor available today. The property of LDR is that its resistance varies inversely proportional to the intensity of the ambient light, i.e. when the intensity of light increases, it decreases the resistance of LDR and vice versa.

Touch Sensor: This is the technology used for the detection of something such as a touch of a finger or stylus. Its name implies that it's used for sensing something. They come in two kinds: (i) Resistive type, (ii) Capacitive type. Nowadays, almost every modern touch sensor is of capacitive type. Because they are more accurate and have a better signal-to-noise ratio.

Range Sensing: Range sensing is concerned with the detection of how near or far a component is from the sensing position, although they can also be used as proximity sensors. Distance or range sensors employ noncontact analog techniques. Short-range sensing, between a few millimeters and a few hundred millimeters, is carried out using electrical capacitance, inductance, and magnetic technique. Longer-range sensing is achieved using transmitted energy waves of all types, e.g. radio waves, sound waves, and lasers.

Mechanical Sensor: Any suitable mechanical or electrical switch can be used, but since a definite amount of force is needed to actuate a mechanical switch, microswitches are generally used.

Pneumatic Sensor: These proximity sensors work by either interrupting or changing an airflow. Pneumatic proximity sensors are contact-type sensors and should not be used in applications where light components may be blown away.

Optical Sensor: Optical proximity sensors work by interrupting a light beam that hits a light-sensitive device, such as a photocell. These are noncontact sensors. The lighting environment must be treated with care because optical sensors can be blinded by flashes from arc welding, and airborne dust or smoke may interfere with light transmission.

Speed Sensor: A speed sensor is a sensor that measures the speed of an object or vehicle in motion. Some examples of speed sensors are wind speed sensors, speedometers, UDAR, and ground speed radar.

Temperature Sensor: Temperature sensors monitor and track temperature and provide electrical signals corresponding to the measured temperature. These electrical signals are usually voltage, and directly proportional to the measured temperature.

PIR Sensor: A PIR (Passive Infrared) sensor, also called a pyroelectric sensor, detects infrared (IR) light radiating from objects in its field of view. It is mainly used for motion and movement detection.
Ultrasonic Sensor: An ultrasonic sensor works on the same principles as SONAR and RADAR, which are based on high-frequency waves that reflect back to determine the properties of a target. The ultrasound is produced by interpreting these high-frequency sounds or radio waves.

7.3.9 What Is 3D Printing?

3D printing is a kind of additive manufacturing that forms an object based on a digital model. Different types of 3D-printing technologies and materials are available but share the same underlying principle: transforming a digital model into a 3D physical solid object through material addition in the form of successive layers.

Understanding 3D printing begins with the understanding that every print starts as a digital file of the 3D design, much like a blueprint of an actual object, which is sliced into super-thin layers and sent to the 3D printer. The method used varies by technology. Desktop printers melt plastic material and lay it down onto a print platform, whereas high industrial machines use lasers to melt metal powder at extremely high temperatures. It takes hours for the procedure, depending on the size, but once the objects are printed, they then have to be processed to achieve the finish that one wants.

Most of the following 3D-printing technologies produce a physical object from digital designs, layer by layer, in a proprietary method for each. Some of the most common are fused deposition modeling (FDM), stereolithography and digital light processing (SLA and DLP), SLS, Material Jetting (PolyJet and MultiJet Modeling), and Binder Jetting.

Essentially, FDM utilizes a solid material termed filament taken from a reel into a heated nozzle that melts the material. The melted material then allows itself to be deposited in layers to build an object. FDM may be used ideally for rapid and relatively low-cost prototyping and could be suited for other applications as well.

SLA and DLP: This process works on projecting a light source into the liquid resin, creating the solidified 3D-printed object. In this process, the build platform is submerged in a tank containing liquid resin, and a light source solidifies each layer of the object one layer after the other until a whole object is produced that has very accurate parts and a very smooth surface finish.

The SLS method uses a laser to melt and solidify powdered material, bit by bit, layer by layer, to create finished objects. This process melts the material in the powder where it is desired, layer after layer, building the object. SLS is widely used for functional prototypes and parts.

Material Jetting, which is also known as PolyJet or MultiJet Modeling, works by jetting droplets of liquid photopolymer onto a build tray and then instantly curing them with UV light. The printer jets the liquid material onto the build tray, and UV light instantly cures the droplets of liquid photopolymer. They have strong detail resolution with down to 16-micron layer thickness.

Binder Jetting uses the principle of fabricating an object layer by layer, starting from a thin powdered material. A binding agent is squeezed out of a nozzle to bind the powder together, and the process is repeated until the desired object is fully formed. After the removal of the object from the print bed, excess powder is cleaned off and coated with adhesive glue to provide the object with strength and resistance against discoloration. The technology is relatively cheap compared to SLS, as the process of printing requires less energy. However, the limitations of 3D printing are higher cost of large production runs, fewer material choices, colors and finishes. Limited strength, endurance, and lower precision.

7.3.10 Case Studies, Retrofitting Process Automation

Automated Dairy Pasteurization Process Components

Sensors:

Temperature Sensors: Monitor milk temperature.

Flow Meters: Measure the flow rate of milk.

Pressure Sensors: Monitor the pressure in the pasteurization tank.

Actuators:

Heating Elements: Control the temperature of the milk.

Cooling System: Rapidly cool the milk after pasteurization.

Valves: Control the flow of milk through the system.

Controllers:

PLC: Manages the automation of the pasteurization process based on sensor inputs.

SCADA System: Supervisory system for monitoring and control, providing an interface for operators.

HMI (Human-Machine Interface): Interface for operators to interact with the SCADA system, view data, and adjust parameters, the figure of the process is represented in Figure 7.13.

HMI/Operator: Operators use the HMI to monitor the system and make adjustments.

SCADA Server: Collects data from the PLC, stores it, and manages alarms/notifications.

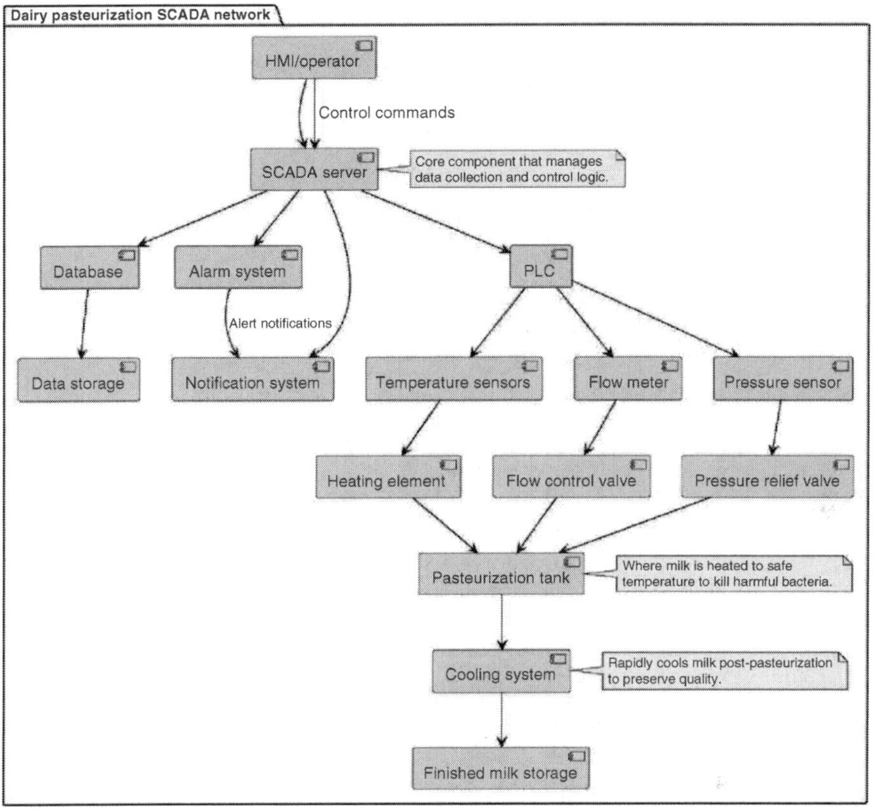

Figure 7.13 Dairy pasteurization SCADA network.

PLC: The core controller that processes inputs from sensors and sends commands to actuators.

Sensors: Continuously monitor conditions (temperature, flow, pressure) and send data to the PLC.

Actuators: Include heating elements for pasteurization and a cooling system for post-pasteurization.

Raw Milk Supply: The input stream that feeds into the pasteurization tank.

Pasteurization Tank: Where the milk is heated to the required temperature.

Finished Milk Storage: The final product is stored after processing.

Component Descriptions:

HHMI: Provides a user-friendly dashboard for operators to monitor system status and control parameters.

SCADA Server:
Collects data from various sensors and PLC.
Manages alarms and control logic to ensure optimal processing conditions.

Database:

Stores historical data for analysis and reporting purposes.
Provides insights into operational efficiency and product quality.
Alarm System: Triggers alert for any deviations from set parameters (e.g. temperature too high or low).

Notification System: Sends alerts via email or SMS to relevant personnel for immediate attention.

PLC: Central control unit that processes input from sensors and manages outputs to actuators.

Sensors:

Temperature Sensors: Measure the temperature of the milk throughout the process.

Flow Meter: Monitors the flow rate of milk entering the pasteurization tank.

Pressure Sensor: Ensures pressure within the tank is maintained at safe levels.

Actuators:

Heating Element: Increases the temperature of the milk to the required pasteurization level.

Flow Control Valve: Regulates the flow of milk based on the processing requirements.

Pressure Relief Valve: Ensures that pressure within the pasteurization tank does not exceed safe limits.

Pasteurization Tank:

Where milk is heated to kill harmful bacteria, it is then cooled rapidly.

Cooling System:

Quickly cools the pasteurized milk to preserve quality before storage.

Finished Milk Storage:

The final product is stored, ready for packaging and distribution.

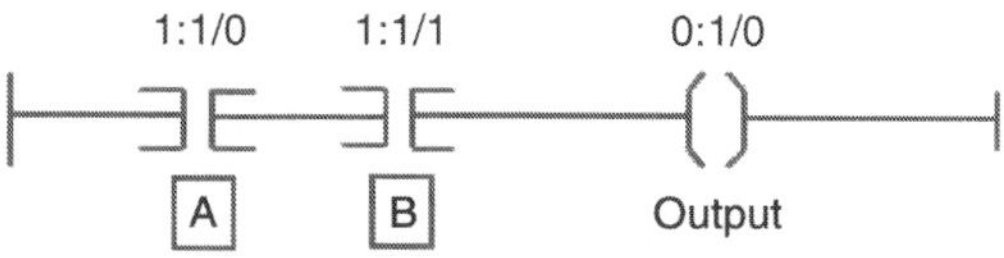

Figure 7.14 Dairy pasteurization SCADA network.

7.4 Review Questions

Q1 What is ladder logic?

Q2 What do the vertical lines in a ladder diagram represent?

Q3 What do the horizontal lines (rungs) in a ladder diagram represent?

Q4 Which type of logic gate is shown in the given ladder programming if A and B are inputs?

Q5 Automate a milk processing plant using SCADA.

Bibliography

M Åkerman. *Implementing Shop Floor IT for Industry 4.0*. Tekniska Hogskola (Sweden): Chalmers, 2018.

JM Aguilera and DJ Park. Texture-modified foods for the elderly: status, technology and opportunities. *Trends in Food Science & Technology*, 57: 156–164, 2016.

L Bruzzone and G Quaglia. Review article: locomotion systems for ground mobile robots in unstructured environments. *Mechanical Sciences*, 3:49–62, 2012.

M Chignoli, D Kim, E Stanger-Jones, and S Kim. The mit humanoid robot: design, motion planning, and control for acrobatic behaviors. In: *2020 IEEE-RAS 20th International Conference on Humanoid Robots (Humanoids)*, 1–8, 2021, IEEE.

I Dankar, A Haddarah, FEL Omar, F Sepulcre, and M Pujolà. 3D printing technology: The new era for food customization and elaboration. *Trends in Food Science & Technology*, 75:231–242, 2018.

W Friedrich and P Lim. Robotic friedrich2001roboticfood applications example: ice cream portioning. *Proceedings of the 2001 Australian Conference on Robotics and Automation*, 56–60, 2001.

K Hirai, M Hirose, Y Haikawa, and T Takenaka. The development of honda humanoid robot. In: *Proceeding of 1998 IEEE International Conference on Robotics and Automation (Cat*, No. *98CH36146)*, vol. 2, 1321–1326, 1998, IEEE.

SH Huang, P Liu, A Mokasdar, and L Hou. Additive manufacturing and its societal impact: a literature review. *The International Journal of Advanced Manufacturing Technology*, 67:1191–1203, 2013.

JAL Galdeano. *3D Printing Food: The Sustainable Future*. (Doctoral dissertation, Universitat Politècnica de Catalunya. Escola Tècnica Superior d\'Enginyeria Industrial de Barcelona, 2014 (Programes de mobilitat "outgoing" (ETSEIB)), 2014.

HH Lund, M Exner, NE Jensen, M Leggieri, M Outzen, G Ravn-Haren, Mv Sehested, A Væring, and R Andersen. Growbot: an educational robotic system for growing food. *Applied Sciences*, 12(11):5539, 2022.

M Molitch-Hou. Branch technology is 3D printing the future of construction one wall at a time. 3D Printing Industry. 2015.

D Periard, N Schaal, M Schaal, E Malone, and H Lipson. Printing food. 2007.

H Rogers, N Baricz, and KS Pawar. 3D printing services: classification, supply chain implications and research agenda. *International Journal of Physical Distribution & Logistics Management*, 46(10):886–907, 2016.

C Severini, D Azzollini, M Albenzio, and A Derossi. On printability, quality and nutritional properties of 3d printed cereal based snacks enriched with edible insects. *Food Research International*, 106:666–676, 2018.

AS Shaikat, MM Rahman, M Khan, MHK Nouman, and S Akter. Development of VFD, PLC and SCADA based fluid temperature, pressure and level control for food manufacturing industry. In: *Proceedings of the 2nd International Conference on Industrial and Mechanical Engineering and Operations Management (IMEOM)*, 12–13, 2019.

J Sun, Z Peng, L Yan, JYH Fuh, and GS Hong. 3D food printing—an innovative way of mass customization in food fabrication. *International Journal of Bioprinting*, 1(1):27–38, 2015.

J Wei, X Ma, and S Zhao. Food messaging: using edible medium for social messaging. In: *Proceedings of the SIGCHI Conference on Human Factors in Computing Systems*, 2873–2882, 2014.

F Yang, M Zhang, B Bhandari, and Y Liu. Investigation on lemon juice gel as food material for 3D printing and optimization of printing parameters. *LWT*, 87:67–76, 2018.

8

PID Control Techniques in Food Processing

After reading this chapter, you should be able to:

- This chapter provides a review of the proportional-integral-derivative (PID) control principles, focusing strictly on techniques applicable to needs in the food industry, and from basic principles of industrial control to such advanced structures as supervisory control and data acquisition (SCADA) PID, control in order to provide an excellent theoretical background and hands-on experience.
- Effective Troubleshooting and Optimization: Process loop checklists, simple PID tuning methods, and common issues in implementation will allow the reader to troubleshoot and optimize control systems effectively in food processing environments.
- Applied Insights Through Case Studies: Case studies, which are presented from the real world, improve comprehension by showing applications of PID control in food processing conditions. Readers will be able to apply the concepts learned directly to the challenges facing them in industrial practice, in order to optimize the processes.

Control algorithms determine the course of action through two primary mechanisms. The open loop system operates on fixed parameters that do not adjust during its operation. Conversely, the closed loop system employs sensors to gauge the discrepancy between the system's desired and actual states, using this error to guide future actions.

Open-loop Control

Consider, for instance, a toaster, which works almost independently. Here, the user introduces slices of bread, sets a timer, and then, with the insertion of a lever, activates the process. But the outcome is always uncertain: if the timer has been set

Food Engineering Automation with Robotics and AI, First Edition. Abir Chakravorty.

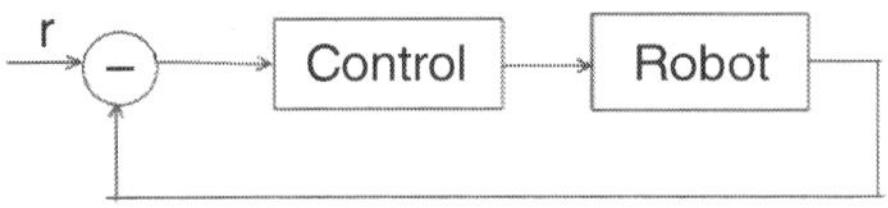

Figure 8.1 Control system-closed loop.

too short, the bread must be toasted again; if it has been set too long, the kitchen smells of burnt toast. This vagueness is because the toaster is an open-loop control system—it does not have the functionality to measure the toasting result in order to decide whether the result is obtained. Open-loop systems are very prevalent. For example, with a washing machine, you adjust the temperature of the water, cycle length, and amount of detergent, but it does not evaluate the cleanliness of the clothes in order to modify its performance. Similarly, the semi-autonomous mobile robot that travels by odometry is an open-loop controller. Based on monitoring motor power and operation duration, distance traveled is calculated, but factors like variation in wheel speed and terrain introduce uncertainty into arriving at the destination. In general, odometry only moves the robot close to its desired location before the sensors are used for positioning. The remaining distance to hit the object precisely can now be measured by using sensors.

Closed-loop Control

Robots have closed-loop controlling systems, which give them autonomy. We have already seen this in Braitenberg vehicles, in which the behavior is determined by details of the interaction, attraction, or repulsion with external objects. Thus, if an object approaches from behind the robot, then it retreats until the object moves out of range. This means the robot must continuously measure the distance to the object and stop moving when the distance is sufficiently large. Power in the motor, and consequently speed in the robot, depends on distance measurements since closed-loop control systems with dynamic adjustments, as shown in Figure 8.1, have been used in order to respond to real-time data.

8.1 Process Control Methods in the Food Industry

Terms used to describe how a control system works with robots, based on the use of a cyclical process for adjustment, are "open" and "closed loop." A control system in which the adjustment process is based on a cyclical process is often called a closed-loop system. It describes how robots operate the performance of a task based on reference values transformed into control values. As an illustration, a warehouse robot can keep its motor operating settings and times alive based on its distance from a storage shelf or the distance of its gripper arm to an object. The system is constantly getting feedback about its output, such as the distance to an object, to make appropriate adjustments that keep it on the correct trajectory, wayward of the desired task parameters. Some of the other themes dealt with in this text are scheduling control algorithms. The authors consider this to be quite vital, as it needs to be carried out periodically for the timely completion of the tasks and for system responsiveness. The algorithms need to be time- and resource-optimally balanced between the need to prevent overloads or wastage of resources, and the ability to maintain accuracy and efficiency in the robot operations.

8.1.1 Methods of Process Control in the Food Sector

The food industry has, over the years, adopted various control approaches. Among these are traditional PID controls, which are employed in virtually all applications (Huang, 2013). These controls have, however, shown limited performance due to their suitability for only low-order linear dynamical processes. Pursuing better performance of controls for high-order nonlinear dynamical processes, which also display time-variant characteristics, led to the development of control approaches that are much more superior. Model-based control strategies are formulated from both predictive process models (be they analytical or empirical) and particular process control requirements. Among techniques used to model the input–output relationship of systems without directly measuring the output are artificial neural networks (ANNs) and fuzzy logic (FL). Harnessing the strengths of both ANNs and FL, neurofuzzy controllers are able to successfully predict the output of processes which are otherwise not measurable with direct methods. Importantly, FL and ANNs fall under soft computing methodologies (Huang et al., 2010), which in previous years have been applied in the field of food science and technology (Kondakci and Zhou, 2017). Other soft computing techniques such as GAs and SVMs are also increasingly being discussed in the development of control procedures.

8.1.1.1 PID Controller

The relevance of the PID controller has been proven through its use within the classic closed-loop control feedback frameworks (Padula and Visioli, 2012). This is a design founded on dynamic analysis of second-order linear systems. The creation of a feedback loop is established by integrating a PID controller at the system's input, thus allowing modification of system behavior according to desired parameters. The operational principle of a PID controller has been encapsulated within its defining Eq. (8.1). A list of symbols, parameters, and their corresponding values is represented in Table 8.1.

Table 8.1 List of symbols, parameters, and their corresponding values.

Symbol	Description	Value
d_f	Diameter of the impeller in the ripple plate (m)	0.4
A_f	Flow area of the ripple plate (m^2)	0.002
A_{RP}	Surface area of the ripple plate (m^2)	2.5
A_B	Surface area of the boiler (m^2)	2.3
A_T	Storage tank surface area (m^2)	2.2
A_{HC}, A_{WT}	Heat transfer area of the heating coil and water tank (m^2)	0.507
V_{HC}, V_{WT}	Volume of the heating coil and water tank (m^3)	4.9
V_B	Boiler volume (m^3)	1.20
V_{RP}	Ripple plate volume (m^3)	0.50
V_{CT}	Cooling tower volume (m^3)	0.05
d_p	Diameter of the holding tube (m)	3.83×10^{-2}
L_p	Length of the holding tube (m)	12
n_w	Daily output of pasteurized milk (kg/day)	1×10^4
$T_{m,i}$	Inlet milk temperature (°C)	72
F_m	Milk volumetric flow rate (m^3/s)	4×10^{-4}
$F_{m,i}$	Inlet milk volumetric flow rate (m^3/s)	5×10^{-4}
$F_{m,o}$	Outlet milk volumetric flow rate (m^3/s)	4×10^{-4}
F_{hw}	Hot water volumetric flow rate (m^3/s)	1.60×10^{-3}
F_{tw}	Tap water volumetric flow rate (m^3/s)	4.80×10^{-3}
F_{iw}	Ice water volumetric flow rate (m^3/s)	1.92×10^{-3}
F_{hw}	Hot water return flow rate to the boiler (m^3/s)	3.20×10^{-3}
T_w	Water makeup temperature (°C)	27
T_a	Ambient air temperature (°C)	30
E	Cooling tower evaporation rate (m^3/s)	1.7×10^{-5}
h_A	Heat transfer coefficient at the water surface (W/K)	2000
λ_v	Latent heat of water vaporization ($kj \cdot kg^{-1}$)	2410
L_D	Cooling tower mechanical drift loss (%)	2
n_s	Rotational speed (rpm)	5000
m_f	Boiler fuel consumption rate (kg/s)	1.80×10^{-3}
LHV	Fuel combustion lower heating value ($kj \cdot kg^{-1}$)	49,888

CL	Ripple plate refrigeration rate (J/s)	5.20×10^{-4}
ρ_m	Milk density ($kg \cdot m^{-3}$)	1015.96
ρ_{tw}	Tap water density ($kg \cdot m^{-3}$)	997
ρ_{iw}	Ice water density ($kg \cdot m^{-3}$)	917
ρ_{hw}	Hot water density ($kg \cdot m^{-3}$)	977.3
$C_{p,m}$	Specific heat capacity of milk ($J\,kg^{-1} \cdot C^{-1}$)	3.855
$C_{p,tw}$	Specific heat capacity of tap water ($J\,kg^{-1} \cdot C^{-1}$)	2.356
$C_{p,hw}$	Specific heat capacity of hot water ($J\,kg^{-1} \cdot C^{-1}$)	108.376
$C_{p,iw}$	Specific heat capacity of ice water ($J\,kg^{-1} \cdot C^{-1}$)	2.108
ΔT_{PHE}	Temperature difference in the heat exchanger (°C)	15.6
$T_{m,i}, T_{m,o}$	Milk temperature inside the holding tube (°C)	76
U_{HC}, U_{WT}	Heat transfer coefficient of heating coil and water tank ($W/m^2 \cdot K$)	490
h_{fg}	Latent heat of vaporization at the ripple plate ($kj \cdot kg^{-1}$)	217

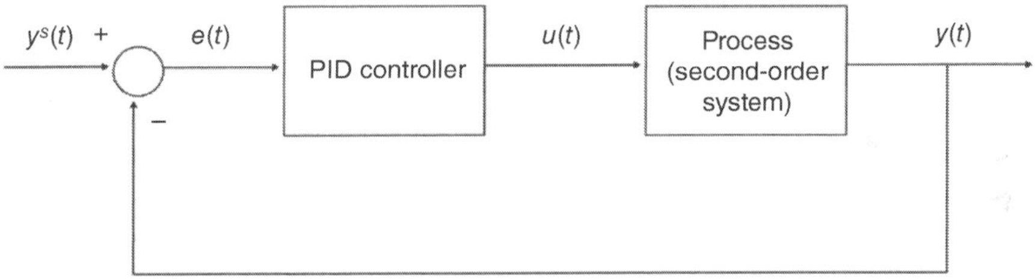

Figure 8.2 A PID control system (Huang [2013]/Elsevier).

$$u(t) = K_p e(t) + K_i \int_0^t e(\tau)d\tau + K_d \frac{d}{dt} e(t) \tag{8.1}$$

In this context, $u(t)$ represents the outcome of the PID controller, and serves as the input to the process at the moment t. Additionally, $e(t)$ denotes the discrepancy between the desired system output $y^s(t)$ and the actual output $y(t)$ represented in Figures 8.2 and 8.3.

At an instant t, y (desired output) is a function of three key parameters in the PID controller system: K_p (proportional gain), K_i (integral gain), and K_d (derivative gain). The role of tuning parameters is depicted in Figure 8.3 in terms of the

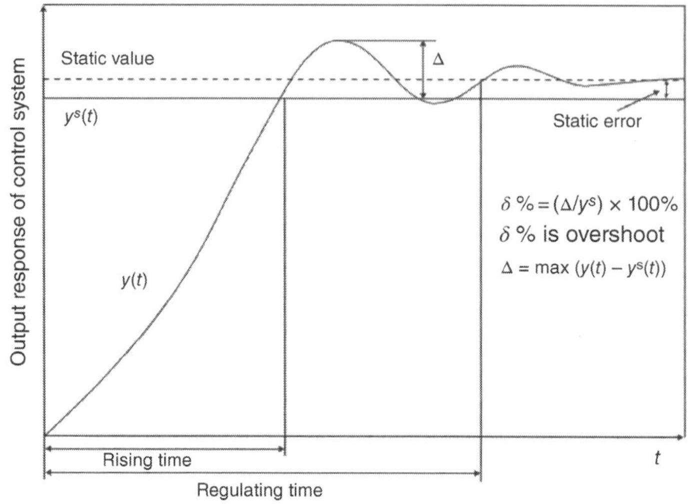

Figure 8.3 A PID control (Huang [2013]/Elsevier).

response to a step input. Proportional control reduces rise time and somewhat reduces but does not eliminate the steady-state error. The steady-state error will be suppressed by the integral control, which may lead to a higher overshoot and a longer time before the system's transient response settles. In comparison, derivative control enhances stability, reduces overshooting, and smoothes out the transient response. By and large, PID controllers encompass all these three modes, wherein K_p, K_i, and K_d are used as the proportionate, integral, and derivative controls, which approximate the cumulative effect of proportional, integral, and derivative actions. In some instances, optimal system control may be achieved by just one or two modes, which is achieved by making the coefficients that are unnecessary zero, and the resulting terms are utilized in the form of proportional-integral (PI), proportional-derivative (PD), P, I, or D controllers, as applicable to the active control actions.

PID controllers are used extensively in a wide range of applications be it in food industry applications, or in the process of process control. The performance varies from satisfactory to downright miserable despite pretuning efforts. System nonlinearity—the way to even greater severity, especially under transient conditions—can weaken the performance of a traditional PID controller by introducing more oscillations, overshoots, and even longer settling times. Some of the worse side effects' elimination and further enhanced closed-loop performance call for even more sophisticated, as well as intricate, nonlinear control techniques. This, in turn, requires an advanced nonlinear process model for a still better process control result.

8.2 Industrial PID Control

8.2.1 Case Study: PID Control in Industrial Processes Introduction

8.2.1.1 Overview of PID Control

8.2.1.1.1 PID Control

It is type of control loop feedback mechanism widely used in industrial control systems. It consists of three parts that, together, keep an output steady: proportional, integral, and derivative.

8.2.1.2 Objective

Bringing the controlled variable (for example, temperature, speed) as close as possible to the setpoint (SP) despite disturbances.
Key Characteristics of PID Controllers:

1. Proportional Control: It measures the present error; it controls the output proportionally with respect to the size of the error.
2. Integral Control: It takes into consideration past errors by cumulating them over a period and helps in removing steady-state errors.
3. Derivative Control: It presages the future errors based on the change rates, thereby creating a damping effect on the system.

8.2.2 Industrial Process Dynamics

Self-regulating Processes: Most industrial processes are inherently self-regulating. That means that a disturbance will cause the process to deviate from its steady-state point but will eventually return to that point after the disturbance. Some common examples include fluid levels in tanks and chemical reactors.
FOPTD Dynamics: First-order plus time delay (FOPTD) models often successfully represent these processes, capturing many of the essentials of many industrial applications.

8.2.3 Tuning PID Controllers

1. Optimization Techniques: Optimality techniques often accompany PID controller tuning that involve optimality over more than one objective; then the controller would work well for all conditions.
2. Conversion Relations: The same PID algorithms are implemented in different ways by different manufacturers. This creates a need for conversion relations to make performance equivalent across various implementations.

8.2.4 Applications of PID Control

1. **Single-Input Single-Output (SISO)**: PID control was first implemented in SISO systems, and in many single-loop applications, such as temperature control in furnaces and speed control in vehicles, it remains in use.
2. **Multiple-Input Multiple-Output (MIMO)**: PID control was later generalized to more complex systems; this is the basis for the management of multiple inputs and outputs simultaneously.

Taking a comprehensive practical case into account, let's discuss the continuously stirred tank reactor (CSTR) operating under isothermal conditions, illustrated in Figure 8.4b, where the series/parallel Van de Vusse reaction occurs, as documented by Rojas et al. (2021a), Daoutidis et al. (1990), and Arrieta et al. (2016). This reaction is characterized by the scheme outlined below and presented in Figure 8.4:

$$\begin{aligned} A &\xrightarrow{k_1} B \xrightarrow{k_2} C \\ 2A &\xrightarrow{k_3} D \end{aligned} \tag{8.2}$$

By applying a mass balance, the system can be described by the following model:

$$\begin{aligned} \frac{dC_A(t)}{dt} &= \frac{F_r(t)}{V}(C_{Ai} - C_A(t)) - k_1 C_A(t) - k_3 C_A^2(t) \\ \frac{dC_B(t)}{dt} &= -\frac{F_r(t)}{V} C_B(t) + k_1 C_A(t) - k_2 C_B(t) \end{aligned} \tag{8.3}$$

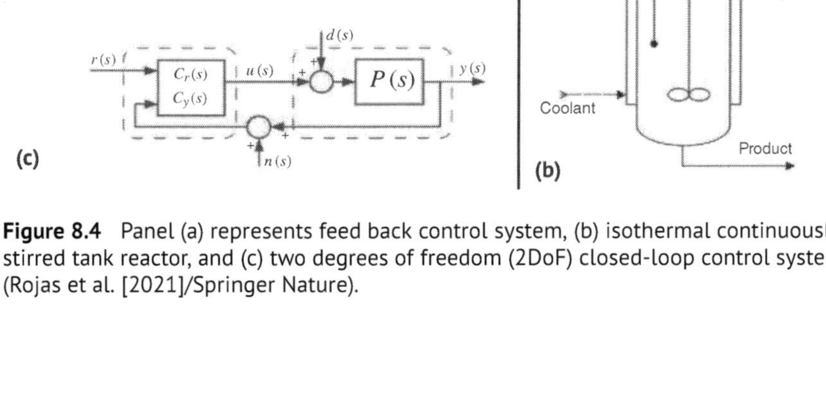

Figure 8.4 Panel (a) represents feed back control system, (b) isothermal continuously stirred tank reactor, and (c) two degrees of freedom (2DoF) closed-loop control system (Rojas et al. [2021]/Springer Nature).

This example considers a case where the feed rate of the product A into the reactor is F_r, which is kept constant, and the volume of the reactor V is also constant; concentration inside the reactor of reactants C_A for the A reactant C_B for B reactant, and so on, values of $k_i (i = 1, 2, 3)$ represent the three rate constants involved.

In Figure 8.5, the concentration of product C_B is the variable under control and the feed flow rate F_r varies as the manipulated variable. The system is expected to experience disturbances from both the input concentration of C_A and the flow rate. The output concentration of C_B in the reactor actually does depend on its two controlling inputs: the feed flow rate F_r and the concentration of C_A. Since the concentration of C_A cannot be manipulated—as it is unmeasured by the control scheme—the feed flow rate, F_r, is instead the variable to be manipulated, and therefore to be controlled to regulate and adjust the reactor. Any change in C_A is regarded as a disturbance, and the closed-loop system has been designed to counteract any such effects so that they do not have an effect on the output concentration C_B.

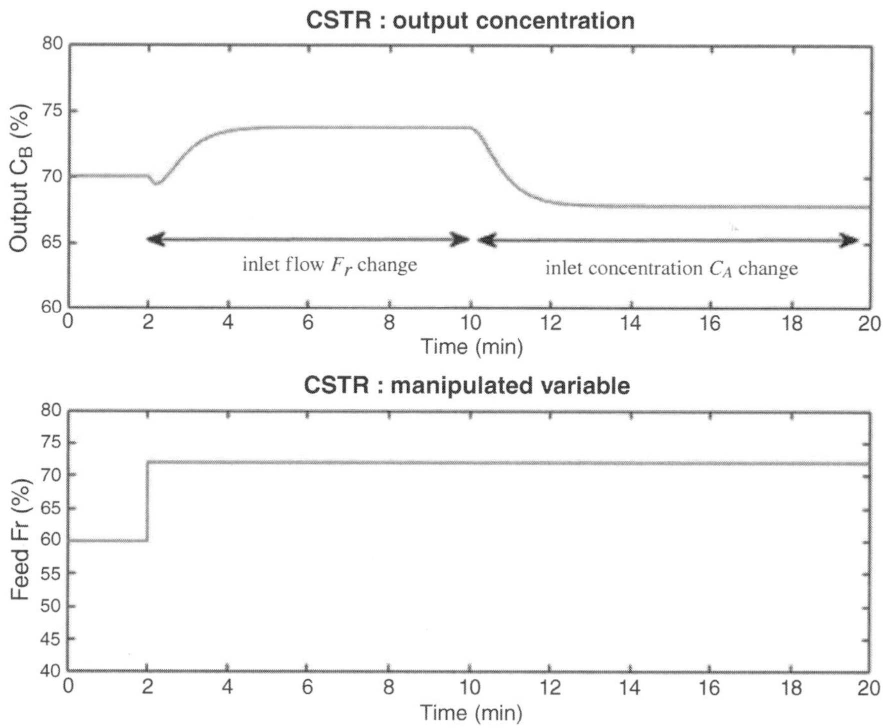

Figure 8.5 CSTR open-loop output to a change in the inlet flow and concentration.

The balance of the discussion concerns the prestigious feedback control structure, which is vested in the PID control algorithm. Since its birth with the advent of the first commercially available PID controller in the 1940s, it has been a stalwart in the industrial process control field.

The performance of a control algorithm is defined by the variation of the parameters until satisfactory performance occurs. The integral part takes into account the accumulated errors over time in the past, whereas the derivative part forecasts the likely future discrepancies. This tuning procedure often occurs in the simulation setup prior to final implementation in the process environment. The following section discusses the tuning of PID controllers specifically through optimization techniques dealing with multiple objectives. However, we will define a clear context for this design effort as a prefatory step.

This leads us to the description of a typical 2DoF closed-loop control system, as shown in Figure 8.4c, consisting of the process control model $P(s)$, and the controller transfer functions represented by $\{C_r(s), C_y(s)\}$. Here, once again, $r(s)$ is the SP, $u(s)$ the controller output signal, $d(s)$ the disturbance factor, $y(s)$ the controlled process variable (PV), and $n(s)$ the measurement noise. The disturbances are said to be loaded in the input stage. This arrangement depicts the control variable output $y(s)$ of the process in terms of a derivative of its inputs $r(s)$, $d(s)$, and $n(s)$.

$$y(s) = M_{yr}(s)r(s) + M_{yd}(s)d(s) + M_{yn}(s)n(s) \tag{8.4}$$

where

$$M_{yr}(s) \doteq \frac{C_r(s)P(s)}{1 + C_y(s)P(s)} \tag{8.5}$$

is the servo-control closed-loop transfer function,

$$M_{yd}(s) \doteq \frac{P(s)}{1 + C_y(s)P(s)} \tag{8.6}$$

is the regulatory control closed-loop transfer function, and

$$M_{yn}(s) \doteq \frac{-C_y(s)P(s)}{1 + C_y(s)P(s)} \tag{8.7}$$

Noise sensitivity in measurements is often significant in process control. Regulatory control primarily acts to counteract load disturbances by returning the controlled variable to a preset SP whenever a disturbance occurs within the system. The goal of servo control is to adapt to a new SP, guiding the controlled variable to the new target value. When applying controllers to these operations, opportunities for undesirable magnification of measurement noise should be considered. For instance, in a sample application of the CSTR, the controller is capable of tracking a step change in the SP with some alleviation of two different types of disturbances: one on the concentration in the feed flow C_{Ai}, and another on the flow rate to the supply.

PID control is the most widely implemented control methodology among all applications of process control since its discovery. Its popularity is due to the simplicity of its configuration, where only three parameters need to be adjusted, along with its transparency through which it executes, thus providing control engineers with an easier to understand perspective compared to more complex strategies of control. This has been driven by continued research into new design methodologies and tuning guidelines toward improvement of the efficiency of PID controller-based control loops. Many papers report the fact that PID controllers continue to be widely used in process industries, albeit now as parts of more sophisticated control schematics rather than as single entities, indeed forming part of overall plant-wide systems.

Therefore, a significant effort remains on PID controller effectiveness in today's process industry.

This PID controller operates on the basis of three special kinds of actions which get equilibrated: current error realized by the term of Proportional action $u_P(t)$, accumulation of past errors coming from the term of the Integral action $u_I(t)$, and the future trend of errors predicted by the term of Derivative action $u_D(t)$. The three are formulated as the basis of the control signal in a PID.

$$u(t) = u_P(t) + u_I(t) + u_D(t) = K_p \left(e(t) + \frac{1}{T_i} \int_0^t e(\tau)\mathrm{d}\tau + T_d \frac{\mathrm{d}}{\mathrm{d}t} e(t) \right) \tag{8.8}$$

which corresponds, when expressed in the form of the transfer function from the error $e(s)$ to $u(s)$, to

$$u(s) = K_p \left(1 + \frac{1}{T_i s} + T_d s \right) e(s) \tag{8.9}$$

8.3 Variations of PID Controller Algorithms

The most commonly referred to PID controller algorithm is known as the standard model, but there are other configurations with several models that combine the three basic control mechanisms in different ways. The control algorithm implementation will normally differ depending on the maker and may not all be packaged in a single controller unit. It is also pretty common for various manufacturers to refer to the same PID algorithm under different names (O'Dwyer et al., 2016). Moreover, there could even be instances where the tuning rule of interest was derived based on a control algorithm other than one utilized in the controller that was being tuned. In such cases, since there is no requirement for an equivalent controller model to exist, it is also apparent that equations for converting controller parameters, which also enable checking the existence of an equivalent controller, should be used. Below, we give descriptions of the basic PID controllers, and for each we use unique parameter designations to allow transformation equations to be written between different models.

Standard PID Model: This model outlines the class of PID control algorithm, described by Åström and Hägglund (2000), for which its output is defined by the equation that follows:

$$u(s) = K_p\left(1 + \frac{1}{T_i s} + \frac{T_d s}{\alpha T_d s + 1}\right)e(s) \tag{8.10}$$

Independent gains PID configuration: The parallel or independent gains approach in PID control algorithm is

$$u(s) = \left(K_p + \frac{K_i}{s} + \frac{K_d s}{\alpha_p K_d s + 1}\right)e(s) \tag{8.11}$$

In this arrangement, the gains for each control action are entirely independent. The relationship between the gains in the parallel form and the standard form is almost very transparent. It is very obvious that the proportional gain happens to be the same, while some specific pattern emerges in the case of the gain for integral action, and the gain for derivative action.

$$K_i = \frac{K_p}{T_i}, \quad K_d = K_p T_d, \quad \alpha_p = \alpha K_p \tag{8.12}$$

In the series or interacting version of the PID control strategy, we see a sequential linkage of a PI controller and a PD controller. The outcome of this configuration is expressed through its transfer function.

$$u(s) = K_p'\left(\frac{T_i' + 1}{T_i' s}\right)\left(\frac{T_d' s + 1}{\alpha' T_d' s + 1}\right)e(s) \tag{8.13}$$

Here, for clarity in separating parameters from standard form, they are carried with a prime. The most important thing that distinguishes a standard from a series PID is that in the series form, complex conjugate zeros can't be used. Assuming the ideal case where $\alpha = \alpha' = 0$, matching a standard with a series PID can only occur if $T_i \geq 4T_d$ so that the Standard PID will have no complex conjugate zeros.

Filtered Ideal PID Form: This approach emerges as an alternative solution to the implementation challenges associated with the ideal derivative. In this scenario, the overall control variable undergoes filtering.

$$u(s) = K_p^*\left(1 + \frac{1}{T_i^* s} + T_d^* s\right)\left(\frac{1}{T_f s + 1}\right)e(s) \tag{8.14}$$

In this instance, parameters are marked with a star to set them apart from those in the standard and series forms. It's important to note that the controller variable filter incorporates the filter time constant T_f as an extra parameter for the controller. This addition complicates the correlation of these controller formulation parameters with the earlier ones. A more comprehensive explanation of their relationship will be provided later in a broader context.

8.3.1 Process Loop Issues—A Summary Checklist

Issues with process loops, mainly MIMO ones, arise during decentralized PID controller tuning for multivariable processes. The key points on these issues are subsequently followed by a case study example.

Process Loop Issues

1. Interaction Between Variables: Most system inputs could influence every system output in MIMO processes, which makes the control of interactions very hard. Consequently, single-loop controllers cannot efficiently handle all the different interactions (Euzébio et al., 2021).
2. Disturbance Rejection: A change in one loop becomes a disturbance to the other loops; this complicates the process control. The problem is more significant in multivariable systems where changes are majorly disturbing.

   It is concerned to be as such for robustness against modeling errors and all the stability over all loops. The proposed method in the paper addresses those concerns by designing a single parameter that balances performance versus robustness.
3. Sensitivity Function: This sensitivity function is to be employed while checking the robustness of each loop. The maximum has to be kept in check so that the stability of the system under various operating conditions is maintained.

8.3.2 Case Study: Shell Heavy Oil Fractionator (Nonlinear Process Modeling)

Case Study Overview: The paper presents a case study related to the Shell heavy oil fractionator. The process is a multivariable one having considerable interactions and large dead times.

Control Loops: There are three input variables in this case: top draw, u1, side draw, u2, and bottom reflux, u3. Interactions among these loops are very pronounced and need a very strong control strategy.

Upper Bounds for Robustness: For the case study, it is decided that the upper bounds of the robustness constraints are MS1 = 2.30, MS2 = 2.30, and MS3 = 1.20. The main motivation for selecting these values was to obtain a faster response in loops 1 and 2 in order to demonstrate another useful application of the proposed tuning method.

Comparative Performance: It is observed that the control response was smoother for the new method compared to the previous methods. It, therefore, states that the proposed method can handle the complexities of the process efficiently.

This case study portrays the problems of industrial process loops and the effectiveness of the proposed decentralized PID method in overcoming these drawbacks to improve performance and robustness in industrial applications.

8.4 Three-term Control, or Proportional, Integral, or Derivative Control

Three-term control, also known as PID control, is the fundamental principle of control systems engineering. The following can be provided as its major elements and features:

Basic Concept: This involves the perception of taking decisions based on past, current, and future information. It simulates the manner of choice-making ability of human beings by considering such things as past experiences and the knowledge of current affairs that have been gathered, and the predictions related to the outcome of such a decision that may be in the future.

8.4.1 Elements of Control

Proportional Control: Here, it considers the error that exists at the present time, that is, the deviation between a set value at which it has to operate, and the concerned actual PV. This kind of controller provides direct action at once without waiting for any future information.

Derivative Control: This predicts future errors by anticipating how fast the error is changing. It further helps in predicting the future performance of the system and might even provide some stability augmentation. It responds based on the rate of change of the error.

Mathematical Representation: The three-term controller can be mathematically written as the sum of these three items, where every term is weighted by a gain factor, with proportional, integral, and derivative gains.

Popularity and Practicality: Three-term control is widely applied in many industries since it is very simple and effective. It proves to be quite useful in environments where the technicians are not much advanced or do not have many hours available. Even though the PID controllers will not be correctly tuned for such a scenario, they still behave quite well and do their intended job of automatic control.

Digital Implementation: In modern applications, three-term control is very much carried out in digital implementation, and it can be given more advanced features of automatic tuning. This adaptation will keep the performance going in more complex systems. The three-term control is a versatile and essential method in control systems combining past, present, and future information, to yield a desirable regulation of processes. Its widespread use justifies its effectiveness and simplicity in numerous applications in industry.

8.5 Parallel PID Controllers in Food Processing

An analytical approach for designing PID controllers for the class of systems controlled in a parallel cascade configuration is presented here. The starting point is a discussion that develops general design concepts for the control systems to improve response to SPs and disturbances. Building on these design concepts, the remainder of the chapter demonstrates the IMC (internal model control) design method to develop a concept for PID controller tuning strategies that can be applied to a more general process model. This method provides a very convenient way to tune parameters of the PID regulator, in systems controlled in parallel cascade, regarding dynamics between primary and secondary loops. Simulation results demonstrate the efficiency of the offered methodology and its superiority over others. An example of parallel cascade systems is given in Figure 8.6.

8.5.1 General Structure for Parallel Cascade Control and Corresponding IMC Structure

The parallel cascade control system is shown in Figure 8.6, where the transfer functions of the process and controller are denoted by p and G, respectively. The output, disturbance, and SP are indicated by the variables r, d, and y. Both disturbance rejection and SP tracking performance in the secondary loop have an impact on how well disturbances are rejected in the main loop. Inadequate SP response, or disturbance rejection, can result from optimal disturbance rejection (or SP tracking) in the secondary loop, which will adversely impact disturbance rejection in the main loop. In order to address this problem, the researchers Lee et al. (2006) used 2-DOF controllers for each loop; in Figure 8.6, q_{f1} and q_{f2} serve as SP filters.

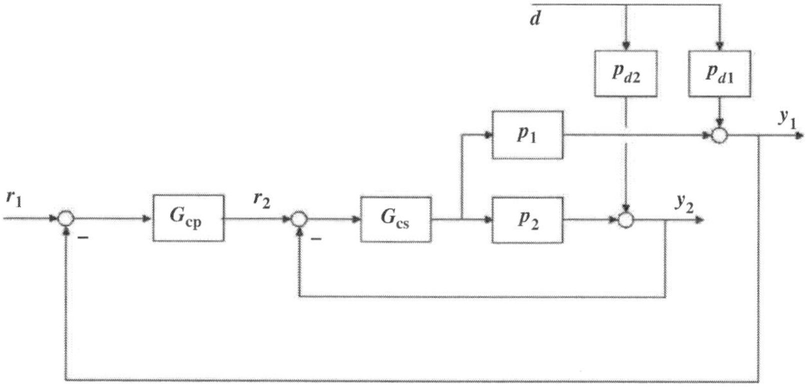

Figure 8.6 Cascade systems in parallel (Lee et al. [2006]/Elsevier).

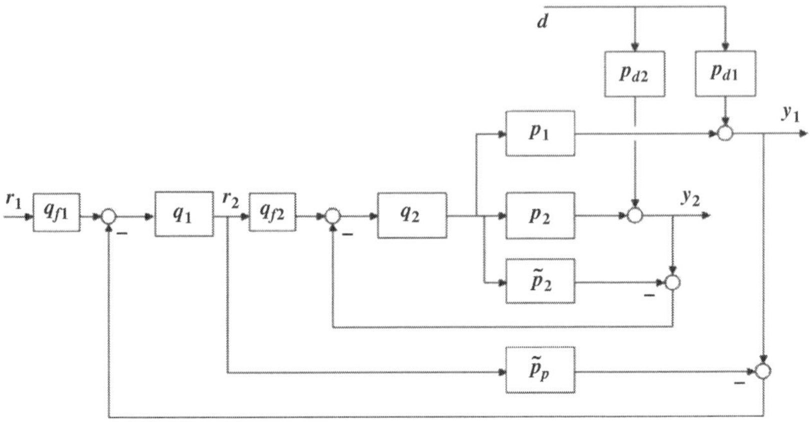

Figure 8.7 IMC implementation of the general structure for a parallel cascade control system (Lee et al. [2006]/Elsevier).

We also present a primary and secondary controller crafting approach based on IMC. Internal Model Control, or IMC, is a method of creating controllers for linear systems by using an internal representation to forecast the behavior of the system.

The typical parallel cascade control structure according to the IMC implementation is shown in Figure 8.7, where $\tilde{p}$ is the transfer function for the process model and q is the transfer function linked to the IMC controller. The outputs from the primary and secondary systems shown in Figure 8.16 are the same in the nominal case, or situation when there is no difference between the plant and the model.

$$
\begin{aligned}
y_2 &= p_2 q_2 q_{f2} r_2 + (1 - p_2 q_2)\, p_{d2} d \\
y_1 &= q_2 q_{f2} p_1 q_1 q_{f1} r_1 + (1 - q_2 q_{f2} p_1 q_1)(p_{d1} - p_{d2} q_2 p_1)\, d
\end{aligned}
\tag{8.15}
$$

Note that the primary process model p_p is equal to $q_2 q_{f2} p_1$. It is assumed the nominal case throughout this chapter.

8.5.1.1 IMC Controller for the Secondary Loop

According to the IMC parameterization, the process model of the secondary loop is decomposed into two parts:

$$
p_2(s) = p_{2m}(s) p_{2a}(s) \tag{8.16}
$$

The model of the control system can be separated into two sub-parts as follows: p_{2m}—this is the portion of the model that the controller explicitly inverts, and p_{2a}—this is the portion of the model for which the controller does not intend to invert. Some general constituents of this part of the noninvertible model are parts with an all-pass configuration and are characterized by the presence of time delays and right-half plane zeros.

In those processes requiring a fast closed-loop response but slow dynamics in an open-loop environment, the q_2 design for the IMC controller must meet the following conditions: (i) It is necessary that q_2 has zeros at the same places $z_1, z_2, \ldots, z_m$ where the process model, p_2, contains slow poles at those places. (ii) The factor $1 - p_2 q_2$ must have zeros at $z_{d1}, z_{d2}, \ldots, z_{dm}$ to have a sufficiently large control if the process model p_{d2} contains slow poles at $z_{d1}, z_{d2}, \ldots, z_{dmn}$.

Obviously, the first requirement is satisfied because the IMC controller is of the form $q_2 = p_{2m}^{-1} f_2$. For the second requirement, the IMC filter has to be set up in such a way that it produces a certain equality:

$$f_2 = \frac{\sum_{i=1}^{m} \alpha_i s^s + 1}{(\lambda_2 s + 1)^{r_2}} \tag{8.17}$$

Within the context, λ_2 is an equable value compromising between robustness and performance in the secondary loop. To maintain the semi-proper characteristics, the IMC controller, q_2, is set with a sufficiently large parameter r_2. The variable m may be one or two for the number of dominant poles to be canceled. Furthermore, it makes use of Eq. (8.18) to calculate the coefficients α_i that cancel the dominant poles in p_{d2}.

$$1 - p_2 q_2 |_{s-z_d \cdots -z_{dz}} = \left| 1 - \frac{p_{2a}\left(\sum_{i-1}^{m} \alpha_s s^i + 1\right)}{(\lambda_2 s + 1)^{\prime 2}} \right|_{s=z_d,\ldots,z_{in}} = 0 \tag{8.18}$$

The designed internal model controller,

$$q_2 = p_{2m}^{-1} f_2 = p_{2m}^{-1} \frac{\sum_{i=1}^{m} \alpha_i s^i + 1}{(\lambda_2 s + 1)^{r2}} \tag{8.19}$$

results in the following closed-loop response of the secondary output:

$$y_2 = \frac{p_{2a}\left(\sum_{i=1}^{m} \alpha_i s^i + 1\right)}{(\lambda_2 s + 1)^{r2}} q_{f2} r_2 + \left(1 - \frac{p_{2a}\left(\sum_{i=1}^{m} \alpha_i s^i + 1\right)}{(\lambda_2 s + 1)^{r2}}\right) p_{d2} d \tag{8.20}$$

The lead term in the transfer function for the closed-loop SP response may cause excessive overshoot and a large settling time, which may lead to degradation of the primary loop performance. To eliminate the overshoot, the SP filter q_{f2} is designed as

$$q_{f2} = \frac{1}{\sum_{i=1}^{m} \alpha_i s^i + 1} \tag{8.21}$$

which gives the following expression for the output of the secondary loop:

$$y_2 = \frac{p_{2a}}{(\lambda_2 s + 1)^{r2}} r_2 + \left(1 - \frac{p_{2a}\left(\sum_{i=1}^{m} \alpha_i s^i + 1\right)}{(\lambda_2 s + 1)^{r2}} \right) p_{d2} d \tag{8.22}$$

We can clearly see that λ_2 has to be greater than the closed-loop time constant in the response of the SP of the secondary loop. A smaller λ_2 results in a higher rate of response from the system but requires a more forceful management measure. Instead, the choice of a larger λ_2 results in a slow closed-loop reaction but a less forceful management method.

Pole cancelation cannot be recommended for Case B processes for which the speed of all open-loop poles will exceed that requested for closed-loop response. Using such cancelations in the IMC design provides no benefit because the open-loop dynamics already provide a faster response than is desired for closed-loop response. In fact, these cancelations add unnecessary complexity to the controller and may introduce more discrepancies if the controller is eventually reduced to a PID format. Hence, the SP filters and the IMC design for Case B are straightforward.

$$f_2 = \frac{1}{(\lambda_2 s + 1)^{r2}}; \quad q_{f2} = 1 \tag{8.23}$$

This is the case of a conventional 1DoF parallel cascade structure, for which the secondary output is obtained by setting $\alpha_i = 0$ in Eq. (8.20).

8.5.1.2 IMC Controller for the Primary Loop

With the designed secondary controller, a primary process model p_p used in the design of the primary IMC controller becomes

$$p_p = q_2 q_{f2} p_1 = \frac{p_{2m}^{-1}}{(\lambda_2 s + 1)^{r2}} p_1 \tag{8.24}$$

which is then factored as

$$p_{\mathrm{p}} = p_{\mathrm{p}m} p_{\mathrm{pa}} \tag{8.25}$$

where the invertible part is

$$p_{\mathrm{p}m} = \frac{p_{2m}^{-1} p_{1m}}{(\lambda_2 s + 1)^{r2}} \tag{8.26}$$

while the noninvertible part p_{pa} is equal to p_{1a}, which is the noninvertible part of p_1.

The primary output, Eq. (8.15), can now be expressed as

$$\begin{aligned} y_1 = & \frac{p_{2m}^{-1}}{(\lambda_2 s + 1)^{r_2}} p_1 q_1 q_{f1} r_1 \\ & + \left(1 - \frac{p_{2m}^{-1}}{(\lambda_2 s + 1)^{r2}} p_1 q_1\right) p_{dp} d \end{aligned} \tag{8.27}$$

where $p_{dp} = p_{d1} - p_{d2} q_2 p_1$.

When compared to the desired closed-loop response speed, case A's process is typified by poles that are far slower. This state arises when the anticipated closed-loop dynamics are significantly outpaced by the dominant open-loop dynamics of the fundamental process. The filter f_1 should be constructed in accordance with the reasoning used in the secondary loop.

$$f_1 = \frac{\sum_{i=1}^{n} \beta_1 s^i + 1}{(\lambda_1 s + 1)^{11}} \tag{8.28}$$

Within the current context, λ_1 denotes a filter parameter used in the primary control loop to balance robustness and performance. In order to guarantee that the principal IMC controller is (semi)proper, the variable r_1 is assigned to a high enough value. The number of poles that need to be canceled, denoted by the variable n, is one or two in many real-world situations. To counterbalance the dominating poles $v_{d1}, v_{d2}, \ldots, v_{dn}$ in p_{dp}, the coefficients β_i are set up in this way. The following equation must be solved in order to obtain these coefficients:

$$1 - p_{\mathrm{p}} q_1 \big|_{s = v_d \ldots\ldots r_{ih}} = \left| 1 - \frac{p_{\mathrm{pa}} \left(\sum_{i=1}^{n} \beta_i s^i + 1 \right)}{(\lambda_1 s + 1)^{r1}} \right|_{s = r_d \ldots\ldots r_{dk}} = 0 \tag{8.29}$$

The primary IMC controller is then equal to

$$q_1 = p_{pm}^{-1} \frac{\left(\sum_{i=1}^{n} \beta_i s^j + 1\right)}{(\lambda_1 s + 1)^{r1}} \tag{8.30}$$

and the closed-loop transfer function describing the response of the primary loop to the SP change becomes

$$\frac{y_1}{r_1} = p_p q_1 q_{f1} = \frac{p_{pa}\left(\sum_{i=1}^{n} \beta_i s^i + 1\right)}{(\lambda_1 s + 1)^{r1}} q_{f1} \tag{8.31}$$

An overshoot caused by the lead term in Eq. (8.31) can be eliminated by designing the SP filter, q_f, as

$$q_{f1} = \frac{1}{\sum_{i=1}^{n} \beta_i i^i + 1} \tag{8.32}$$

The primary output of the closed-loop system is then given by

$$y_1 = \frac{p_{pa}}{(\lambda_1 s + 1)^{r1}} r_1 + \left(1 - \frac{p_{pa}\left(\sum_{i=1}^{n} \beta_i s^i + 1\right)}{(\lambda_1 s + 1)^{r1}}\right) p_{dp} d \tag{8.33}$$

In case B, if all the open-loop poles are greater than the desired speed of the closed-loop response:

In reality, the maximum possible speed of the closed-loop response would be limited by slow sampling and nonzero time-delays for the measurement of main outputs. This is because the dominant process time constant dictates the response speed. Because of this, both IMC and SP filters need to be chosen with this limitation in mind.

$$f_1 = \frac{1}{(\lambda_1 s + 1)^{r1}}; \quad q_{f1} = 1 \tag{8.34}$$

The expression for the primary process output is obtained by setting $\beta_i = 0$ in Eq. (8.33).

8.5.1.3 PID Controller Design

The real work is better done by PID controllers than all other complicated solutions. This section finds analytical PID tuning rules and uses PID controllers to approximate the suggested IMC structure for the parallel cascade control system.

8.5.1.3.1 Equivalent Feedback Controller

The proposed IMC control system described in the previous section can be represented by the classical feedback structure as shown in Figure 8.6. Here, the equivalent feedback controllers, G_c, are related to the IMC controllers as shown in Eq. (8.35):

$$G_c = \frac{q}{1 - pq} \tag{8.35}$$

Substitution of Eqs. (8.16) and (8.19) into Eq. (8.35) gives the equivalent feedback controller of the secondary loop

$$G_{cs}(s) = \frac{p_{2m}^{-1}(\sum_{i=1}^{m} \alpha_i s^i + 1)}{(\lambda_2 s + 1)^{r2} - p_{2a}(\sum_{i=1}^{m} \alpha_i s^i + 1)} \tag{8.36}$$

Similarly, using Eqs. (8.25) and (8.30) in Eq. (8.35) obtains the equivalent feedback controller of the primary loop:

$$G_{cp}(s) = \frac{p_{pm}^{-1}(\sum_{i=1}^{n} \beta_i s^i + 1)}{(\lambda_1 s + 1)^{r1} - p_{pa}(\sum_{i=1}^{n} \beta_i s^i + 1)} \tag{8.37}$$

If either a secondary or primary loop belongs to case B, then the equivalent feedback controller is simplified by setting $\alpha_i = 0$, or $\beta_i = 0$.

8.5.1.4 PID Approximation of the Equivalent Feedback Controllers

Using the Lee et al. (1998) and Lee et al. (2006) technique, the PID approximation of controller Eqs. 3.5–3.7 is derived. The secondary and primary loops are indicated by the subscripts s, p, which are removed from the notation to simplify it. Given that the noninvertible element of the process model's steady state gain equals 1, it is simple to confirm by examination that the controllers provided by Eqs. (8.36) and (8.37) have a pole at the origin. Consequently, both controllers can be written as

$$G_c \equiv g(s)/s \tag{8.38}$$

The PID approximation of G_c can now be obtained using the Maclaurin series approximation of g as a function of s:

$$G_c(s) = \frac{1}{s}\left(g(0) + g'(0)s + \frac{g''(0)}{2!}s^2 + \cdots + \frac{g^i(0)}{i!}s^i + \cdots\right) \tag{8.39}$$

The series approximation truncated to the first three terms can be interpreted as the standard PID controller given by

$$G_c(s) = K_c\left(1 + \frac{1}{\tau_1 s} + \tau_D s + \cdots\right) \tag{8.40}$$

where the analytical expressions for the PID constants are given by

$$K_c = g'(0);\ \tau_1 = g'(0)/g(0); \quad \tau_D = g''(0)/2g'(0) \tag{8.41}$$

When the process has a dominant lead term, the PID tuning according to Eq. (8.41) may lead to negative values of integral and derivative time constants irrespective of the selected filter time constant, λ. To address this problem, the equivalent feedback controller, G_ε, can be approximated as a lagged PID controller

$$K_c\left(1 + \frac{1}{\tau_1 s} + \tau_D s\right)/(\tau_F s + 1)$$

To obtain the parameters of the PID·lag controller, we rewrite G_c as

$$G_c \equiv [g(s)(\tau_F s + 1)]/s(\tau_F s + 1) \tag{8.42}$$

By expanding $[g(s)(\tau_F s + 1)]$ into the Maclaurin series, one obtains:

$$G_c(s) = \frac{1}{s(\tau_F s + 1)}\{g(0) + [g'(0) + \tau_F g(0)]s + [g''(0) + 2\tau_F g'(0)]s^2/2! + [g'''(0) + 3\tau_F g''(0)]s^3/3! + \cdots\} \tag{8.43}$$

The third-order term in Eq. (8.43) is eliminated by setting the lag parameter, τ_F, equal to

$$\tau_F = -g'''(0)/3g''(0) \tag{8.44}$$

which gives the following tuning parameters for the lagged PID controller:

$$K_c = g'(0) + \tau_F g(0); \quad \tau_1 = K_c/g(0); \quad \tau_D = [g''(0) + 2\tau_F g'(0)]/2K_c \tag{8.45}$$

After obtaining the PID controller for the process model by ignoring the lead term, an alternative method to approximate the equivalent feedback controller for the process with a dominant lead term is to add a lag with a time constant tau_F equal to the time constant of the dominating lead term that is ignored. The resulting PID *l*ag controller will be almost identical to the one obtained with equations, although the less complex and compact tuning rule may be derived in addition to Eq. (8.44).

8.5.1.4.1 PID Controller Tuning for FOPDT Process

Consider a specific scenario where both p_1 and p_2 are represented as first-order plus dead time (FOPDT) processes in order to demonstrate how the established method can be applied:

$$p_2(s) = p_{d2}(s) = \frac{K_2 e^{-\theta_2 s}}{\tau_2 s + 1} p_1(s) = p_{d1}(s) = \frac{K_1 e^{-\theta_1 s}}{\tau_1 s + 1} \tag{8.46}$$

Beginning with the secondary loop, factor the process model as

$$p_{2m} = \frac{K_2}{\tau_2 s + 1}; \quad p_{2a} = e^{-\theta_2 s} \tag{8.47}$$

For case A, the IMC filter is

$$f_2 = \frac{\alpha s + 1}{(\lambda_2 s + 1)^2} \tag{8.48}$$

Therefore, the equivalent feedback controller of the secondary loop becomes

$$G_{cs} = \frac{(\tau_2 s + 1)(\alpha s + 1)}{K_2[(\lambda_2 s + 1)^2 - e^{-\theta_2 s}(\alpha s + 1)]} \tag{8.49}$$

For case B, the IMC filter is simply set to

$$f_2 = \frac{1}{\lambda_2 s + 1} \tag{8.50}$$

and the equivalent feedback controller of the secondary loop is equal

$$G_{cs}(s) = \frac{\tau_2 s + 1}{K_2(\lambda_2 s + 1 - e^{-\theta_2 s})} \tag{8.51}$$

The tuning rules for the secondary are obtained by using Eq. (8.41) and expanding Eq. (8.49) into a Maclaurin series.

PID controller in the A (or B) scenario. Table 8.2 provides a summary of the outcome.

Now let's look at the main PID controller. After designing the secondary controller, the primary model is provided by Eq. (8.24), where

$$p_p = \frac{p_{2m}^{-1}}{(\lambda_2 s + 1)^{r2}} p_1 = \frac{\tau_2 s + 1}{K_2(\lambda_2 s + 1)^{r2}} \frac{K_1 e^{-\theta_1 s}}{\tau_1 s + 1}$$

which we factor into

$$p_{pw} = \frac{K_1(\tau_2 s + 1)}{K_2(\lambda_2 s + 1)^{r2}(\tau_1 s + 1)}; \quad p_{pa} = e^{-\theta_1 s} \tag{8.52}$$

Table 8.2 PID controller tuning rules for the FOPDT model in parallel cascade control.

Secondary loop	Case A	Case B
Process model	$p_2 = \frac{K_2 e^{-\theta_2 s}}{\tau_2 s+1}$	
Reference trajectory	$\frac{y_2}{r_2} = \frac{e^{-\theta_2 s}}{(\lambda_2 s+1)^2}$	$\frac{y_2}{r_2} = \frac{e^{-\theta_2 s}}{\lambda_2 s+1}$
K$_c$	$K_{cs} = \frac{\tau_1 S}{K_2(2\lambda_2+\theta_2-\alpha)}$	$K_{cs} = \frac{\tau_I S}{K_2(\lambda_2+\theta_2)}$
τ_1	$\tau_1 s = \tau_2 + \alpha + \frac{\lambda_2^2+\theta_2\alpha-\frac{1}{2}\theta_2^2}{2\lambda_2+\theta_2-\alpha}$	$\tau_1 S = \tau_2 + \frac{\frac{1}{2}\theta_2^2}{\lambda_2+\theta_2}$
τ_D	$\tau_D s = \frac{\tau_2\alpha - \frac{\frac{1}{6}\theta_2^3-1-\theta_2^2\alpha}{2\lambda_2+\theta_2-\alpha}}{\tau_I S} - \frac{\lambda_2^2+\theta_2\alpha-\frac{1}{2}\theta_2^2}{2\lambda_2+\theta_2-\alpha}$	$\tau_D S = \frac{\theta_2^2}{2(\lambda_2+\theta_2)}\left(1 - \frac{\frac{1}{3}\theta_2}{\tau_1 s}\right)$
Variable	**Case A**	**Case B**
Setpoint filter	$q_{f2} = \frac{1}{\alpha s+1}$; $\alpha = \tau_2\left(1-\left(1-\frac{\lambda_2}{\tau_2}\right)^2 e^{\frac{-\theta_2}{\tau_2}}\right)$	$q_{f2} = 1$
Primary loop	**Case A[a]**	**Case B[b]**
Process model	$p_1 = \frac{K_1 e^{-\theta_1 s}}{\tau_1 s+1}$	$p_1 = \frac{K_2 e^{-\theta_1 s}}{\tau_1 s+1}$
Reference trajectory	$\frac{y_1}{r_1} = \frac{e^{-\theta_1 s}}{(\lambda_1 s+1)^2}$	$\frac{y_1}{r_1} = \frac{e^{-\theta_1 s}}{\lambda_1 s+1}$
K$_c$	$K_{cp} = \frac{\tau_{Ip}}{\frac{K_1}{K_2}(2\lambda_1+\theta_1-\beta)}$	$K_{cp} = \frac{\tau_{Ip}}{\frac{K_1}{K_2}(\lambda_1+\theta_1)}$
τ_I	$\tau_{Ip} = \tau_1 + \beta + \frac{\lambda_1^2+\theta_1\beta-\frac{1}{2}\theta_1^2}{2\lambda_1+\theta_1-\beta}$	$\tau_{Ip} = \tau_1 + \lambda_2 - \tau_2 + \frac{\frac{1}{2}\theta_1^2}{\lambda_1+\theta_1}$
τ_D	$\tau_{Dp} = \frac{\tau_1\beta - \frac{\frac{1}{\theta_1^3}\theta_1^3-\frac{1}{2}\theta_1^2\beta}{2\lambda_1+\theta_1-\beta}}{\tau_{Ip}} - \frac{\lambda_1^2+\theta_1\beta-\frac{1}{2}\theta_1^2}{2\lambda_1+\theta_1-\beta}$	$\tau_{Dp} = \frac{\frac{1}{2}\theta_1^2}{\lambda_1+\theta_1} + \frac{\tau_2^2+\tau_1\lambda_2-\lambda_2\tau_2-\tau_2\tau_1-\frac{l_0^{\rho_1^3}}{\lambda_1+\theta_1}}{\tau_{Ip}}$
Setpoint filter	$q_{f1} = \frac{1}{\beta s+1}$; $\beta = \tau_1\left(1-\left(1-\frac{\lambda_1}{\tau_1}\right)^2 e^{\frac{-\theta_1}{\tau_1}}\right)$	$q_{f1} = 1$

a In case A, a PID · lag controller is always used; the lag filter is set as $\frac{1}{\tau_2 s+1}$.

b In case B, when a PID · lag controller is used, the lag filter is set as $\frac{1}{\tau_2 s+1}$. The tuning rule is then modified simply by setting $\tau_2 = 0$.

Secondary loop	Case A	Case B
Process model	$p_2 = \frac{K_2 e^{-\theta_2 s}}{\tau_2 s+1}$	$p_2 = \frac{K_2 e^{-\theta_2 s}}{\tau_2 s+1}$
Reference trajectory	$\frac{y_2}{r_2} = \frac{e^{-\theta_2 s}}{(\lambda_2 s+1)^2}$	$\frac{y_2}{r_2} = \frac{e^{-\theta_2 s}}{\lambda_2 s+1}$
K_c	$K_{cs} = \frac{\tau_1 S}{K_2(2\lambda_2+\theta_2-\alpha)}$	$K_{cs} = \frac{\tau_I s}{K_2(\lambda_2+\theta_2)}$
τ	$\tau_1 s = \tau_2 + \alpha + \frac{\lambda_2^2+\theta_2\alpha-\frac{1}{2}\theta_2^2}{2\lambda_2+\theta_2-\alpha}$	$\tau_1 S = \tau_2 + \frac{\frac{1}{2}\theta_2^2}{\lambda_2+\theta_2}$
τ_D	$\tau_D s = \frac{\tau_2\alpha-\frac{\theta_2^3-1-\theta_2^2\alpha}{2\lambda_2+\theta_2-\alpha}}{\tau_I S} - \frac{\lambda_2^2+\theta_2\alpha-\frac{1}{2}\theta_2^2}{2\lambda_2+\theta_2-\alpha}$	$\tau_D s = \frac{\theta_2^2}{2(\lambda_2+\theta_2)}\left(1-\frac{\frac{1}{3}\theta_2}{\tau_1 s}\right)$
Setpoint filter	$q_{f2} = \frac{1}{\alpha s+1}$; $\alpha = \tau_2\left(1-\left(1-\frac{\lambda_2}{\tau_2}\right)^2 e^{\frac{-\theta_2}{\tau_2}}\right)$	$q_{f2} = 1$

Primary loop	Case A[a]	Case B[b]
Process model	$p_1 = \frac{K_1 e^{-\theta_1 s}}{\tau_1 s+1}$	$p_1 = \frac{K_2 e^{-\theta_1 s}}{\tau_1 s+1}$
Reference trajectory	$\frac{y_1}{r_1} = \frac{e^{-\theta_1 s}}{(\lambda_1 s+1)^2}$	$\frac{y_1}{r_1} = \frac{e^{-\theta_1 s}}{\lambda_1 s+1}$
K_c	$K_{cp} = \frac{\tau_{Ip}}{\frac{K_1}{K_2}(2\lambda_1+\theta_1-\beta)}$	$K_{cp} = \frac{\tau_{Ip}}{\frac{K_1}{K_2}(\lambda_1+\theta_1)}$
τ_I	$\tau_{Ip} = \tau_1 + \beta + \frac{\lambda_1^2+\theta_1\beta-\frac{1}{2}\theta_1^2}{2\lambda_1+\theta_1-\beta}$	$\tau_{Ip} = \tau_1 + \lambda_2 - \tau_2 + \frac{\frac{1}{2}\theta_1^2}{\lambda_1+\theta_1}$
τ_D	$\tau_{Dp} = \frac{\tau_1\beta-\frac{\frac{1}{\theta_1^3}\theta_1^3-\frac{1}{2}\theta_1^2\beta}{2\lambda_1+\theta_1-\beta}}{\tau_{Ip}} - \frac{\lambda_1^2+\theta_1\beta-\frac{1}{2}\theta_1^2}{2\lambda_1+\theta_1-\beta}$	$\tau_{Dp} = \frac{\frac{1}{2}\theta_1^2}{\lambda_1+\theta_1} + \frac{\tau_2^2+\tau_1\lambda_2-\lambda_2\tau_2-\tau_2\tau_1-\frac{{}_{l_0}\rho_1^3}{\lambda_1+\theta_1}}{\tau_{Ip}}$
Setpoint filter	$q_{f1} = \frac{1}{\beta s+1}$; $\beta = \tau_1\left(1-\left(1-\frac{\lambda_1}{\tau_1}\right)^2 e^{\frac{-\theta_1}{\tau_1}}\right)$	$q_{f1} = 1$

a In case A, a PID · lag controller is always used; the lag filter is set as $\frac{1}{\tau_2 s+1}$.

b In case B, when a PID · lag controller is used, the lag filter is set as $\frac{1}{\tau_2 s+1}$. The tuning rule is then modified simply by setting $\tau_2 = 0$.

For the main controller, the PID-lag configuration is recommended when the time constant τ_2 in the lead term is large (dominant). As mentioned for the process model with the neglected lead term in Eq. (8.31), and defining τ_F to be τ_2, provides a reasonable tuning for the PID-lag controller.

Now consider case A: The outer (primary) loop should be far faster than the open-loop system. Using the simplified PID - lag controller design method, with τ_2 set to zero in the tuning rule, we have a neat tuning rule. Finally, the controller design is also simplified since the secondary loop, namely $\lambda_2 \ll \lambda_1 \ll \tau_1$, typically needs to be sufficiently faster than the primary loop.

$$p_{\text{pm}} = \frac{K_1}{K_2 (\tau_1 s + 1)} \tag{8.53}$$

The IMC filter is set to

$$f_1 = \frac{\beta s + 1}{(\lambda_1 s + 1)^2} \tag{8.54}$$

Therefore, the equivalent feedback controller of the primary loop becomes

$$G_{\text{cp}} = \frac{K_2 (\tau_1 s + 1) (\beta s + 1)}{K_1 \left[(\lambda_1 s + 1)^2 - e^{-\theta_1 s}(\beta s + 1)\right]} \tag{8.55}$$

In case B, which often occurs when the primary output is measured with a long time delay, the IMC filter is simply set to

$$f_1 = \frac{1}{(\lambda_1 s + 1)} \tag{8.56}$$

and the equivalent feedback controller for the primary loop is equal to

$$G_{\text{cp}} = \frac{K_2 (\lambda_2 s + 1)(\tau_1 s + 1)}{K_1 (\tau_2 s + 1)(\lambda_1 s + 1 - e^{-\theta_1 s})} \tag{8.57}$$

When a PID · lag controller is used, the equivalent feedback controller is modified simply by setting $\tau_2 = 0$ in Eq. (8.57).

Using the tuning parameters provided by Eq. (8.41), the PID controller is produced which approximates G_{cp} given by Eqs. (8.55) or (8.57). Table 8.3 contains a list of the analytical results. The intended closed-loop time constants for the primary and secondary loops, λ_1 and λ_2, are necessary for the suggested tuning rules to function. It could take multiple iterations to choose λ_1 and λ_2, which manage the tradeoff between robustness in plant-model mismatch and possible closed-loop performance. Seldom are the closed-loop time constants chosen in process control practice to be less than a tenth of the dominating open-loop time constant. Process time delays should also be taken into consideration while choosing λ. Based on our experience, λ should be selected initially as follows. For case A, $\lambda_1/\theta_1 \geq 1$ and $\lambda_2/\theta_2 \geq 1$; for case B, $\lambda_1/\theta_1 \geq 0.5$ and $\lambda_2/\theta_2 \geq 0.5$.

Example 1. Consider the process with

$$p_1 = p_{d1} = \frac{e^{-4s}}{(20s + 1)} p_2 = p_{d2} = \frac{1}{(10s + 1)}$$

Table 8.3 PID tuning values for Example 1.

	Secondary loop				Primary loop				
	K_{cs}	τ_{Is}	τ_{Ds}	q_{f2}	K_{cp}	τ_{Ip}	τ_{Dp}	Lag	q_{f1}
Yu	1	10	0		3.24	22	0		
Proposed (case B)	10	10	0		2.75	22	1.85	$\frac{1}{10s+1}$	
Proposed (case A)	19	1.9	0	$\frac{1}{1.9s+1}$	4.41	10.9	1.24	$\frac{1}{10s+1}$	$\frac{1}{9.52s+1}$

Despite the lag dominance of both the primary and secondary processes, we use rules for both situations A and B when designing PID controllers and we compare their performance. Because the secondary process's time constant is important, the PID·lag controller is employed as the primary controller. We have chosen $\lambda_1 = 4$ and $\lambda_2 = 1$. Table 8.3 contains a list of the tuning parameters that were obtained. The application of the unit step disturbance occurs at $t = 0$ in computer simulations. Figure 8.8 displays, together with the PID approximations for the IMC controllers in examples A and B, the closed-loop response of the principal output. It also displays the response using the design that was created. The IMC controller with case A design exhibits the best performance, as predicted. The response is still superior to the alternatives even when it becomes more oscillatory when approximated by the PID controller. The case B design's IMC controller responses and its PID approximation are nearly the same.

Table 8.3 represents PID tuning values for example 1.

Again, this 1DoF parallel cascade structure is better than the remaining design. In the frequency domain, the quality of the PID approximations of the comparable feedback controllers was also described. The approximation for the case B design is excellent over the whole frequency range, as shown in Figure 8.9. Only at low frequencies can the PID controller accurately simulate the more sophisticated version of the IMC controller that results from the case A design.

We further assume that for predicting process model parameters, there is a $\pm 40\%$ error for both the primary and secondary loops in order to probe the robustness of the proposed controller. For example, the actual process gain, time constant, and time delay are off by 40% from the nominal model values. Figure 8.8 illustrates the response of the proposed controller, case A compared to the one of case B. The suggested controller maintained the high stability of the control system for heavy process uncertainties also, as the one evidenced from Figure 8.8.

Example 2. The following process model was used by Semino and Brambilla (1996):

$$p_1 = p_{d1} = \frac{1.24e^{-33s}}{(30s+1)}$$

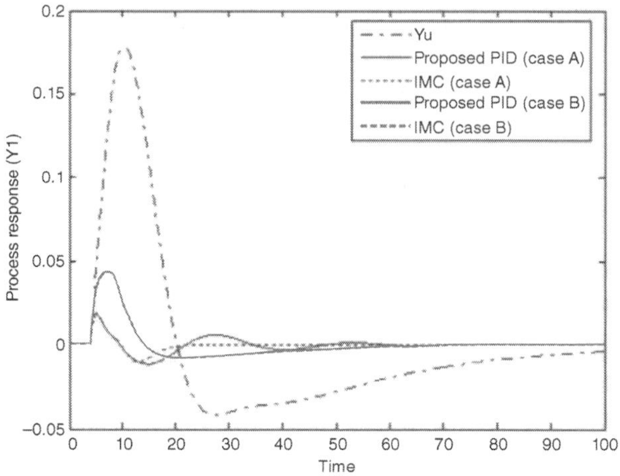

Figure 8.8 Comparison of the closed-loop responses to a load change in case of example 1 (Lee et al. [2006]/Elsevier).

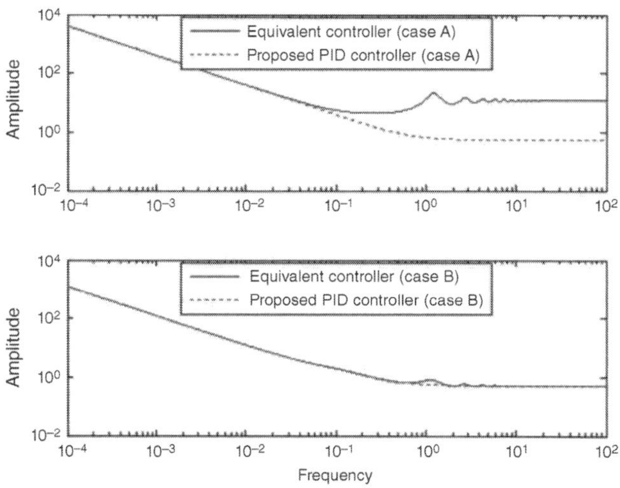

Figure 8.9 Comparison of frequency (example 1) plots for the equivalent feedback controller and the proposed PID controllers (Lee et al. [2006]/Elsevier).

$$p_2 = p_{d2} = \frac{3.1e^{-9s}}{(30s + 1)}$$

The closed-loop time constants are selected as $\lambda_2 = 5$ and $\lambda_1 = 17$. Since the secondary process is obviously dead-time dominated, it falls under case A. The main process is in case B and is dead-time dominated. The primary loop's intended closed-loop dynamics should not be selected fast enough compared to the open-loop dynamics due to large dead time. Thus, it is expected that the removal of the process pole will lead to a more complicated controller with little advantage.

The development of parallel cascade controllers was executed for three distinct scenarios: (i) the procedure for case *B* was utilized to design both the primary and secondary controllers; (ii) the primary loop was designed following the case *A* strategy, while the secondary controller followed the case *B* design method; (iii) the case *A* approach was employed for designing both the primary and secondary controllers. Figure 8.10 represents the comparison of frequency plots for the equivalent feedback controller and the proposed PID controllers. A summary of the outcomes for each scenario is presented in Table 8.4, with a comparative analysis through computer simulations of the closed-loop reaction to a unit step

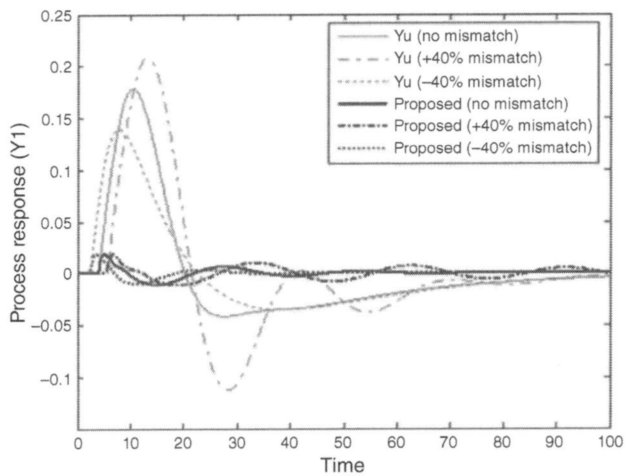

Figure 8.10 Comparison of frequency (example 1) plots for the equivalent feedback controller and the proposed PID controllers (Lee et al. [2006]/Elsevier).

Table 8.4 PID tuning values for example 2.

	Secondary loop				Primary loop				
	K_{cs}	τ_{Is}	τ_{Ds}	q_{f2}	K_{cp}	τ_{Ip}	τ_{Dp}	Lag	q_{f1}
Semino and Brambilla	0.77	34.5	0		0.78	12	0		
Proposed (case 1)	0.76	32.9	2.63		2.30	45.9	11.6	$\frac{1}{30s+1}$	
Proposed (case 2)	1.35	18.5	3.27	$\frac{1}{14.6s+1}$	2.30	45.9	11.6	$\frac{1}{30s+1}$	
Proposed (case 3)	1.35	18.5	3.27	$\frac{1}{14.6s+1}$	2.63	40.8	9.24	$\frac{1}{30s+1}$	$\frac{1}{28.1s+1}$

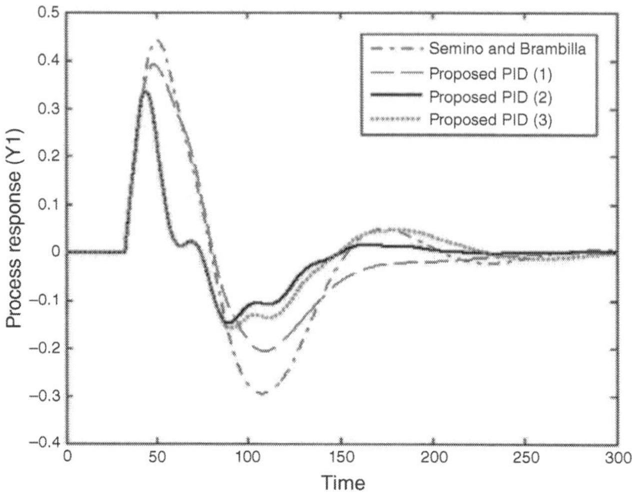

Figure 8.11 Example 2: Comparison of the closed-loop responses to a load change (Lee et al. [2006]/Elsevier).

disturbance introduced at $t = 0$. Figure 8.11 illustrates the performance of each design and benchmarks them against the controller performance. As anticipated, the PID controller configuration for scenario (ii) delivers superior efficiency, with all our PID designs surpassing the alternative.

8.5.2 Development of a Nonlinear PID Controller and Adjustment Guidelines for First-order Plus Time Delay Systems

8.5.2.1 FOPTD Model and PID Controller

8.5.2.1.1 First-order Plus Time Delay Model

Typically, processes characterized by significantly damped dynamics can be depicted through a model known as first-order plus time delay (FOPDT). This model serves as a simplified representation for a variety of complex processes, effectively capturing a wide spectrum of process behaviors. Its application is particularly prevalent in the area of controller tuning, where it forms the basis for several renowned tuning approaches, including Ziegler–Nichols (ZN), Cohen–Coon, Tyreus–Luyben, and Chien–Hrones–Reswick tuning, among others. These methods rely on the specific parameters defined within the FOPDT model (O'Dwyer, 2006). In terms of mathematics, the FOPDT model is expressed through a specific time domain formula.

$$\tau \frac{dy(t)}{dt} + y(t) = Ku(t - L) \tag{8.58}$$

where K denotes the process gain, τ is the process time constant, and L is the process time delay.

8.5.2.2 Conventional PID Controller and Tuning Rules

In general, a parallel form of conventional PID controllers is given as follows:

$$u(t) = K_p \left[e(t) + \frac{1}{T_i} \int e(t)dt + T_d \frac{de(t)}{dt} \right] \tag{8.59}$$

where u is the control input, e is the error between the SP and the PV, and Kp, Ti, and T_d, denote the proportional gain, the integral time, and the derivative time, respectively.

It is when the magnitude of a SP change or a disturbance is large enough that both the proportional and derivative actions are active. The proportional action will respond to the size of the error, whereas the derivative action responds to the rate at which it is changing. Integral action will cancel out any persistent error by accumulating both the current value and its history. On the other hand, an integral action far too active causes problems in terms of overshoots, integral windup in the transient phase, and a loss in stability. Therefore, integral action must be handled with quite a lot of care, especially when enormous errors are in question. In such cases, minimizing the integral of the error comes handy in enhancing the transient response. Conversely, when the error is small, maximizing the integral of the error brings in quick offsetting corrections. This means the integral action needs to be constantly readjusted to fulfill both needs effectively. To counter this, it is proposed that a variation of the usual PID controller, called the nonlinear PID controller, be used. In the current model, all the proportional and derivative actions are in a linear nature, but the integral action features a nonlinear function. This function scales the error input to the integral action by multiplying the error with a nonlinear gain. The structure of this innovative NPID controller, which incorporates this special integral action, is illustrated in a block diagram (see Figure 8.12), marking the introduction of the NPID controller. Table 8.5 represents the tuning formulas for the SP tracking proposed.

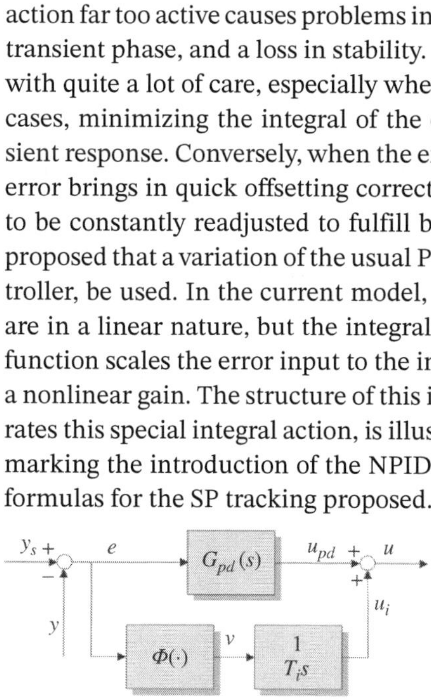

Figure 8.12 The proposed NPID controller configuration (Lee et al. [2006]/Elsevier).

Table 8.5 Tuning formulas for step SP tracking proposed by Tavakoli and Tavakoli (2003).

Performance index	Dimensionless parameters		
	KK_p	T_i/τ	T_d/τ
ISE	$\frac{0.3\frac{L}{\tau}+0.75}{\frac{L}{\tau}+0.05}$	$\frac{2.4}{\frac{L}{\tau}+0.4}$	$\frac{1}{90\frac{L}{\tau}}$
IAE	$\frac{1}{\frac{L}{\tau}+0.2}$	$\frac{0.3\frac{L}{\tau}+1.2}{\frac{L}{\tau}+0.08}$	$\frac{1}{90\frac{L}{\tau}}$
ITAE	$\frac{0.8}{\frac{L}{\tau}+0.1}$	$0.3+\frac{1}{\frac{L}{\tau}}$	$\frac{0.06}{\frac{L}{\tau}+0.04}$

Source: Adapted from Tavakoli and Tavakoli (2003).

8.5.2.3 Proposed Nonlinear PID Controller

8.5.2.3.1 Structure of the Nonlinear PID Controller

This typically yields reactions from both the proportional action and the derivative action, reacting to the magnitude of error and how rapidly it is changing. The integral action works by summing accumulated errors present both currently and in history in order to eliminate any persistent error. On the other hand, over-integral action leads to some drawbacks, such as overshoots, integral windup in transients, and loss of stability.

Therefore, integral action should be controlled and this is essential particularly at large errors. At this point, it is desirable to minimize the integral of the error to improve the transient response. However, during small errors, optimizing the integral of the error will rapidly correct some offsetting. For that reason, this integral action must at all times be readjusted only in a way that can balance between these two opposing needs.

These challenges are addressed by what has been termed the nonlinear PID (NPID) controller—a modified version of the classical PID controller. Here, the proportional and derivative actions remain linear, but a nonlinear function is added to the integral action. This function scales the error input to the integral action by multiplying the error with a nonlinear gain. The structure of this innovative NPID controller, which incorporates this special integral action, is illustrated in a block diagram (see Figure 8.12), marking the introduction of the NPID controller.

In Figure 8.12, y_s and y denote the SP and PV, respectively. u_{pd} and u_i denote the outputs of the PD action and the I action, respectively. The equation of the NPID controller is of the following form:

$$v(t) = \Phi(e) = k(e)e(t)$$
$$u(t) = K_p\left[e(t) + \frac{1}{T_i}\int v(t)dt + T_d\frac{de(t)}{dt}\right] \quad (8.60)$$

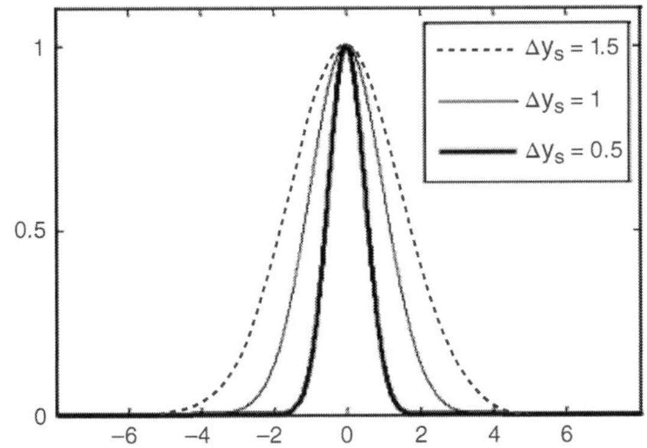

Figure 8.13 Shapes of $k(e)$ versus e according to the change of y_s (Jin and Son [2019]/ICI Bucharest).

where $v(t)$ is a scaled error, $\Phi(e)$ is a nonlinear function which provides a nonlinearly-scaled error according to e, and $k(e)$ is a nonlinear gain. $k(e)$ is described as:

$$k(e) = \exp\left(-\frac{e^2}{2\Delta y_s^2}\right) \tag{8.61}$$

In the given context, $\Delta y_s(\neq 0)$ represents the variation between the current SP, denoted as $y_s(n)$, and the preceding SP, $y_s(n-1)$, with n symbolizing the discrete time instance.

Referring to Figure 8.13, we observe the characteristic curves of $k(e)$ versus e as influenced by alterations in Δy_s. It is important to note that $k(e)$ is symmetric with respect to the y-axis. The form of the nonlinear gain function $k(e)$ is determined by both e and the change in Δy_s. A smaller Δy_s results in a narrower curve, whereas a larger Δy_s leads to a broader curve. Furthermore, $k(e)$ tends towards 0 as e approaches infinity, indicating a lower boundary, and approaches 1, an upper boundary, as e nears 0. $k(e)$ modulates the error to prevent undue control actions. This characteristic of $k(e)$ is advantageous for achieving desirable transient behavior.

8.5.2.3.2 Dimensional Analysis

Equation (8.58) reveals that the FOPTD model comprises three distinct parameters. The units for τ and L are measured in time, whereas the unit for K varies, depending on the process's input and output. Finding the optimal controller settings based on these parameters proves to be a challenging task. Consequently, a dimensionless method is employed to simplify the optimization of settings by minimizing the number of parameters involved and to analyze the process's behavior without relying on physical dimensions.

By introducing a dimensionless variable $t' = t/\tau$ and redefining $y(\tau t') = \bar{y}(t')$ to streamline the notation, the derivative $dy(t)/dt$ can then be expressed in a dimensionless form.

$$\frac{dy(t)}{dt} = \frac{dy(\tau t')}{dt'}\frac{dt'}{dt} = \frac{1}{\tau}\frac{d\bar{y}(t')}{dt'} \tag{8.62}$$

Also a delay of L can be expressed by a delay of L/τ in the new time variable t' by defining $u(\tau t') = \overline{u}(t')$:

$$u(t-L) = u\left[\tau\left(t' - \frac{L}{\tau}\right)\right] = \bar{u}\left(t' - \frac{L}{\tau}\right) \tag{8.63}$$

Then, Eq. (8.58) can be written in the dimensionless form

$$\frac{d\bar{y}(t')}{dt'} + \bar{y}(t') = K\bar{u}\left(t' - \frac{L}{\tau}\right) \tag{8.64}$$

Similarly, by defining variables and applying the dimensional analysis to the controller equation, Eq. (8.60) can also be rewritten in terms of t' as

$$\bar{u}(t') = K_p\left[\bar{e}(t') + \frac{\tau}{T_i}\int \bar{v}(t')\,dt' + \frac{T_d}{\tau}\frac{d\bar{e}(t')}{dt'}\right] \tag{8.65}$$

Taking the Laplace transform for Eqs. (8.64) and (8.65), it yields the overall block diagram with unity feedback, as depicted in Figure 8.14.

As illustrated in Figure 8.14, the behavior of the controlled object is defined solely through the dimensionless ratio L/τ, which results from the integration of the process gain and the proportional gain from the controller. A significant benefit of the newly suggested NPID controller lies in its characterization by three distinct parameters: KK_p, $T_i\tau\tau$, and $T_d\tau$, akin to the parameters used in the traditional PID controller.

Tuning of the NPID Controller

We introduce the NPID controller in the above discussion, which might be utilized for all types of systems. In the above discussion, it was mentioned that numerous industrial processes can be represented by FOPTD, and hence, we would be

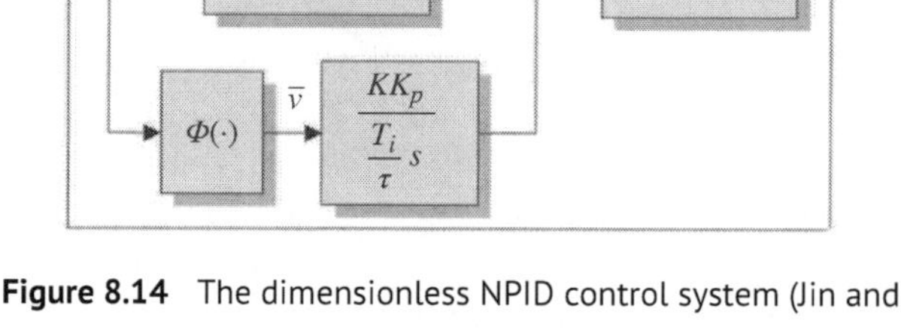

Figure 8.14 The dimensionless NPID control system (Jin and Son [2019]/ICI Bucharest).

focusing on tuning the NPID controller specifically for the FOPTD system. For the parameterization of the NPID controller, we are using a dimensionless representation as shown in Figure 8.14 along with a genetic algorithm (GA) approach. The optimization strategy is pursued by minimizing the three main performance metrics portrayed in dimensionless forms, namely integral square error (ISE), integral absolute error (IAE), and integral time absolute error (ITAE).

$$\begin{aligned} \text{ISE} &= \int_0^{\infty} \bar{e}^2(t')\,dt' \\ \text{IAE} &= \int_0^{\infty} |\bar{e}(t')|\,dt' \\ \text{ITAE} &= \int_0^{\infty} t'\,|\bar{e}(t')|\,dt' \end{aligned} \tag{8.66}$$

ISE is commonly utilized in optimal control for its ease of analysis. However, it disproportionately penalizes errors, treating errors above one more severely and those below one less so, which results in a lack of sensitivity to parameter changes around the optimal solution. In contrast, IAE improves sensitivity by considering the absolute size of the error, while ITAE introduces a penalty for prolonged transient periods, offering better distinction than both IAE and ISE.

This study focuses on cases where the ratio L/τ of the controlled objects falls within the 0.01 to 3 range, consistent with prior research (O'Dwyer, 2006). Thus, the proposed configurations apply solely to processes that fit the FOPTD model with a L/τ ratio within this specified range. For each performance metric, the algorithm is executed 10 times using different random seeds to derive solution sets (KK_p, T_i/τ, and T_d/τ) that minimize the performance metric in response to SP changes, while adjusting the L/τ ratio from 0.01 to 3. The outcomes are then averaged.

Consideration is given to both linear and nonlinear regression models for curve fitting in order to derive the tuning rules.

$$\begin{aligned} KK_p &= a_1\left(\frac{L}{\tau}\right)^{b_1} \\ \frac{T_i}{\tau} &= a_2 + b_2\left(\frac{L}{\tau}\right) \\ \frac{T_d}{\tau} &= a_3\left(\frac{L}{\tau}\right)^{b_3} \end{aligned} \tag{8.67}$$

In the expression where parameters $a_1, b_1, a_2, b_2, a_3, b_3$ remain undefined, Eq. (8.67) is interpreted as a variation of the rules established by O'Dwyer, A. in

Table 8.6 ISE, IAE, and ITAE tuning rules for step setpoint tracking ($0.01 \leq L/\tau \leq 0.9$).

Performance index	Dimensionless parameters		
	KK_p	T_i/τ	T_d/τ
ISE	$1.2886\left(\frac{L}{\tau}\right)^{-0.9182}$	$0.9217 + 0.2375\frac{L}{\tau}$	$0.4302\left(\frac{L}{\tau}\right)^{0.9645}$
IAE	$1.0350\left(\frac{L}{\tau}\right)^{-0.9327}$	$0.9465 + 0.1398\frac{L}{\tau}$	$0.3527\left(\frac{L}{\tau}\right)^{0.9406}$
ITAE	$1.0019\left(\frac{L}{\tau}\right)^{-0.9457}$	$0.8723 + 0.1848\frac{L}{\tau}$	$0.3401\left(\frac{L}{\tau}\right)^{0.9799}$

Table 8.7 ISE, IAE, and ITAE tuning rules for step setpoint tracking ($1 \leq L/\tau \leq 3$).

Performance index	Dimensionless parameters		
	KK_p	T_i/τ	T_d/τ
ISE	$1.3362\left(\frac{L}{\tau}\right)^{-0.5488}$	$0.8419 + 0.3190\frac{L}{\tau}$	$0.4137\left(\frac{L}{\tau}\right)^{0.75}$
IAE	$1.0822\left(\frac{L}{\tau}\right)^{-0.5495}$	$0.8237 + 0.2692\frac{L}{\tau}$	$0.3331\left(\frac{L}{\tau}\right)^{0.6831}$
ITAE	$1.0093\left(\frac{L}{\tau}\right)^{-0.5198}$	$0.7813 + 0.2781\frac{L}{\tau}$	$0.2878\left(\frac{L}{\tau}\right)^{0.7317}$

2006. Utilizing specific software to fit curves based on the given datasets and each corresponding model is recommended. By applying the logarithmic transformation to both sides, the nonlinear regression Eq. (8.67) is transformed into a linear regression framework. The derived tuning rules for each performance metric are detailed in Tables 8.6 and 8.7. It should be noted that the curve fittings depicted in part (b) serve as a continuation of those shown in part (a).

8.5.2.4 Nonlinear PID Controller for FOPTD Systems Through Particle Swarm Optimization

Industry-wide, the most common type of this kind of controller found is the PID controller. Traditionally, a PID controller becomes cumbersome while handling nonlinear systems of sufficient complexity since it may not be able to provide adequate response to the system. Moreover, the conventional PID controllers have a trade-off between performance and robustness, where neither of the two can actually be made up without sacrificing either quickness or stability. It introduces a novel nonlinear gains proportional to a feasible control method known as the numerical, linear, and derivative, which communicates advancements in both disturbance rejection and SP tracking while being concomitantly accepted into nonlinear systems. The paper demonstrates the effectiveness and robustness of the suggested regulator for the very common FOPTD scenario frequently found in business. Application of the particle swarm optimization (PSO) technique charges is utilized to tune the proposed NLPID. The performance of the proposed NLPID controller through simulation is evaluated and compared against well-known controllers, like the 2DoF PID and conventional PID as well as Smith Predictor PID controllers that are applied to a FOPTD system in MATLAB/Simulink with varying levels of uncertainties and disturbances. This analysis shows that the proposed NLPID controller yields faster rise and settling time with excellent disturbance rejection without any overshoot. The result is perfect stability and a lack of delay without sacrificing speed, and it resists parametric, additive and multiplicative doubts.

Three gains are represented by the PID controller, which combines the previous errors (integration), present errors (proportional), and future error estimates (derivative) in a linear fashion. The following are the time-domain and transfer function representations of the traditional PID control, respectively:

$$K_{\mathrm{PID}}(s) = k_{\mathrm{p_c}} + k_{\mathrm{i_c}}\frac{1}{s} + k_{\mathrm{d_c}} s u_{\mathrm{PID}}(t) = k_{\mathrm{p_c}}\epsilon(t) + k_{\mathrm{i_c}}\int_0^{t_\mathrm{f}} \epsilon(t)dt + k_{\mathrm{d_c}}\dot{\epsilon}(t) \quad (1)$$

where $k_{\mathrm{pc}}, k_{\mathrm{i_c}}$, and $k_{\mathrm{d_c}}$ are the proportional, integral, and derivative gains, respectively, $\epsilon(t)$ is the feedback error and t_f is the integration time.

This is one of the most widely used feedback-type control systems of all the existing control systems. In the last few years, researchers and engineers have been interested in this controller because of its linearity and simplicity, which are helpful for industrial applications. It is a known fact that nonlinear systems with modeling uncertainty and process disruptions are frequently encountered in industry. A 1DoF control mechanism, the PID controller is associated with some trade-off between robustness, and performance. The SP response is likely to overshoot once the PID controller has been tuned to reject a disturbance. The same would be likely to occur after it has been adjusted to remove the overshoot in terms of disturbance rejection. By creating a controller that strikes a compromise between reducing overshoot and generating quick disturbance rejection, this can be maximized. On the other hand, this necessitates the formation of overshoot with significant input costs in nonlinear and higher-order systems. Gain scheduling and real-time techniques have been presented to solve the issue of employing a 1DoF PID controller in order to simplify tuning algorithms. However, these methods may use a lot of memory on the controller's hardware. Thus, costs will increase. Another method is to prevent overshoot without sacrificing disturbance rejection capability by using a 2DoF PID control structure. One can get robust rejection of disturbances through such a framework and retain PID controller tuning in the other loop for the elimination of overshoot at the (SP) response. This requires complicated, multi-objective tuning algorithms, a larger search space for the parameters, significant memory, and a long time for calculation. In order to decrease the tuning effort while maintaining the effectiveness of the 2DoF structure, an advanced control methodology that utilizes linear functions for representing time- and feedback-error-varying PID gains is being developed. This allows one to use a function to compute new tuning by mistake from response instead of taking up too much computer memory as different values of tuning are required. Though it enhances systems and eliminates many of the problems related to PID, industry suffers from nonlinear dynamics that linear function gains cannot sufficiently compensate for the PID controllers. Among the frequently suggested solutions in research, one of the solutions is to express PID gains with nonlinear functions known as a NLPID control structure. For some nonlinear systems, a large number of NLPID controllers were offered as an effective control option. NLPID control has been the subject of intense and ongoing study for creating industrial NLPID controllers that can operate for a class of nonlinear systems. The application of passivity-based theory with a more extensive set of nonlinear functions has recently inspired new research and industrial applications for modern NLPID controllers, which naturally created a wealth of new research opportunities. More recent research introduced different sets of nonlinear functions where one method explains the proportional, integral, and derivative functions. For example, if the Gaussian error function is used to compute the benefits, the integral gain decreases with increasing error but the proportional and derivative gains increase. For a much larger class of nonlinear functions, including exponential and error power functions, an extension of the Popov criterion-based NLPID control design is proposed. It is worth noting that an enhanced nonlinear tracking differentiator was also presented and demonstrated together with Han's nonlinear PID controller. In this paper, the tracking differentiator is applied at both the feedback and SP. A second nonlinear PID controller was also proposed; in this case, the integral gain of the controller is zero for high errors but increases at steady state. Specifically, the controller achieves a large proportional gain for large error values but decreases for small errors. When used in conjunction with a nonlinear derivative function that is a function of both error and error rate, this controller has already demonstrated the potential to improve performance over traditional approaches. The stability and appropriate responses of both linear and delay-type systems have been achieved using a nonlinear PID controller that is based only on a scaled integral nonlinearity. To improve the transient responsiveness and rejection of external disturbances, the innate nonlinear unstable dynamics of a magnetic levitation system were compensated for by using a traditional PID controller with derivative filtering in conjunction with nonlinear functions in the feedforward and feedback loops. Recently developed was a nonlinear PID controller using an online internal model control tuning PID in conjunction with fuzzy fusion, applied for temperature control in a nonlinear CSTR. Likewise, nonlinear proportional gain has also been used in the enhancement of nonlinear PID, which has been applied in an excavator's hydraulic drive control for improving stability. In one of the most recent papers, nonlinear gains from a set through a nonlinear PID controller were also adapted with the MIT rule. This resulted in better performance than those obtained when only conventional and nonlinear PID techniques were used. Recently, a bright nonlinear PID controller was developed in order to regulate the temperature of a CSTR system. That includes (NN) as weights, which describe the gains along with the use of PSO to optimize the weighting parameters (Jin and Son, 2019). A cross-coupled nonlinear PID controller is utilized in the most recent study in this report, which involves the application in a highly coupled twin rotor MIMO system to achieve noticeably superior performance.

8.5.2.4.1 Novel NLPID Controller

This paper presents a novel nonlinear controller. The time-varying profits of the controller are quantified based on the rate of mistake and the magnitude of error. To improve the transient response and maintain robustness, a set of nonlinear functions is employed to express these gains. The equation of the proposed NLPID controller is structured in a similar form to a parallel linear PID controller. The following is the time-domain NLPID controller equation:

$$u_{\mathrm{NLPID}}(\epsilon(t), \dot{\epsilon}(t), r(t)) = k_{\mathrm{p}}(\epsilon(t), r(t))\epsilon(t) + k_{\mathrm{i}}(\epsilon(t), r(t)) \int_{0}^{t_{\mathrm{f}}} \epsilon(t)\,\mathrm{d}t + k_{\mathrm{d}}(\epsilon(t), r(t))\dot{\epsilon}(t) \tag{2}$$

For such a circumstance, it is guaranteed that the suggested NLPID controller provides fast (SP) tracking and fast disturbance rejection without overrun. In this scenario, the maximum negative effect on the overshoot of these PID gains is the proportional and integral gains. Because larger proportional and integral gains cause an oscillatory response that occurs along with a considerable overshoot, this results in prompt reaction and rejection of disturbances. This would balance out the overshoot and a huge proportional signal at high error may come from a small integral signal at high error. This allows the necessary quick adjustment. Once again, as the output approaches steady-state, the gain of the proportional part must fall swiftly while the gain of the integral part must rise sharply to compensate for any steady-state errors. There is a similarity in the structure of the integral and derivative gains. Here, the derivative gain considers the error rate to eliminate noise and derivative kicks. The benefit is 0 if the error rate is increasing. According to the documented PID control behavior in the literature, the proposed NLPID controller architecture must have nonlinear functions. The nonlinear function processing such a feature is the mollifier function, which originates from distribution theory and has never been used in the NLPID control literature. The mollifier has the following mathematical definition:

$$M(x(t)) = \begin{cases} e^{\left[\frac{1}{|x(t)|^2-1}\right]} & \text{if}|x(t)| < 1 \\ 0 & \text{if}|x(t)| \geq 1 \end{cases} \quad (3)$$

Besides, to maximize the effect of nonlinearity in minimizing overshoot, the mollifier is utilized in this research in an application that applies nonlinearity in the transient response area. Therefore, the following is a description and illustration of the adopted nonlinear gains for the proposed NLPID controller:

The proportional gain

The proportional nonlinear gain is represented by the following function:

$$k_p(\epsilon(t), r(t)) = \begin{cases} ak_0 - k_0 e^{\left[\frac{1}{\left|\frac{\epsilon(t)}{r(t)}\right|^2 - 1}\right]} & \text{if}\left|\frac{\epsilon(t)}{r(t)}\right| < 1, r(t) \neq 0 \\ ak_0 & \text{if}\left|\frac{\epsilon(t)}{r(t)}\right| \geq 1,\ r(t) \neq 0 \\ ak_0 - k_0 e^{\left[\frac{1}{|\epsilon(t)|^2-1}\right]} & \text{if}\ |\epsilon(t)| < 1,\ r(t) = 0 \\ ak_0 & \text{if}\ |\epsilon(t)| \geq 1,\ r(t) = 0 \end{cases} \quad (4)$$

where a is the mean or shift value of the nonlinear function that sets the higher gain bounds at either higher or lower values directly related to a and k_0, and k_0 is the proportional constant gain. In order to have the controller act nonlinearly within the proper error range, the function also depends on the SP function $r(t)$, which increases and decreases the nonlinearity. An example of a proportional gain k_p produced using this function and tuned for specific values of a,k_0, and $r(t)$ is shown in Figure 8.15. In order to compensate for the error with a quick transient response, the proportional nonlinear gain is made to generate the maximum proportional signal at big errors.

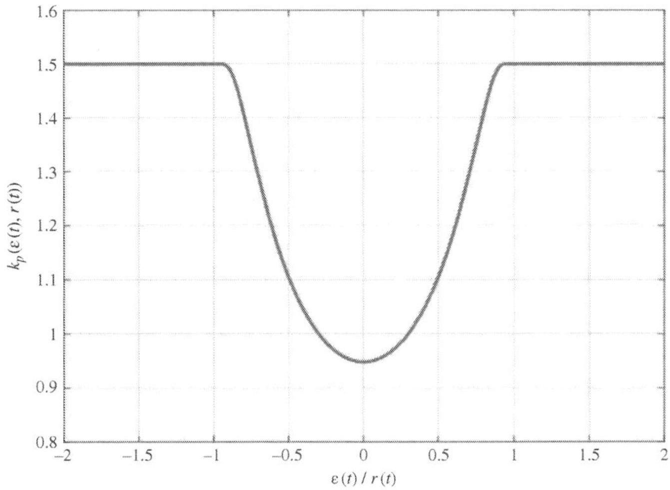

Figure 8.15 An example of tuned nonlinear proportional gain shown for a step SP function $r(t) = 1$, for values $a = 1$, and $k_0 = 1.5$ (Kara-Mohamed [2023]/Elsevier/CC BY 4.0).

The proportional gain

The proportional nonlinear gain is represented by the following function:

$$k_p(\epsilon(t), r(t)) = \begin{cases} \dfrac{1}{\left|\dfrac{\epsilon(t)}{r(t)}\right|^2 - 1} \\ ak_0 - k_0 e^{\left[\frac{\epsilon(t)}{\left|\frac{\epsilon(t)}{r(t)}\right|^2 - 1}\right]} \\ \text{if}\left|\dfrac{\epsilon(t)}{r(t)}\right| < 1,\ r(t) \neq 0 \\ ak_0 \\ \text{if}\left|\dfrac{\epsilon(t)}{r(t)}\right| \geq 1,\ r(t) \neq 0 \\ ak_0 - k_0 e^{\left[\frac{1}{|\epsilon(t)|^2 - 1}\right]} \\ \text{if}\,|\epsilon(t)| < 1,\ r(t) = 0 \\ ak_0 \\ \text{if}\,|\epsilon(t)| \geq 1,\ r(t) = 0 \end{cases}$$

where a is the mean or shift value of the nonlinear function that causes higher bounds set at either higher or lower values directly proportional to a and k_0, and k_0 is the proportional constant gain. This function further relies on SP function $r(t)$, which expands or contracts nonlinearity so that the controller behaves nonlinearly over the proper error range. It will allow constructing a proportional gain k_p. This constructed k_p is later tuned for desired values of a, k_0, and $r(t)$. In order to compensate for the error with a quick transient response, the proportional nonlinear gain is made to generate the maximum proportional signal at big errors. In order to minimize the overshoot, the value then rapidly and smoothly drops. The requirements of a quick transient reaction without overshoot are thereby partially met by this proportional gain. The limitations of traditional PID control can thus be improved.

The Integral Gain

The integral nonlinear gain is represented by the following function:

$$k_i(\epsilon(t), r(t)) = \begin{cases} k_1 e^{\left[\frac{1}{\left|\frac{c(t)}{r(t)}\right|^2 - 1}\right]} & \text{if} \left|\frac{c(t)}{r(t)}\right| < 1,\ r(t) \neq 0 \\ 0 & \text{if} \left|\frac{c(t)}{r(t)}\right| \geq 1,\ r(t) \neq 0 \\ k_1 e^{\left[\frac{1}{|\epsilon(t)|^2 - 1}\right]} & \text{if} \left|\epsilon(t)\right| < 1,\ r(t) = 0 \\ 0 & \text{if} \left|\epsilon(t)\right| \geq 1,\ r(t) = 0 \end{cases} \tag{5}$$

where k_1 is the integral constant that determines the largest value of the integral nonlinear gain. In this case, the set-point function $r(t)$ also affects the gain where it enlarges and shrinks the integral nonlinearity in order to adapt to the error range, so that the nonlinearity is active throughout the transient response.

The integral gain, shown in Figure 8.16, is built according to Eq. (8.5) and it is designed so that it starts from a value of zero and increases as the error approaches steady-state, approaching its maximal bounded value. This allows for the integral to error-correct the system during steady-state while keeping a low integral value during the transient response, which helps maintain low overshoot.

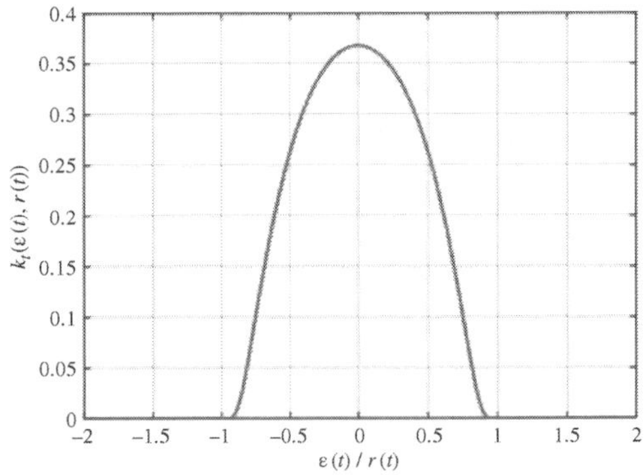

Figure 8.16 An example of tuned nonlinear integral gain shown for a step SP function $r(t) = 1$, for value $k_1 = 1$ (Kara-Mohamed [2023]/Elsevier/CC BY 4.0).

The derivative gain

Finally, the derivative nonlinear gain is represented by the following function:

$$k_d(\dot{\varepsilon}(t), r(t)) = \begin{cases} k_2 e^{\left[\frac{1}{\left|\frac{\dot{\varepsilon}(t)}{r(t)}\right|^2 - k_3^2}\right]} \\ \text{if} \left|\frac{\dot{\varepsilon}(t)}{r(t)}\right| < k_3,\ r(t) \neq 0 \\ 0 \\ \text{if} \left|\frac{\dot{\varepsilon}(t)}{r(t)}\right| \geq k_3,\ r(t) \neq 0 \\ k_2 e^{\left[\frac{1}{|\dot{\varepsilon}(t)|^2 - k_3^2}\right]} \\ \text{if}\, |\dot{\varepsilon}(t)| < k_3,\ r(t) = 0 \\ 0 \\ \text{if}\, |\dot{\varepsilon}(t)| \geq k_3,\ r(t) = 0 \end{cases} \quad (6)$$

where $r(t)$ is the set-point function that can increase and decrease the nonlinearity appropriately in a fashion similar to the earlier nonlinear gains, and k_2 is the derivative constant that raises the maximum derivative value. As can be noted, the error rate, instead of the error, is the input to the nonlinear derivative function for the derivative gain. This makes it easier for the controller to determine the steady-state position, where the derivative gain is zero at error rate values greater than the filter constant k_3 and maximized for enhanced damping and minimized overshoot.

The filtering constant, k_3, is a design parameter defined by the designer based on the quantity of derivative that must be included in the controller. This constant is used to resolve some of the well-known PID constraints. It reduces the effect of high-frequency noise that may have an impact on the controller input and derivative kick. It identifies the locations where the noise and derivative kick cancelation takes place and alters the range where the nonlinearity operates. This means that the noise signals one wants to cancel can be conveniently modified by the control designer.

Figure 8.17 shows a plot of the derivative gain tuned. It is similar in form to the integral gain, with the design filtering constant k_3 controlling the difference between the two gains.

The nonlinear derivative gain is meant to minimize the derivative kicks and noise effects on the feedback response of the system. This boosts the input signal

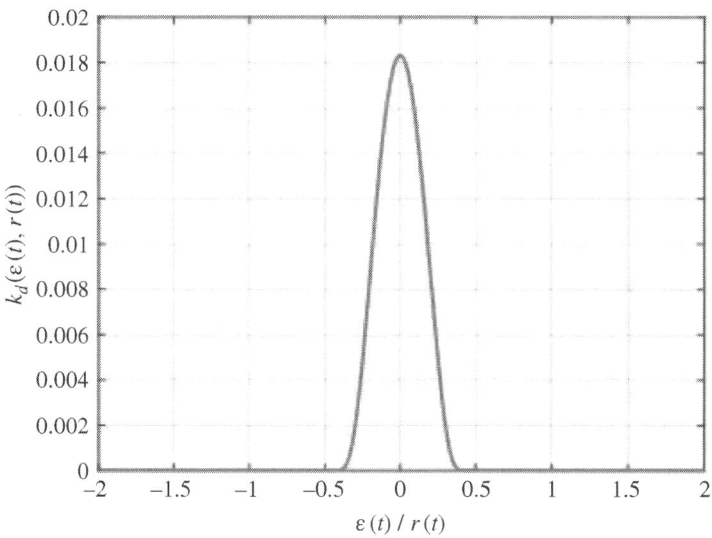

Figure 8.17 An example of tuned nonlinear derivative gain shown for a step setpoint function $r(t) = 1$, for values $k_2 = 1$ and $k_3 = 0.5$ (Kara-Mohamed [2023]/Elsevier/ CC BY 4.0).

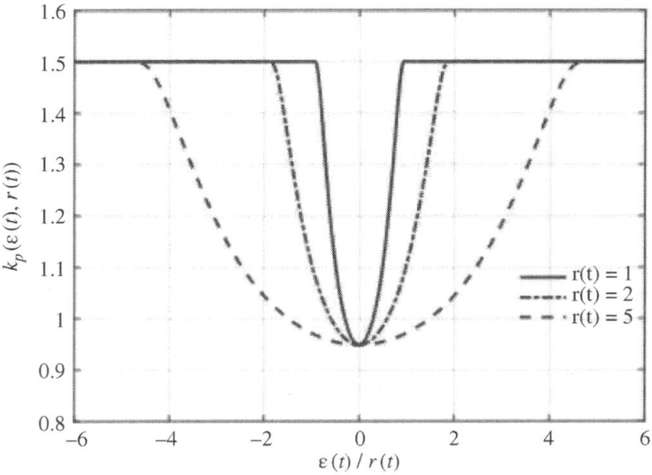

Figure 8.18 The tuned nonlinear proportional gain as it adapts to new step SP values of *r*(*t*) = 1, 2, and 5 (Kara-Mohamed [2023]/Elsevier/CC BY 4.0).

to the system, thus eradicating the traditional drawbacks of the PID controller. To finally minimize overshoot and increase the response speed, the derivative gain is increased to a value close to the steady state for better transient performance.

Figure 8.18 depicts how adjusting the SP changes the recommended nonlinear proportional gain. The wider is the SP, the broader are the nonlinearities are that occur, but retaining the design values of constants; maximum and minimum gains values. Therefore, the nonlinearities would be active within the range: $-\epsilon_{max} \leq \epsilon \leq \epsilon_{max}$. As explained previously, this also affects the integral and derivative gains. The reference function indicates that the three gains operate in synchronization when $r(t)$ is included in their definition. In order to minimize overshoot and enable the gains to stabilize at steady-state, this engages the nonlinearities during the transient response. As demonstrated subsequently, this design enhances the rejection of any unmodeled uncertainty, enhances disturbance rejection, and generates the same transient response for any given SP function.

In the following section, the tuning methodology of the controllers is shown to ensure fair comparisons with established controllers in the literature so that the improvements are shown and are isolated to be due to the proposed nonlinear function gains.

8.5.2.4.2 Particle Swarm Optimization Algorithm

PSO is among those stochastic and evolutionary optimization techniques that are very simple but yet very efficient to solve challenging optimization problems.

Indeed, the algorithm of PSO is one of the commonly chosen algorithms for engineers and researchers in control systems because of simplicity, efficiency, and quick convergence without derivatives. Yet, apart from this optimization algorithm, several others exist in the literature which are applied to particular applications. This means that the suggested NLPID controller is, in this research work, fine-tuned with PSO. However, its major drawback is that PSO tends to fall very easily in a local minimum. Two stages have been presented in order to eradicate the problem, that could be lessened if present. For starters, during set times, particles are restricted from dropping within a selected range of values. Second, the personal best value of each particle is excluded, and cognition is not used in this situation. Here, social or global intelligence is applied. To make falling at a local optimum more difficult, the social best objective value is required. The objective function and the optimization problem established with the ITAE performance measure and the settling time of the system, which is mathematically expressed as follows, are utilized to tune the proposed NLPID parameters k_0, k_1, k_2, and a.

$$\underset{k_0,k_1,k_2,a}{\text{minimize}} \quad f(t, \epsilon(t), t_s) = \int_0^{t_f} t|\epsilon(t)| \quad dt + t_s \tag{8.68}$$

$$k_{min} \leq k_0, k_1, k_2 \leq k_{max} \tag{8.69}$$

$$a_{min} \leq a \leq a_{max} \tag{8.70}$$

where the ultimate time is t_f, the feedback error is $\epsilon(t)$, and the settling time is t_s. To guarantee stability and reduce the likelihood of becoming trapped inside local optima, the optimization problem is defined with the parameter constraints falling within a given range. The goal function is to minimize rise-time and overshoot in order to minimize the feedback error in the least amount of time. Furthermore, incorporating the settling time results in a quicker approach to steady-state, which aids in fulfilling the suggested design specifications. The filtering parameter k_3 is chosen by the designer based on the system and required derivative amount. This implies that the PSO technique need not be used to adjust this parameter because the designer can modify it once the controller has been adjusted. Using the subsequent stages, the PSO algorithm iterates until the ultimate iteration is achieved: (i) In the range $[k_{min}, k_{max}]$, create n random location particle vectors X_0^n, for $[a_{min}, a_{max}]$, and k_0, k_1, k_2 for a. (ii) Let $V_0^n = 0$ be the starting velocity vector. (iii) Simulate the control system in Simulink. (iv) Compute $f(t, \epsilon(t), t_s) = \int_0^{tf} t|\epsilon(t)|dt + t_s$. (v) If values surpass the defined range, re-initialize a random number in range $[k_{min}, k_{max}]$ for k_0, k_1, k_2, and re-initialize a random number in range $[a_min, a_{max}]$ for a. (vi) Then compute the new velocity and position values, which are modified to be as:

$$
\begin{aligned}
{}_kV^j_{i+1} &= |{}_kV^j_i +_k r^j_i c_1({}_kGbest_i -_k P^j_i)| \\
{}_kX^j_{i+1} &= |{}_kX^j_i +_k V^j_{i+1}|
\end{aligned} \tag{8.71}
$$

(vii) Re-iterate.

where k is a natural number between 1 and 4 which cycles through the four parameters of the parameter set; ${}_kV^j_i$ represents the velocity vector for each cycle i, particle j; tuning parameter $k,{}_kX^j_i$ represents the position vector for each cycle i, particle j; and tuning parameter $k,{}_kr^j_i$ is the stochastic variable that differs with every iteration and ranges from [0,1]. Gbest is the minimum objective function value for all particles during iterations, with each particle being one specific tuning parameter. The particular new particles are reinitialized in the pre-specified range if the new position ${}_kX^j_{i+1}$ lies outside the defined range of values. A tuning parameter obtained from PSO research surveys is $c_1 = 1.3$. In this process, only the social best values are taken into account, which is a variant of PSO, in which history for the minimal objective value was taken for each particle. By applying the modified PSO method, the nonlinear gain parameters k_0, k_1, k_2, and a that would reduce the settling time, overshoot, and transient response to satisfy the design requirements have been found. Benchmark simulations for SP tracking and distaste rejection in the next chapter, establish the following example for FOPTD; PID, 2D-PID and SP-PID algorithms against the developed NLPID.

8.5.3 Series PID Controllers in Food Processing

The most widely used controller in the industry is the PID controller. The great advantage of this controller is that it does not require dynamic process modeling. People who are not experts in automatic control can calibrate the settings. Figure 8.19 shows the different kinds of controllers.

8.5.3.1 Series and Parallel PID

The two structures are described by the following equations:

Series:

$$C_{\text{series}}(s) = k(1 + {}^1/_{(sT_i)})(1 + sT_d)$$

Parallel:

$$C_{\text{parallel}}(s) = k_p + {}^{k_i}/_s + k_d s$$

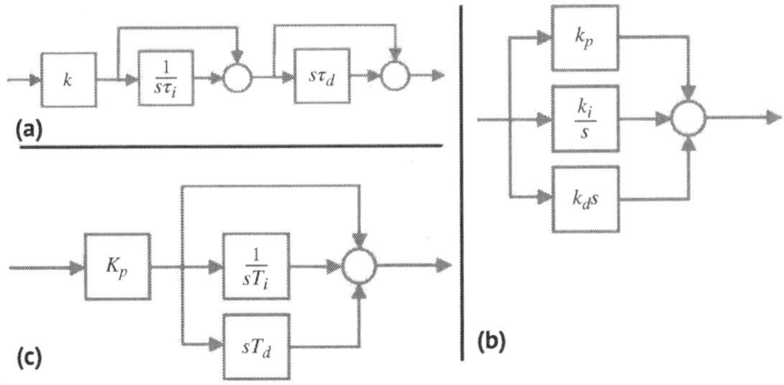

Figure 8.19 The PID controllers in (a) series, (b) parallel, (c) mixed form.

As you can see, they are mathematically equivalent. From a serial controller, it is always possible to switch to a parallel form, thanks to these formulas:

$$k_p = k(1 + {}^{T_d}/_{T_i})$$

$$k_i = {}^{k}/_{T_i}$$

$$k_d = kT_d$$

8.5.3.1.1 A third way to Implement a PID

Another way to build PID exists! It is called "standard", "mixed," or sometimes "ideal". It is defined by the equations below:

$$C_{\text{standard}}(s) = K_p(1 + 1/sT_i + sT_d)$$

This new form and a regular parallel form are a lot alike. So, transformation equations are easy to obtain:

$$K_p = k(1 + T_d/T_i)T_i = T_i + T_dT_d = T_dT_i/(T_d + T_i)$$

Ultimately, PID can be implemented in three nearly mathematically similar methods. In general, there are formulas for changing a controller's form (with the exception of k_i=0).

8.5.4 Simple PID Tuning in Food Industry

The Cohen–Coons and ZN method stands out as the most recognized traditional approach for PID tuning. Given the system's nonlinear nature, tuning becomes a challenging endeavor for researchers. Additionally, system parameters are prone to variations over time. In such scenarios, adopting an adaptive and smart control strategy might be more advantageous. Lately, bio-inspired intelligent computing techniques such as FL (FIS), NN, and GA have been effectively utilized in tackling intricate issues. The PID controller utilizes a basic yet robust method of control. It is the predominant choice in industrial settings, with its adoption rate standing at approximately 90%. A PID controller is made up of three basic gain parameters (K_p, K_i, K_d). Figure 8.20 displays the PID controller schematic. Both the rising time and the steady-state error diminish as the proportional gain increases. The rise time (t_{rt}) reduces if the K_i increases, but the settling time (t_{st}) and the overshoot increase, which can be compensated for by increasing the derivative gain. In this case, $e(t)$ is the input to the PID controller, and $u(t)$ is the output. The following equation illustrates the relationship between input and output.

$$u(t) = k_p e(t) + k_i \int e(t) + k_d \frac{de(t)}{dt}$$

Taking Laplace transformation:

$$u(s) = k_p E(s) + k_i \frac{E(s)}{s} + s k_d E(s)$$

$$\frac{u(s)}{E(s)} = k_p + \frac{k_i}{s} + s k_d$$

$$C(s) = k_p + \frac{k_i}{s} + s k_d$$

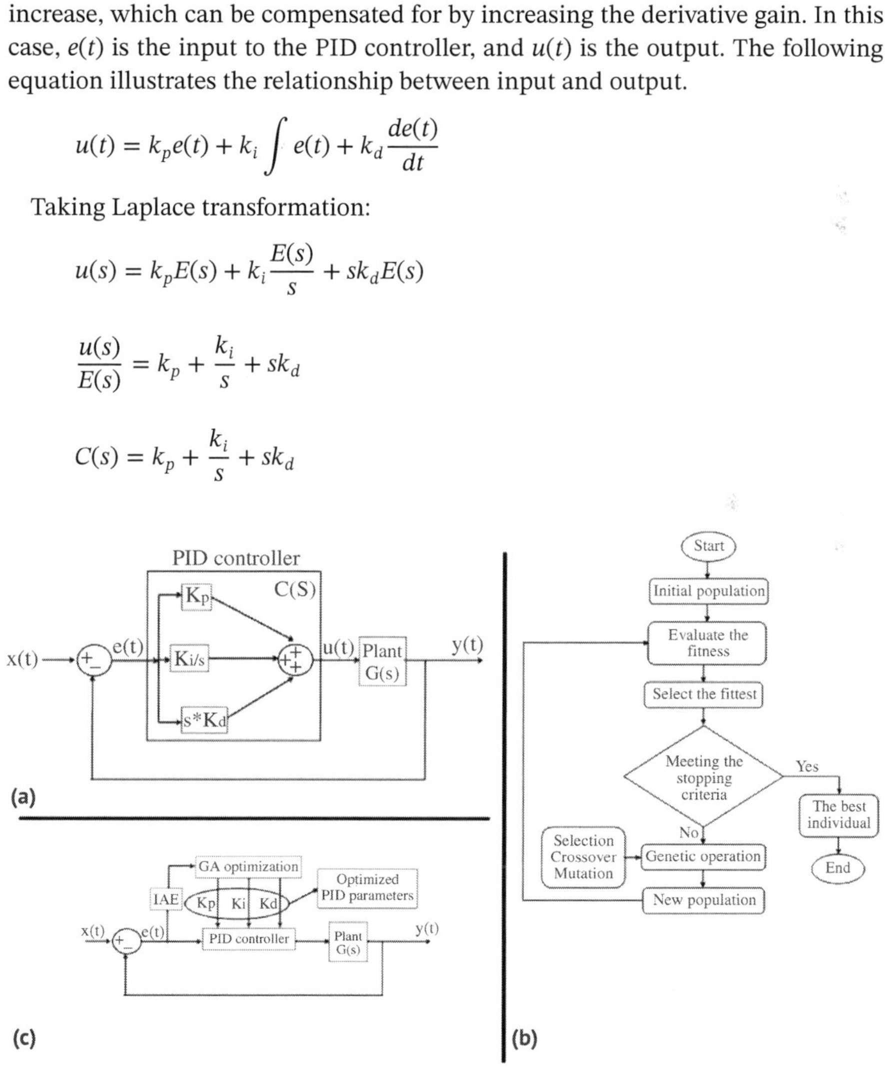

Figure 8.20 (a) PID controller, (b) genetic algorithm, and (c) proposed GA-based PID controller.

8.5.4.1 Genetic Algorithms

Optimization algorithms mimic natural evolutionary procedures. Influenced by the mechanisms of natural selection and genetics, they actually belong to an extensive field called evolutionary computing. This is one aspect of artificial intelligence. An essential thing in solving GAs is finding the right objective functions which set a value for them as optimal. GAs begin a blind search for the best solution by examining a population of possible solutions. Each possible solution within this population is called a chromosome. The chromosomes are ranked and compared according to their fitness value, which ensures that only the fittest member survives. This GA technique uses three primary operations: mutation, crossover, and reproduction. The process continues until such time that the member chosen satisfies the predetermined stopping criterion.

The algorithm continuously applies its mechanisms to generate a new generation of solutions. This cycle repeats until a particular stopping condition is satisfied. The workflow of the GA process is illustrated in step detailed as follows:

Step 1: Initial Population: The GA kicks off by initializing the population, which comprises potential solutions to the problem at hand.

Step 2: Fitness Evaluation: At this juncture, the fitness of each member in the population is assessed. The potential solutions, i.e. individuals, are ranked according to their fitness scores.

Step 3: Genetic Operations: Mimicking natural evolution, genetic operations are employed to evolve the population towards an optimal solution. This involves three fundamental steps:

1. Selection: This step involves picking individuals (chromosomes) from the current population for further processes. The selection criteria are based on Darwin's survival of the fittest principle, where individuals with higher fitness scores are chosen as parents for the next step.
2. Crossover: During the crossover, selected parents are paired to produce offspring (new solutions), introducing new combinations of solutions. This is achieved through a predefined crossover probability, facilitating the exchange of genetic information between parents, Figure 8.20 (Gani et al., 2019) represents the PID controller and the proposed GA-based PID controller.

8.5.5 PID Controller Implementation Issues

Implementing PID controllers can present several challenges, particularly in tuning and performance optimization. Here are some key issues associated with PID controller implementation:

Tuning Complexity: One of the primary challenges in implementing PID controllers is the tuning process. Properly setting the proportional, integral, and derivative gains is crucial for achieving desired system performance. The tuning process can be complex and time-consuming, often requiring trial and error or advanced techniques to find optimal values (Chaturvedi and Kumar, 2023).

Nonlinear Systems: Many real-world systems are nonlinear, which complicates the tuning of PID controllers. Traditional tuning methods may not yield satisfactory results in such cases, necessitating the use of advanced techniques like artificial intelligence or metaheuristic algorithms for better performance.

Objective Function Selection: The choice of objective function for optimization can significantly impact the performance of the PID controller. Different objective functions, such as ITAE, ISE, and IAE, can lead to varying results. Selecting the most appropriate function is essential for effective tuning.

Performance Comparison: When implementing a PID controller, it is important to compare its performance against conventional methods. This includes evaluating the controller's response time, overshoot, and stability. The paper highlights that optimized PID controllers, particularly those tuned using techniques like PSO, can outperform traditional methods.

Integration with Other Techniques: Many researchers have explored hybrid approaches that combine conventional PID tuning methods with evolutionary algorithms. This integration can enhance performance but also adds complexity to the implementation process, requiring a deeper understanding of both methodologies.

System Modeling: Accurate modeling of the system is crucial for effective PID controller implementation. The mathematical model must reflect the system's dynamics accurately, as any discrepancies can lead to suboptimal performance. The paper discusses the development of a mathematical model for a cruise control system, emphasizing the importance of this step.

In summary, while PID controllers are widely used due to their simplicity, their implementation can be fraught with challenges, particularly in tuning, system modeling, and performance evaluation. Addressing these issues is essential for achieving optimal control in various applications.

8.5.6 Bandwidth-limited Derivative Control

Constraints on actuator bandwidth can have adverse effects on torque-based control system stability and performance within robotics. This constraint especially plays a significant role in operational space control, wherein an even restricted bandwidth in one actuator limits the efficiency of tasks related to it simultaneously (Lee and Park, 2021). It proposes a new strategy to deal with this problem by selectively penalizing actuators with lower performance for better overall functionality of the operational space controller. The basic idea is based on a strategy of allocating joint torques primarily to actuators demonstrating high performance when control bandwidth is insufficient to meet the desired levels across all actuators. This strategy may be used recursively, hence allowing the near-optimal performance of the controller, even though a large number of actuators are jammed due to actuator bandwidth limitations. If this measure is inadequate, then further torques are directed toward actuators of even superior performance.

8.5.6.1 Case Study: Operational Space Control Under Actuator Bandwidth Limitation

8.5.6.1.1 Overview

The paper addresses the challenges posed by actuator bandwidth limitations in torque-based robot controllers, particularly in operational space control. It proposes a method to enhance performance by penalizing low-performance actuators, ensuring that control bandwidth is effectively utilized.

Key Concepts

1. Actuator Bandwidth Limitation: This limitation can significantly degrade the stability and performance of robotic systems, affecting all related tasks simultaneously.
2. Operational Space Control: A control strategy that focuses on the robot's end-effector performance rather than individual joint movements.
3. Proposed Method Penalization of Low-Performance Actuators: The method involves adding joint torques primarily to high-performance actuators when the control bandwidth is insufficient. This approach can be executed recursively to maximize performance under constraints.
4. Experimental Validation: The proposed method was tested on the Franka Emika Panda robot manipulator, demonstrating its effectiveness in real-world applications.
5. Technical Details Torque Commands: The torque commands are modeled to account for actuator limitations, using a first-order lag model to represent the cut-off frequency of each actuator.
6. Control Gains: The paper discusses the adjustment of control gains based on actuator performance, with specific examples illustrating how the maximum feasible gain can be increased by penalizing lower-performing actuators.
7. Performance Improvement: By selectively applying torques to higher-performing actuators, the method significantly enhances the overall system performance, as evidenced by experimental results.
8. Root-locus Analysis: The paper includes a root-locus analysis to visualize the impact of actuator selection on system stability and performance.

8.5.7 Proportional and Derivative Kick

The PID controller is the most well-known in industry for stability and quick responses across a very wide range of conditions. This classic controller model is based on three fundamental parameters: the proportional gain (KP), the integral gain (KI), and the derivative gain (KD). Changes to these parameters, known colloquially as tuning, can be undertaken either conventionally or more advanced. Traditionally, methods like ZN give optimum adjustment. On the other hand, better tuning is possible through more advanced methods, especially through GA, which draw inspiration from genetic and natural selection processes. These algorithms select for the most fit "chromosomes" or solutions in evolving a population toward optimal PID parameter settings, balancing broad search spaces with good solutions.

While traditional PID controllers have been in existence for decades, literature often focused on their ability to minimize errors in controlling plants rather than the interactions of control signal peaks. The important trade-off is between overshoot, settling time, and the derivative kick, which is what people call a voltage spike resulting from derivative action in the controller. The derivative kick is often the result of sudden SP changes and can produce a signal which is impulsive in nature—a really bad condition for control elements such as motors, electronics, or valves.

The following studies have researched the following parallel controller configurations to overcome the derivative kick: PI-PD, I-PD, and, more recently, fractional-order PID (FOPID). The PI-PD setup has demonstrated exceptional workability in the management of unstable and resonant systems' responses to setpoint changes. The I-PD is superior, like the PI-PD, in terms of stabilizing an unstable system, but has been designed without being focused directly on removing the derivative kick, which can then be seen as an indirect result of its structure. The derivative kick is specifically targeted in the design of the FOPI-D controller as part of its configuration adjustment of the derivative term inside the controller structure. The feedback loop contains a block that alters some qualities of the control system. The FOPID controller, containing fractions, has been researched recently for its performance to be superior to the conventional PID controller. In FOPID controllers, the integral and derivative parts are based on fractional calculus. Conventional methods used to design FOPID controllers both in the time and frequency domains may be found. This makes the FOPID controller adjustable in five parameters: Kp, Ki, and Kd, and the fractional orders of integration, λ and differentiation μ. FO(PID)s are known for their increase in energy efficiency in control systems. Their fractional operators, which possess memory capabilities, use historical data to make filtering decisions, which are more efficient in managing the plants. Apart from these, they provide much more flexible regulation than the typical PID controllers through fractional components, thus allowing gain and phase shift. This flexibility is effectively exploited in applications such as permanent magnet direct current (PMDC) motors. DC motors are used wherever constant or variable speed has to be maintained, and two major types fall into self-excited and externally excited motors. The speed of the paper is controlled using an externally excited DC motor model by changes in armature voltage. The mathematical model for such a DC motor is described. Optimization algorithms often play an optimal role in the synthesis of FOPID controllers. In addition to the algorithm mentioned above, GA, there is the PSO, which is another popular method used for the design of the FOPID controller.

8.5.7.1 The Derivative Kick Issue

Although the FOPID controller is well known to enhance the performance of closed-loop control systems, the derivative kick issue remains a major issue. The derivative kick manifests as an immediate response of the controller and a big control signal U(s), which occurs instantaneously at the step input. Although several research works focused on the reduction of overshoot and an improvement in settling times in the framework of FOPID controller design, the problem of the spike in the control signal is rarely discussed and considered something negligible Ülšetřcite (Lendek and Tan, 2021). In some instances, the spike might bring about a catastrophe and spoil parts of the system. In this regard, as a way to further optimize the implementation of control systems, it is necessary to mitigate the effects of the derivative kick. In the following paper, we outline one method of eliminating the voltage spike in a control signal by exploiting the application of a time-varying derivative FOPID (TVD-FOPID) controller. As described below, this new TVD-FOPID controller is implemented such that it first opposes the derivative action and gradually restores it.

8.5.8 Reverse-acting Controllers

1. The outflow control valve is controlled by the level controller LC. In this experiment, I have attached a level controller to the bottom valve. As the tank level rises, the controller in this setup must increase the signal and consequently the flow.
2. The inflow control valve is controlled by the level controller LC. In this case, when the tank level rises, the controller must decrease the flow.

Although both setups may control the level just as well, they necessitate completely different actions from the controller. This is what control action direction entails.

When the measurement signal rises, the output of a direct-acting controller tends to rise as well.

When the measurement signal rises, the output of a reverse-acting controller tends to fall.

Because all control systems are programmable (often designated DIR/REV), the controller can be changed from direct-acting to reverse-acting, or vice versa.

It is very important to determine the correct direction of the control action. A well-configured feedback control system will always experience negative feedback—that is, it will attempt to make output errors decrease.

The system is likely to become unstable if the direction of control action is incorrect, since it will experience positive feedback and act to reinforce output faults. First of all, if you are having trouble setting up a reliable controller, ensure you have established the correct direction of control action.

1. Direct-acting Controller/Fail-Open Control Valve: In such a case where a direct-acting controller is implemented with a fail-open control valve, the following is the process performance: The valve closes as the process level goes up, as the controller output increases. If the process level dropped below the lower limit, this would leave the controller completely open. Aside from the valve/controller response, this would have been similar to that of a reverse-acting regulator.
2. Reverse-acting Controller/Fail-Open Control Valve: In this scenario, the valve tends to open the controller output because the process level increases. This valve closes the valve when the process level goes lower than the marked point. For example, there is a recycling compressor valve, which falls into this category of use.
3. Direct-acting Controller/Fail-Closed Control Valve: This is equivalent to a reverse-acting controller or fail-open valve. During the rise in the process level, the valve opens as a result of controller output.

The valve is closed if the process level drops below the (SP). A common application is backpressure control Here, the valve controller acts like a PSV.

The PID controller can be set for either DIRECT or REVERSE acting modes. Different conditions present on a closed data flow path that is seen within the control loop would include the process having either a positive or negative gain (Niu and Xiao, 2022), the control valve being air-to-open or air-to-close, and an AO signal being direct or inverted, all with a related impact on the loop gain. This means the loop gain could either be positive or negative and originates from a total of 16 different combinations. For control to be stable, it has to use negative feedback, which means overall that the loop dynamics have to have negative gain. If the action direction is wrong, then control will simply not work because the controller's output will deviate severely and promptly in an undesirable direction. This was discussed in the previous lecture. The general guideline is that the 0% output should be a fully closed valve, and the 100% output should be a fully opened valve. With this setup, the AO block and the control valve dynamics will always present a positive gain. An air-to-close valve needs to be inverted from an AO as shown in order to present a positive gain. In this approach, the system is effectively a simplification of the AO and control valve being ruled out of consideration. The gain for a controller is then simply an issue of selecting which direction to act; this must have an opposite sign by definition, and so give negative feedback, as follows.

8.5.9 Industrial PID Control

The integration of model predictive control (MPC) with PID controllers is one of the efforts of the optimization techniques for industrial processes. While MPC is known for the management of system constraints, it is yet to be widely adopted because of the complexity and familiarity of field engineers with advanced control techniques. The control system is defined hierarchically as the proposed MPC Controller. It sits on a higher plane in a SCADA or supervisory control and data acquisition system. PID Controller: lower-level implementation on a programmable logic controller (PLC).

The system is tested using LabVIEW software, where the MathScript Module is utilized to develop the algorithm.

Advanced Process Control (APC): Enhances system performance and product quality. Includes techniques like MPC and PID control. Requires advanced automation technology for effective implementation.

MPC: A promising advanced control algorithm suitable for various industrial applications. Handles system constraints effectively, but is not widely adopted due to complexity.

PID Controller: Commonly used in industrial settings for its simplicity and robustness. Parameter adaptation is challenging, especially under operational constraints.

8.5.9.1 System Architecture Hierarchical Control Structure

MPC Controller: Implemented on a PC using LabVIEW as the higher control level.

PID Controller: Executed on a Siemens PLC (S7-1200) as the lower control level.

Communication between PC and PLC is facilitated through an OPC server.

8.5.9.2 Implementation Details Simulation Setup

The proposed control system is tested on a three-tank benchmark system, representing a MIMO system. The system is linearized around the operating point using a Taylor expansion.

8.5.9.3 Operational Benefits

The integration of MPC and PID aims to improve industrial process performance while being easily implementable with existing automation technology.

Challenges and Solutions Operator Familiarity: Many operators are more accustomed to traditional control techniques, making the transition to advanced methods challenging.

The study emphasizes simplifying advanced control theory for better understanding and execution by field engineers.

8.5.10 Traditional Industrial PID Terms

PID control remains the main technique for feedback control in the process industry, with ongoing interest in modifications to enhance performance. Major strides in the improvement of PID technology occurred during the last decade, especially in distributed control systems. Routine analysis of the usage of control is important in highlighting measurement and control difficulties, which can then be resolved in operation when management is made aware. These tools allow proper analysis and diagnosis of control performance with minimal manpower. It most significantly affects the throughput of a plant and the quality of the products produced.

8.5.10.1 Case Study

Petrochemical Plant Initial Survey: A survey conducted in a petrochemical plant showed that control utilization was very low. Management decided to commit resources for further analysis.

Formation of an Instrumentation Team: Depending upon this survey outcome, an instrumentation team was put in place to work on those control loops which were not operating in their normal mode of design.

Process improvements: The work yielded high levels of reduction in variability, hence higher plant throughput with quality products. Two years later, a new production record was achieved. Performance monitoring tools include robust report generation capabilities and are usually important for securing management support and using control utilization as an issue. New technologies, such as model switching to perform the identification process, have improved ease of use for online and adaptive tuning in tools.

The paper talks about certain changes in PID controllers with improvements in recovery from saturation conditions and enabling effective control when measurements are updated non-periodically. In the future work of the process control industries, it is planned to consider more utilization of PID control in light of process nonlinearity and operating constraints.

8.5.11 Industrial PID Structures and Nomenclature

Robotics would efficiently improve operations in food processing, become free from the hassles of doing the job manually, and pass the mark for being able to upgrade from old setups. In this regard, current food handling and packaging setups can be said to have a few limitations to their capacity and even less satisfactory in terms of output due to manual processing. It gives a clear preview of the technological parameters that will be essential in the future operation of food processing, not only with regard to Industry 4.0 but also with respect to structural configurations, commercial adaptation, and guidelines on the adaptation of commercially available robots to applications in the food industry (Khan et al., 2018).

It categorizes the robots used in the food industry according to their kinematics and applications in sectors like handling, processing, palletizing, packing, and serving. Classification can bring understanding of the various roles that the robots might assume in producing food.

Challenges and Opportunities: It describes challenges in the food processing industry, mainly in energy consumption and strong human–machine interfaces. However, it also finds many opportunities for technological advancement such as embedded control, sensor fusion, and machine learning.

CPS: Introduction to cyber-physical systems in food production, where food production will be designed as a closed-loop system that can optimize output under varied constraints and risks while improving the safety of food manufacturing processes.

Future Research Directions: The paper ends on an optimistic note about the present research and development ongoing in food robotics. According to the paper, the bright future with regard to the introduction of advanced technologies in the food industry assures a very promising expectation. So much emphasis lies in the coherence of various technologies toward producing innovative solutions to the challenges in food processing.

8.5.12 The Process Controller Unit

The clinical manifestations of the disease include symptoms of gastroenteritis, abdominal cramps, bloody diarrhea, fever, myalgia, headache, nausea, and vomiting. In 2018, Salmonella induced over half of foodborne outbreak illnesses reported in the European Union. Contamination with these bacteria often happens with products such as poultry, cattle, and feed products. Moreover, however, dried foods, infant formula, fruits, vegetables, and even pets are increasingly identified as major sources (Ehuwa et al., 2021). Managing Salmonella encompasses legislative interventions and practices that successfully reduced hospitalization rates from 2014 to 2015. However, a considerable rise in hospitalization was seen in 2016, indicating that controls must be tightened more both by the governmental and private agencies. Food handlers, particularly in the meat industry and when preparing ready-to-eat foods, therefore play a critical role in controlling Salmonella from spreading. This review gives a comprehensive update on the global epidemiology of salmonellosis. While putting emphasis on official interventions, the role of food handlers in the transmission of disease, and the critical nature of food safety measures.

8.5.12.1 Case Study: Salmonella, Food Safety, and Food Handling Practices

Overview of Salmonellosis Prevalence: Salmonellosis is the second most reported gastrointestinal disorder in the European Union, primarily caused by consuming Salmonella-contaminated foods. In 2018, it accounted for over half of foodborne outbreak illnesses reported in the European Union.

Symptoms: Common symptoms include gastroenteritis, abdominal cramps, bloody diarrhea, fever, myalgia, headache, nausea, and vomiting

Sources of Contamination: Common Sources—Salmonella contamination is often linked to poultry, cattle, and their feeds. However, it can also be found in dried foods, infant formula, and fruits, and vegetables.

Role of Food Handlers: Food handlers, especially in meat processing and ready-to-eat food sectors, play a crucial role in the spread of Salmonella.

Training and Knowledge Transfer Knowledge versus Behavior: While knowledge-based training is beneficial, it does not always translate into proper food safety behavior. Studies indicate that increased knowledge does not guarantee improved practices.

Effective Training Models: A combination of knowledge-based and behavior-based training yields the best results. Behavior-based training has been shown to maintain good practices longer than knowledge-based training alone.

Tailored Training: Training should be customized based on the employee's role and background, considering language and education levels to enhance understanding.

Future Perspectives: Regulatory Measures—Stricter government regulations and control systems are essential for reducing salmonellosis cases. This includes both domestic production and international importation requirements.

Public Awareness: There is a need for effective communication strategies to raise awareness about salmonellosis across all population strata.

Consumers should be educated on basic food safety practices to prevent contamination.

Conclusion: Ongoing Challenges—Despite existing training and regulations, outbreaks of salmonellosis continue to occur, highlighting the need for improved training methods and public awareness campaigns. Continuous evaluation and adaptation of food safety practices are crucial for reducing the incidence of foodborne illnesses related to Salmonella.

8.5.13 Supervisory Control and the SCADA PID Controller

In practice, actual traditional PID controllers, as made by applying the ZN technique, typically exhibit large overshoots and longer settling times. A good deal of research efforts have been made to enhance the performance of PID controllers in closed-loop operations, specifically in real-time temperature control applications utilizing SCADA systems, such as, in this case, the HC-900 Honeywell controller. The paper discusses the basics, mathematical backgrounds, and some design issues of PID controllers as well. This study examines the performance of PID controllers through various criteria like the ISE, IAE, ITAE, and integral time square error (ITSE) with simulations which have been prepared to justify these results. In this paper, control strategies based on PID are developed for temperature monitoring and control in the $UT_3 21$ laboratory system via in-depth analysis of a plant process. The tests carried out clearly illustrate satisfactory control performance within a wide range of dynamics. For example, some are low-order, and some are high-order systems; the dead-time is negligible in some, while it could be as high as 40 s. Some show monotonic responses; others show oscillatory responses.

The dynamic model of the system to be controlled is to be established in order to design and tune a controller that meets the required level of performance. Specify desired closed-loop performance by known physical limitations. By using appropriate controllers, you attain the performance that you require. Utilize an appropriate platform for finalizing the controllers. Test the controllers and modify their performance as needed. The transfer function for the PID controller is

$$G_c(s) = K_c\left(1 + \frac{1}{T_I s} + T_D s\right) = K_P + \frac{K_I}{s} + K_D s$$

where

K_P is the proportional gain

T_I is the integral time

T_D is the derivative time

Clearly, this transfer function is improper and practically it can not be used as frequency increases, its gain also increases. Therefore, practical PID controller has the following transfer function:

$$Gc(s) = Kc\left(1 + \frac{1}{Tis} + \frac{Tds + 1}{\lambda s + 1}\right)$$

where λ is a small parameter and it may be set as 10% of derivative term.

In the context of this study, we used methodologies developed by ZN, Cohen–Coon, and Wang for real-time analysis of the performance of the UT_321 laboratory temperature control system. By application of the system identification toolbox in MATLAB 7.1, a model of the UT_321 temperature control system along with SCADA was built to optimize the configuration. Efficacy was determined through frequency domain specifications, time domain specifications, and performance error indices of the controller. In this paper, a model-based PID controller is proposed with its design intent on temperature regulation applications. Control is generally applicable for SISO systems, aside from flow, pressure, and liquid level, and for multivariable systems. The rest of the document proceeds in this manner: Section 1 will introduce the background of PID controllers. Section 2 defines the problem statement. Section 3 states the basics of controller design for FOPDT/SOPDT-type plants, and the UT_321 lab setup for temperature control.

8.5.14 Case Study

8.5.14.1 Problem Statement

This section explains a design methodology for PI/PID controllers based on a model-based control strategy for performance assessment and real-time application to temperature control systems. The method is realized by means of models of the FOPDT/SOPDT variety, using closed-loop pole placement through root loci diagrams, establishing the system model by means of a reaction curve experiment. Its purpose is to calculate PID tuning parameters for performance improvement. The developed controller is executed on a SCADA in addition to documentation of a temperature control simulation performed with a PID controller.

As seen in Figure 8.21, the controller $G_c(s)$ is a PID controller. It is assumed that the system model $G(s)$ is known or has been found through experimentation. Systems of the type shown in the equation as follows are used for FOPTD as follows:

$$G(s) = \frac{Ke^{-t_d s}}{1 + \tau s} \tag{8.72}$$

where $G(s)$ is the open loop transfer function, K is the steady state gain, t_d is the dead time, and τ is the time constant.

8.5.14.2 PID Controller Design Method for SOPDT Model

It is based on the second-order plus time delay model and closed-loop pole allocation through root locus for the design method. Additionally, this method is applicable to the class of FOPTD systems that take the form of a PI controller with $K_D = 0$. First, let us define some terms that are used in the design. The equivalent time constant τ_0 of a process which is inversely proportional to the speed of response. According to equivalent time constant principles, it is given by,

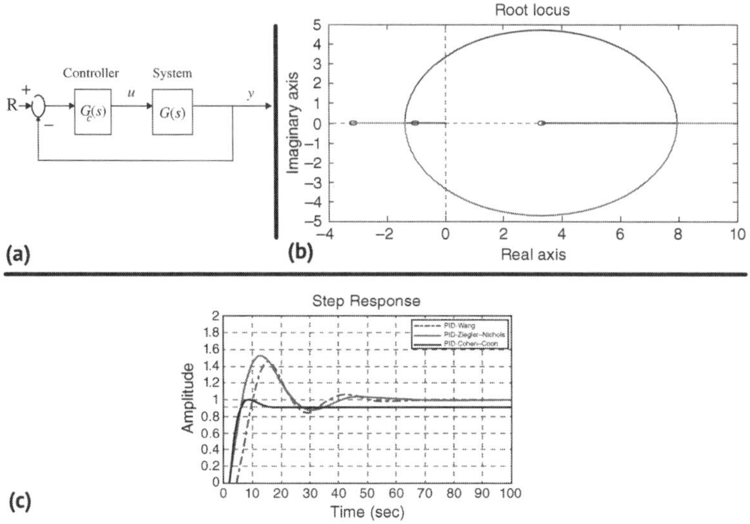

Figure 8.21 (a) A single loop controller configuration, (b) root locus of open loop transfer function, and (c) simulation result of system model (Laware et al. [2013]/International Journal of Application or Innovation in Engineering & Management).

$$\tau_0 = \frac{\sqrt{b^2 - 2ac}}{c}, \qquad b^2 - 4ac \geq 0$$
$$= \frac{2a}{b}, \qquad b^2 - 4ac < 0 \tag{8.73}$$

where a, b, c are system parameters in (4). Another variable of interest is the damping ratio ξ_0 of open loop system model which is given by

$$\xi_0 = 1, \qquad b^2 - 4ac \geq 0$$
$$= \frac{b}{2\sqrt{ac}}, \qquad b^2 - 4ac < 0 \tag{8.74}$$

The PID controller can be written in the form (Laware et al., 2013)

$$G_c(s) = k\left(\frac{As^2 + Bs + C}{s}\right) \tag{8.75}$$

where

$$A = \frac{K_D}{k}, B = \frac{K_P}{k} \text{ and } C = \frac{K_I}{k} \tag{8.76}$$

In the design procedure, if we choose the controller zeros such that they cancel out the poles of system model in Eq. (8.73), we get $A = a, B = b, C = c$. Then the resultant open loop transfer function becomes

$$G(s)G_c(s) = \frac{ke^{-t_d s}}{s} \tag{8.77}$$

Using its root locus, the roots of Eq. (8.77) can be chosen to obtain desired dynamics in a closed loop. The root locus indicates that the desired closed loop poles selected could be real in nature or complex in nature. This choice will depend upon the model under consideration. In highly oscillatory systems, roots come to be the real ones in order to damp these resulting oscillations in the response obtained in closed loop. In contrast, for lightly oscillatory systems, the complex poles are selected with the intent of introducing some amount of overshoot and increased speed of response.

On this basis, the two cases are considered for pole selection as follows: Case I: $\xi_0 > 0.707$ or $0.05 < (t_d/\tau_0) < 0.15$

$$\text{Or}\,(t_d/\tau_0) > 1$$

Since the system in this instance is either non-oscillatory or mildly oscillatory, complex closed loop poles ought to be chosen from the open loop transfer function's root locus in Eq. (8.77). Assume that the damping ratio ξ_n and natural frequency ω_n, which are determined by, are the ideal pair of complex poles on the root locus.

$$8.64s = -\xi_n\omega_n \pm j\omega_n\sqrt{1-\xi_n^2} \tag{8.78}$$

For the pole pair in Eq. (8.64) to be on root locus, it must satisfy the angle and magnitude conditions given by

$$\begin{aligned} &\angle G(s)G_c(s) = -\pi \\ &|G(s)G_c(s)| = 1 \end{aligned} \tag{8.79}$$

Angle condition in Eq. (8.79) results into

$$-\omega_n t_d\sqrt{1-\xi_n^2} - (\pi - \cos^{-1}\xi_n) = -\pi \tag{8.80}$$

From Eq. (8.80) we get ω_n as,

$$\omega_n = \frac{\cos^{-1}\xi_n}{t_d\sqrt{1-\xi_n^2}} \tag{8.81}$$

Using magnitude condition in Eq. (8.79) and ω_n in Eq. (8.81), the value of k can be computed as

$$k = \omega_n e^{-\omega_n\xi_n t_d} \tag{8.82}$$

For the desired damping ratio $\xi_n = 0.707$, the resulting k is given by

$$k = \frac{0.5}{t_d} \tag{8.83}$$

Case II: $\xi_0 \leq 0.707$ and $0.15 \leq (t_d/\tau_0) \leq 1$ Real double closed-loop poles should be chosen from the root locus of the open-loop transfer function, since the system is extremely oscillatory, as can be shown from the considerations in this instance. The location of the poles at $s_{1,2} = -1/\tau_0$ and the poles at the breakaway point on the root locus are used to determine the double real pole on the root locus. Either before the breakaway point or at the breakaway point, the poles at $s_{1,2} = -1/\tau_0$ are chosen. The goal is for the closed-loop reaction speed to be at least as quick as the system model's open-loop response. We obtain the expression for k from the magnitude condition as follows, for the poles at $s_{1,2} = -1/\tau_0$ to be on the root locus:

$$k = \frac{1}{\tau_0} e^{-t_d/\tau_0} \tag{8.84}$$

Now to calculate the breakaway point on root locus, put $\xi_n = 1$ in Eq. (8.81) to get

$$\omega_n = \frac{\cos^{-1}\xi_n}{t_d\sqrt{1-\xi_n^2}} = \frac{1}{t_d} \tag{8.85}$$

From Eqs. (8.82) and (8.85) we get

$$k = \frac{1}{e\tau_0} \tag{8.86}$$

The selection of k from Eqs. (8.85) and (8.86) is done as

$$k = \min\left\{\frac{1}{\tau_0} e^{-t_d/\tau_0}, \frac{1}{et_d}\right\} \tag{8.87}$$

Finally, the PID tuning parameters are obtained as:

$$\begin{bmatrix} K_P \\ K_I \\ K_D \end{bmatrix} = k \begin{bmatrix} b \\ c \\ a \end{bmatrix} \tag{8.88}$$

8.5.14.3 Summary of PID Controller Design

For the given system model, (i) Obtain a, b and c from system model. (ii) Compute τ_0 from Eq. (8.73) and ξ_0 from Eq. (8.74). (iii) For $\xi_0 > 0.707$ or $0.05 < (t_d/\tau_0) < 0.15$ or $(t_d/\tau_0) > 1$, calculate k from Eq. (8.83). (iv) Otherwise, for $\xi_0 \leq 0.707$ and $0.15 \leq (t_d/\tau_0) \leq 1$, calculate k from (5). (v) Obtain PID tuning parameters from Eq. (8.84) (Laware et al., 2013).

8.5.14.4 Determination of System Model

The system model is an important part of the design of a model-based controller. All physical information about the system in question must be correct and full in order to find a white box model. The system in question does not have access to such data. Therefore, system identification is used to define the system model. We used the data from the step test in time domain to calculate the model using the system. We considered the FOPDT model. This step response locates the system parameters such as steady state gain, time delay, and the process's time constant, and the model produced has the general form:

$$G(s) = \frac{k_p e^{-t_d s}}{1 + \tau s}$$

where

k_p is steady state gain of system

τ is time constant of system

t_d is dead time of system

Hence, we get FOPDT model as

$$Gp(s) = 2.2 \times \frac{e^{-6s}}{(1 + 40.484s)}$$

(Water flow through Rota meter is kept at 40 LPH)

Wang technique controller design: the system parameters for the desired system model Eq. (8.90) are $a = 0$, $b = 40.484$, and $c = 1$. Using Eqs. (8.73) and (8.74), the values of τ_0 and ξ_0 are calculated to be 40.484 and 1, respectively. Because $\xi_0 > 0.707$, $k = 0.0833$ is calculated using Eq. (8.83) instead. Thus, the PID parameters, using Eq. (8.90), are $K_p = 3.374$, $K_i = 0.0833$, and $K_d = 0$.

Hence, controller transfer function by Wang method is given by

$$Gc(wn) = \frac{3.374s + 0.0833}{s}$$

PI controller is so obtained. ZN tuning technique controller design From the system model's Bode diagram, we get phase crossover frequency $(\omega_{pc}) = 0.3572$ rad/ sec and gain margin $(GM) = 16.38$ dB. Finally, the final gain k_u and period T_u are obtained as follows:

$$\text{Ku} = \text{GM} = 16.38 \text{ dB}$$

$$Tu = \frac{2\pi}{wpc} = 17.6$$

The controller settings for PID controller as per ZN method are $K_p = 0.6K_u = 4.0$, $T_i = 0.4545$, and $T_d = 0.6$ Hence, ZN method gives following model:

$$Gc(zn) = \frac{0.6s^2 + 4s + 0.4545}{s}$$

Controller design by Cohen–Coon tuning method: from delay time and time constant of (20), we get, $K_p = 4.203$, $T_i = 13.91$, and $T_d = 2.124$. Hence, controller transfer function by Cohen–Coon method is

$$Gc(cc) = \frac{0.51s^2 + 4.203s + 0.3022}{s}$$

Process rate and reaction rate for actual temperature system is

$$\left(\frac{57.5 - 39.3}{40.484 - 3}\right) = 0.49\%/s \left(\frac{57.5 - 39.3}{40.484 - 3}\right) = 0.49\%/s$$
$$\frac{\text{process Rate}}{mv2 - mv1} = \frac{0.49}{8} = 0.061\%/s \tag{8.89}$$

Tables 8.8–8.10 represent comparison of controller on the basis of frequency domain specification, comparison of controller on the basis of time domain specification, and comparison of controller on the basis of performance indices.

Table 8.8 Comparison of controller on the basis of frequency domain specification.

Model	Gain margin in dB			Phase margin in degree		
Model given	W	Z-N	C-C	W	Z-N	C-C
by Eq. (8.90)	5.193	6.514	5.99	32.38	27.71	48.90

Table 8.9 Comparison of controller on the basis of time domain specification.

Model	Peak overshoot (%)			Settling time (s)		
Model given	W	Z-N	C-C	W	Z-N	C-C
by Eq. (8.90)	4.4	5.2	1.6	51	65	13

Table 8.10 Comparison of controller on the basis of performance indices.

Model	ISE			IAE		
Model given	W	Z-N	C-C	W	Z-N	C-C
by Eq. (8.90)	188.7	178.1	148.7	151.99	279.83	295.6
Model	**ITSE**					
Model given by Eq. (8.90)	W	Z-N	C-C	W	Z-N	C-C
	11.58	38.48	42.87	11.4	20.8	21.9

8.6 Review Questions

Q1 A PID controller is employed to stably control a plant with transfer function,

$$P(s) = \frac{1}{(s+1)(s+2)}$$

Now, the proportional gain is increased by a factor of 2, the integral gain is increased by a factor of 3, and the derivative gain is left unchanged. Given that the closed-loop system continues to remain stable with the new gains, the steady-state error in tracking a ramp reference signal.

Q2 In a unity-gain feedback control system, the plant

$$P(s) = \frac{0.001}{s(2\,s+1)(0.01\,s+1)}$$

is controlled by a lag compensator

$$C(s) = \frac{s+10}{s+0.1}$$

The slope (in dB/decade) of the asymptotic Bode magnitude plot of the loop gain at $\omega = 3$ rad/s is (in integer).

Q3 Find the values of K_p, K_I, and K_D for PID controller 8.3 whose characteristic equation has real roots (see Figure 8.3) at -10, $\xi = 0.8$, and $\omega_n = 2$ rad/s.

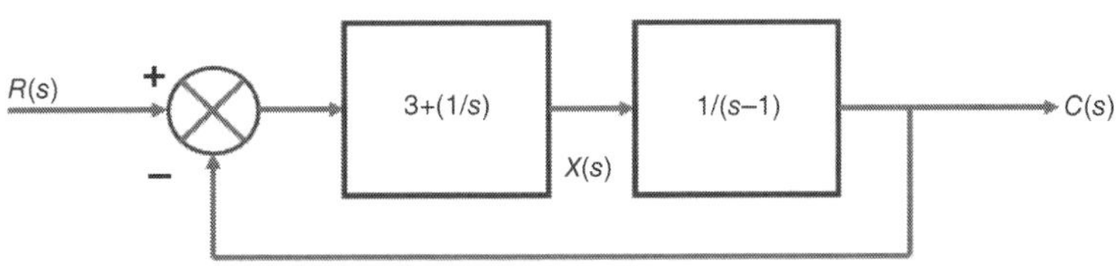

Q4 Consider a unity-gain negative feedback system consisting of the plant $G(s)$ (given below) and a proportional-integral controller. Let the proportional gain and integral gain be 3 and 1, respectively. For a unit step reference input, the final values of the controller output and the plant output, respectively, are

$$G(s) = \frac{1}{s-1}$$

Q5 A plant has a nominal given by

$$G_o(s) = \frac{1}{(s-1)^2}$$

Prove that this system cannot be stabilized with a PI controller.

Bibliography

O Arrieta, R Vilanova, JD Rojas, and M Meneses. Improved PID controller tuning rules for performance degradation/robustness increase trade-off. *Electrical Engineering*, 98:233–243, 2016.

KJ Åström and T Hägglund. Benchmark systems for PID control. *IFAC Proceedings Volumes*, 334:165–166, 2000.

S Charkoutsis and M Kara-Mohamed. A particle swarm optimization tuned nonlinear PID controller with improved performance and robustness for first order plus time delay systems. *Results in Control and Optimization*, 12:100289, 2023.

S Chaturvedi and N Kumar. Design and implementation of an optimized PID controller for the adaptive cruise control system. *IETE Journal of Research*, 6910:7084–7091, 2023.

P Daoutidis, M Soroush, and C Kravaris. Feedforward/feedback control of multivariable nonlinear processes, *AIChE Journal*, 3610:1471–1484, 1990.

O Ehuwa, AK Jaiswal, and S Jaiswal. Salmonella, food safety and food handling practices. *Foods*, 105:907, 2021.

TAM Euzébio, MTD Silva, and AS Yamashita. Decentralized PID controller tuning based on nonlinear optimization to minimize the disturbance effects in coupled loops. *IEEE Access*, 9:156857–156867, 2021.

MM Gani, MS Islam, and MA Ullah. Optimal PID tuning for controlling the temperature of electric furnace by genetic algorithm. *SN Applied Sciences*, 1:1–8, 2019.

Y Huang. Automatic process control for the food industry: an introduction. In: *Robotics and Automation in the Food Industry*, 3–20, 2013, Woodhead Publishing.

Y Huang, Y Lan, SJ Thomson, A Fang, WC Hoffmann, and RE Lacey. Development of soft computing and applications in agricultural and biological engineering. *Computers and Electronics in Agriculture*, 712:107–127, 2010.

GG Jin and YD Son. Design of a nonlinear PID controller and tuning rules for first-order plus time delay models. *Studies in Informatics and Control*, 282:157–166, 2019.

ZH Khan, A Khalid, and J Iqbal. Towards realizing robotic potential in future intelligent food manufacturing systems. *Innovative Food Science & Emerging Technologies*, 48:11–24, 2018.

T Kondakci and W Zhou. Recent applications of advanced control techniques in food industry. *Food and Bioprocess Technology*, 10:522–542, 2017.

AR Laware, VS Bandal, and DB Talange. Real time temperature control system using PID controller and supervisory control and data acquisition system (SCADA). *International Journal of Application or Innovation in Engineering & Management (IJAIEM)*, 2(2), 2013.

H Lee and J Park. Operational space control under actuator bandwidth limitation. In: *2021 IEEE International Conference on Robotics and Automation (ICRA)*, 12847–12852, 2021. IEEE.

Y Lee and S Park. M Lee, C Brosilow. "PID controller tuning for desired closed-loop responses for SI/SO systems". *AICHE Journal*, 441:106–115, 1998.

Y Lee, M Skliar, and M Lee. Analytical method of PID controller design for parallel cascade control. *Journal of Process Control*, 16(8):809–818, 2006.

A Lendek and L Tan. Mitigation of derivative kick using time-varying fractional-order PID control. *IEEE Access*, 9:55974–55987, 2021.

SS Niu and D Xiao. *Process Control: Engineering Analyses and Best Practices*. Springer Nature. 2022.

A O'Dwyer. Performance improvement using simple PID controller tuning formulae. In: *3rd IET International Conference on Power Electronics, Machines and Drives (PEMD 2006)*, 276–280, 2006. Stevenage UK: IET.

E O'Dwyer, L De Tommasi, K Kouramas, M Cychowski, and G Lightbody. Modelling and disturbance estimation for model predictive control in building heating systems. *Energy and Buildings*, 130:532–545, 2016.

F Padula and A Visioli. Optimal tuning rules for proportional-integral-derivative and fractional-order proportional-integral-derivative controllers for integral and unstable processes. *IET Control Theory & Applications*, 66:776–786, 2012.

JD Rojas, O Arrieta, and R Vilanova. Industrial PID control. In: *Industrial PID Controller Tuning: With a Multiobjective Framework Using MATLAB®*, 5–19, 2021a. Springer International Publishing.

JD Rojas, O Arrieta, and R Vilanova. PID tuning as a multiobjective optimization problem. In: *Industrial PID Controller Tuning: With a Multiobjective Framework Using MATLAB®*, 91–113, 2021b. Springer International Publishing.

D Semino and A Brambilla. An efficient structure for parallel cascade control. *Industrial & Engineering Chemistry Research*, 356:1845–1852, 1996.

S Tavakoli and M Tavakoli. Optimal tuning of PID controllers for first order plus time delay models using dimensional analysis. In: *2003 4th International Conference on Control and Automation Proceedings*, 942–946, 2003. IEEE.

9

Online Spectroscopy Techniques to Access Food Quality and Safety, SCADA and HMI Examples

After reading this chapter, you should be able to:

- Innovative Quality and Safety Assessment: How online spectroscopy, optical sensors, and nondestructive inspection methods enhance the quality control of food
- Automation in Food Packaging and Processing: The role of robotics, artificial intelligence/machine learning (AI/ML), and sensors in the automation of food handling, packaging, and safety systems.
- Practical Applications and Industry Case Studies: Real applications of spectroscopy, imaging, and sensing techniques for the monitoring of food safety and quality.

There are ML techniques that improve monitoring and forecasting food safety. The main points in the paper that can be listed are as follows:

Importance of ML in Food Safety: Combining information types using ML improves food safety surveillance quality and cost. It incorporates prior experience and historical data to prioritize monitoring efforts along the food supply chain.

Of course, much research has been done on the application of ML in food-borne diseases and deep learning in food science; however, until this day, a specific overview of the topic has not been done specifically on ML for the monitoring and prediction of food safety. This paper aims to fill the gap by summarizing the relevant studies.

Types of Data and Algorithms: The data types that may occur in ML applications can be divided into structured and unstructured data. It states that Bayesian networks (BNs) and neural networks (NNs) are the most frequently used algorithms for structured and unstructured data, respectively. The third point deals with the importance of model validation in the assurance of the reliability and

Food Engineering Automation with Robotics and AI, First Edition. Abir Chakravorty.

trustworthiness of predictions made by ML. This benefits the assessment of the role of ML in food safety and its capability for enhanced practice monitoring and better forecasts of food safety hazards.

9.1 AI/ML Techniques for Food Safety Case Studies

The section "Application of ML to the monitoring and prediction of food safety: A review" makes several important contributions to the field of food safety and ML. The key contributions include:

The paper gives a systematic review of the existing literature on the use of ML in the surveillance and monitoring of food safety with respect to prediction. In this manner, various findings culled from diverse studies are brought together in one place, making it easier for both researchers and practitioners to understand the present state of the area and spot gaps in knowledge.

Categorization of the Studies: This study actually classifies the reviewed studies into the fields of application, types of data used, and the type of ML model applied. It, therefore, is helpful in understanding the diversity of approaches and methodologies used in the field as further research work may be pursued.

Data Types and Algorithms: The paper focuses on the types of data used within ML applications; these include structured and unstructured data. Work indicates the most frequently used algorithms in the chapter, which comprise BNs, NNs, and support vector machines (SVMs), among others, based on which it can be deduced that one is effective or not for a particular kind of data.

Flowchart Structure of RC-MF Model: The figure shows the structure of the RC-MF model as defined by its flow chart diagram, as depicted in Figure 9.1.

Identification of Future Research Avenues: A number of directions may be used to derive future research, such as the searching and input of more diverse data sources, in order to elicit even better predictions by ML models on food safety. This will help researchers to develop ideas to advance this field.

Validating the Predictive Ability Using Model Validation: To bring out the proper issue of model validation, the job is done since it is very important for validation in terms of the accuracy and reliability of such predictions developed by ML models. All the studies which were reported and included in the reviews indicated high accuracy in making predictions. This thus establishes the potential use of ML applications that have to do with food safety.

Overall, the article remains seminal up to this day; it reviews and puts into perspective relevant literature but also organizes findings and points out opportunities for further research with a view to validation of ML applications toward food safety.

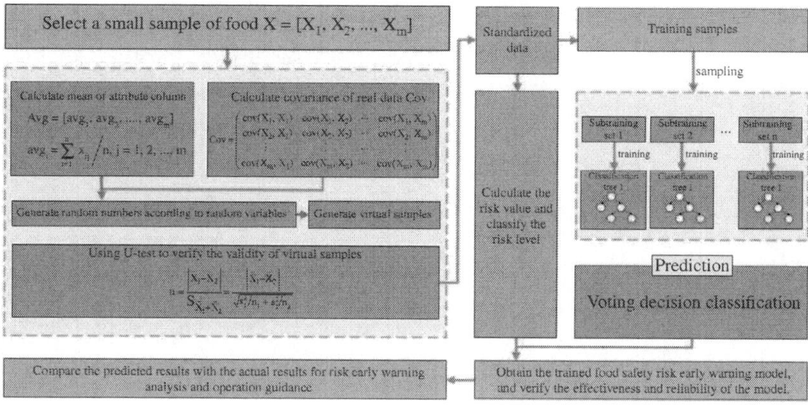

Figure 9.1 The flowchart structure of the RC-MF model (Geng et al. [2022]/Elsevier).

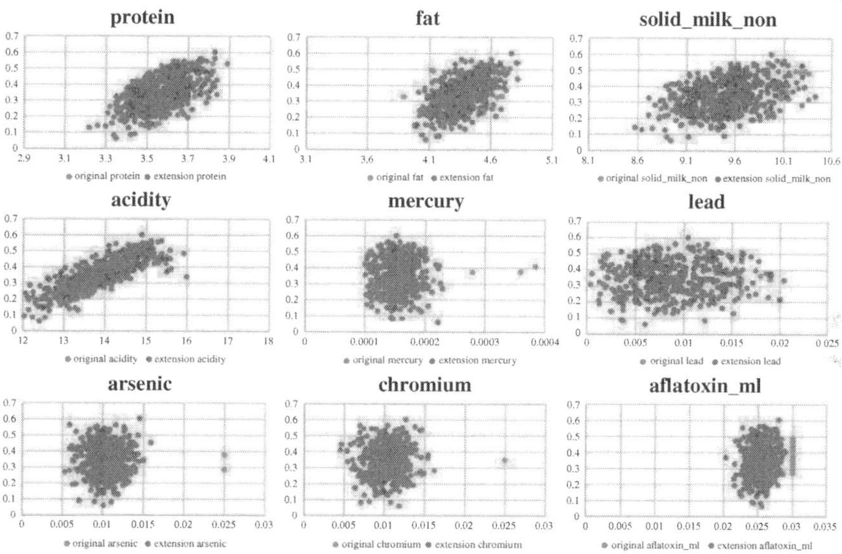

Figure 9.2 Comparison of the original data with the extended data (Geng et al. [2022]/ Elsevier).

This case study addresses the application of ML in the monitoring and prediction of food safety, as mentioned in the paper "Application of ML to the monitoring and prediction of food safety: A review." It enlightens how ML is increasingly being applied to the safety of food through data analysis and modeling. In the figure, the original data is compared with the extended data when compared to Figure 9.2.

Major Contributions: This paper integrates multiple studies with good clarity over the application of ML in food safety.

Some parameters based on which studies were classified are as follows:

Field of Application: It focuses upon the specific food safety hazards, general food safety events, food fraud, and traceability.

Data Types: This helps in the differentiation between structured and unstructured data applied within the ML models.

Common Algorithms: This section identifies the most common ML algorithms employed.

Common ML algorithms: BNs, NNs, SVMs, decision trees, several types of neural networks, data insights, and types of food safety hazard so that it revolves around the following study areas by Wang et al. (2022a).

Biological hazards: 36

Chemical hazards: 22

Physical hazards: 2

Food fraud: 16

Growth in Research: Thus, applications involving ML models in food safety monitoring are growing fast and reflect the advanced sophistication of data analytical techniques in food safety in general.

To explain better, categorizing the studies into their respective fields of application, the types of data used, and the kind of ML model applied will lead to a better comprehension of the diversity in approaches and methodologies. This is a precursor to making future research relatively easier.

Data Types and Algorithms: Discusses types of data applied within an ML context. Distinguishes between structured and unstructured data, reminding the reader about the most popular algorithms: BNs, NNs, and SVMs. It is how effectively these work for different types of data.

Besides giving specific suggestions on further directions for future research, the author provides further data sources and other input variables that can be integrated into an ML model to enhance the predictive capability of such models in food safety. This goes along with guiding researchers with some feasible ways forward to develop the field further.

Paper on Validation of ML Models: Presented the case of model validation for the predictions from the ML models to have reliability and accuracy. All the studies reviewed here had achieved very high prediction accuracy; therefore, it proved that ML was of high potential for application in food safety.

ML Algorithms with Unstructured Data: Wang et al. (2022a) reported that various types of NNs are the most frequently used ML algorithms for data analysis in unstructured data. For example, CNN is used in image data analysis. RNN is used in text data analysis. Recently, Zhou et al. (2024) reviewed ML algorithms applied to the analysis of food safety-related unstructured data. The interested reader is referred to that review for more detailed information (see Wang et al., 2022a).

Design of the AHP-ELM ML model was focused on developing a predictive and optimization model for food safety inspection. For the suggested model, the input layer, output layer, and the number of hidden layer nodes were set to 11, 3, and 20, respectively. ELM is one form of NNs. Analyzed through the suggested model, the food safety inspection data feature complexities that include being highly complex, high in dimension, nonlinear, and discontinuous. Training time and accuracy are validated with 10% randomly selected unused data. The training takes 468 ms, which is about 10 times faster than the ANN technique, and the relative generalization error is as low as 0.5%, nearly 20 times faster than the traditional ANN method. According to recent updates, in 2018, an ELM model was studied with the objective of predicting food safety hazards in dairy products. In order to show that the prediction accuracy of the ELM network is up to 86%, the model is going to be tested using the remaining unusable half of the same dataset.

9.1.1 Support Vector Machine

Using separators of "large margin," which simply refers to distance to the nearest data in that class, categorize new samples into one or the other class to build up an SVM model.

Training Data: Two equal-sized datasets set up as a half-space or hyperplane and create two well-separated areas which have significant margins. The better hyperplane in the SVM has a lower functional margin—larger margins are connected to lower generalization errors—the SVM is thus a nonprobabilistic binary linear classifier. Direct decision functions that were originally created for binary classification but were subsequently generalized to multiclass by a number of different authors, are utilized by SVM. It does well even for high-dimensional issues, since it can result in lower sample complexity if the model algorithm is constrained to produce an extended margin separator.

9.1.2 Decision Tree

A decision tree (DT) is a tree-like model in which every internal node represents a feature or attribute, each branch represents a test on this attribute, and each leaf node represents the class to which an instance is assigned based on the outcome of the test. Each node is a class label, and each node is associated with a branch to represent the relationship of an attribution as well as each internal node to a "test" of an attribute. The paths from the root to the leaf represent the classification rules. A DT is a supervised model for both regression and classification problems and

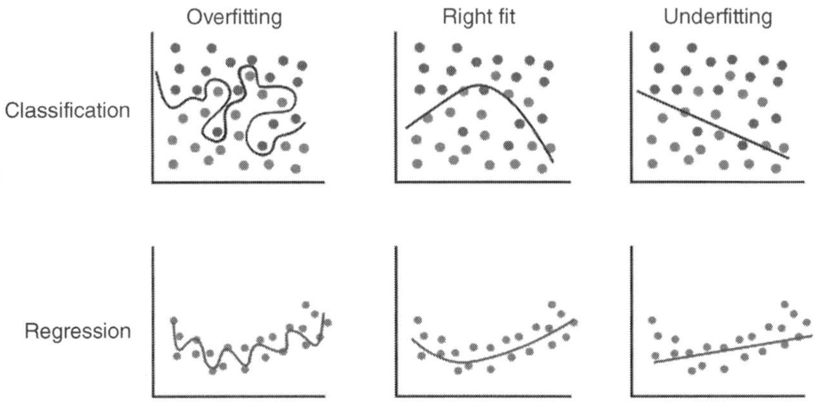

Figure 9.3 Comparison of the overfitting, underfitting, and right fitting [https://www.mathworks.com/discovery/overfitting.html]/The MathWorks, Inc.

can be easily understood. Although it delivers the most optimal result, it doesn't necessarily give the globally optimal solution. The tree structure may turn out to be very complicated when training with complex datasets. Results indicate that the result obtained from DT is very accurate, that is, 98.8%. An RF is an ensemble method of learning that is based on a number of DT classifiers. In model training, it builds a number of DT over different subsets of the dataset, and then outputs the class that is the average prediction of the individual trees. An RF is intended to minimize over-fitting-see boxes. In ML, overfitting happens when the algorithm fits too closely or exactly to its training data; hence, in this case, a model can predict or conclude anything other than the data used for training. As an illustrative example, please see the comparison of overfitting and underfitting, as depicted in Figure 9.3.

9.1.3 ML Algorithm Validation

Model validation in the ML paradigm refers to the process of testing the performance of a trained model using a test dataset, which is an independent data collection used in training the model. Model repeatability and model transportability are what these translate into. For instance, since internal validation, through the characterization of a model's reproducibility by applying the model on unused data from the same target population, can be implemented, model evaluation becomes a combination of both. External validation characterizes the transportability of the model by using underused data from other but related populations. In this case, the mix of training and test data is very similar; external validation also explains how to achieve reproducibility. If cross-validation is applied, the performance of the model is tested both on the test and full datasets. Techniques for this task include bootstrapping, k-fold cross-validation, and random subsampling. Subsampling at random consists of a choice between a training set and an independent test, or validation, set; the dataset will be randomly split into two partitions (e.g. 80/20). In k-fold cross-validation, the training set is partitioned into k folds. Aside from the kth fold, the model learns on every other k-1 fold. Then validates, that is, uses the k-fold to compute the error in its outputs. Repeat the experiment k times, with k changing its value each time. The accuracy of validation of the supplied model will be defined as the average of those validation accuracies. The possibility of information leakage could be an issue. The researcher must select whatever knowledge leaks from the test data to the training set or from the future to the past so that the model does not perform poorly or overestimatedly. Most of the selected articles use internal validation, which is similar to random subsampling validation or cross-validation.

Besides the generalization ability of the model through the work of the ML, it is in the validation stage that distinguishes between the ML models and the statistical models. As much as the statistical models are conceived for inferences, both prediction and classification models have evolved through ML models. It is difficult to determine where exactly the borderline goes between the statistical and ML approaches. However, if we exclude statistical models from our references, like Bayes inference, since it entails the validation step of the ML model as a criterion of selection.

9.1.4 Risk Prediction Models Using ML

Another study by Geng et al. (2022) proposed the development of an advanced random forest prediction method together with the Monte Carlo algorithm for food safety risk prediction to secure and reduce food safety risk. The Monte Carlo method is able to simulate virtual samples using data from very small sample sets, and an extended input into the structures of method types of the random forest approach can be facilitated. A new risk prediction model for food safety is proposed to predict food danger levels and ensure worker safety. Finally, the sterilized milk data is used to validate the accuracy and efficacy of the suggested approach. Experimental data indicate that the newly developed risk prediction model can outperform the state-of-the-art in terms of prediction accuracy and stronger generalization capacity for efficient risk early warning. In addition, the proposed model could offer technical support and decision-making support to the relevant departments to prevent and control the risk event of food.

A new food safety risk prediction model, as represented in Figure 9.4, is proposed based on the random forests integrating Monte Carlo, namely the MC-RF, which will better assess and forecast any food safety hazards and safeguard consumers' lives. The Monte Carlo approach expands the input and separates it into the RF

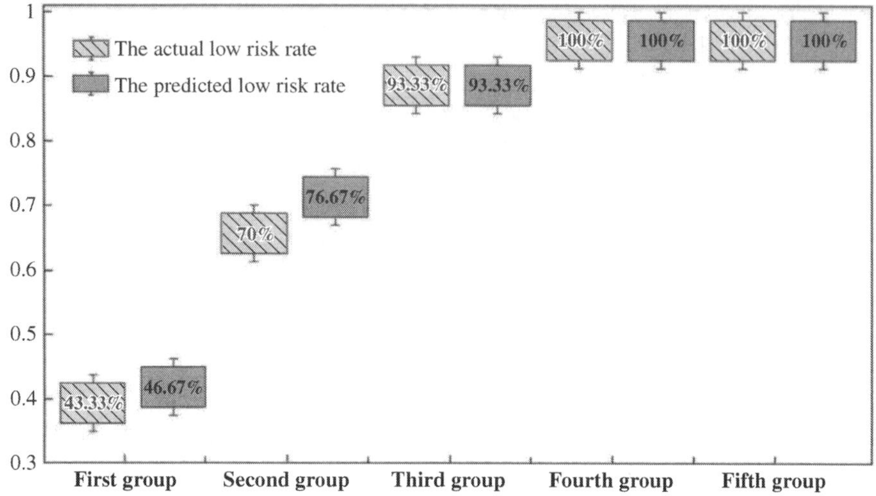

Figure 9.4 The safety of sterilized milk changes with time (Geng et al. [2022]/Elsevier).

model's training and test sets. Lastly, the data from actual sterilized milk samples is used to apply the suggested revolutionary food safety risk prediction model.

Products and Procedures

The MC algorithm, which is a numerical approach to solving structural reliability based on stochastic simulation and statistical testing is also known as the random sampling method/probability simulation method, the numerical simulation method, or the statistical test method. In order to calculate the statistical properties of the parameters, the MC mostly begins with the creation of the probability model or a sampling test. Ultimately, a computer simulation yields an approximate solution.

The basic steps of using the MC to expand data are shown as follows:

1. Select a small sample $X = [X_1, X_2, \ldots, X_m]$ (m is the number of attributes) to be expanded.

$$X = \begin{bmatrix} x_{11} & x_{12} & \cdots & x_{1m} \\ x_{21} & x_{22} & \cdots & x_{2m} \\ \vdots & \vdots & \vdots & \vdots \\ x_{n1} & x_{n2} & \cdots & x_{nm} \end{bmatrix}$$

where n stands for the number of samples.

2. The correct probability model is built, along with the average of the attribute column and the covariance (Cov) of the real data. $[\text{avg}_1, \text{avg}_2, \ldots, \text{avg}_\text{m}]$ *yieldsAvg*. are calculated (avg a_i is the arithmetic mean of the observation in each column), so that the main characteristic parameters of the built model coincide with the real problem.

$$\text{avg}_i = \sum_{i=1}^{n} x_{ij}/n, j = 1, 2, \ldots, m$$

$$\text{cov}\left(X_i, X_j\right) = \sum_{k=1}^{n} \left(\left(x_{ik} - \text{avg}_i\right) * \left(x_{jk} - \text{avg}_j\right)\right)/m - 1$$

$$\text{Cov} = \begin{pmatrix} \text{cov}(X_1, X_1) & \text{cov}(X_1, X_2) & \cdots & \text{cov}(X_1, X_m) \\ \text{cov}(X_2, X_1) & \text{cov}(X_2, X_2) & \cdots & \text{cov}(X_2, X_m) \\ \vdots & \vdots & \cdots & \vdots \\ \text{cov}(X_m, X_1) & \text{cov}(X_m, X_2) & \cdots & \text{cov}(X_m, X_m) \end{pmatrix}$$

where k is the value of the observed variable in each column.

3. Generate random numbers based on the distribution of the random variables in the relevant probability model.
4. The designed and selected sampling method for appropriate sampling, every random variable sampling, and statistical analysis simulation test results generate the virtual samples.

MC is not sensitive to dimension problems and is not constrained by any assumption, and its solution is a relatively accurate solution. So the MC has strong applicability and problem-solving ability and can better solve small samples.

9.1.5 The RF Algorithm

The RF is the combinatorial classifier technology introduced by Breiman after the bagging algorithm, which is an ensemble learning algorithm. Multiple DTs were trained and predicted samples. There are many DTs in the composition of the RF, and each DT is unrelated to each other. By employing a voting method for the DTs about which class the example should have, the class with the most votes will be selected as the final classification result. Key to DT learning is to choose the optimal partitioning attributes and strive for higher and higher "purity" of the nodes with progressive partitioning. The base partition index consists of three elements: information gain, information gain rate, and Gini index.

9.1.5.1 The Information Entropy

The information entropy $E(D)$ (where D is the sample set) is a widely used metric for assessing the purity of a sample set, as defined in Eq. (9.1):

$$E(D) = -\sum_{k=1}^{|y|} p_k \log_2 p_k \tag{9.1}$$

Here, $\mathbf{p_k}$ denotes the fraction of class k inside the sample set. When $E(D) = 0$. Sample set D comprises a singular category. A lower value of E indicates a higher purity of the sample set.

Information gain $\text{Gain}(D, A)$ is the difference in information entropy before and after sample set D partition, namely, the value by which the purity increases. We suppose that the attribute A has m possible values $\{a^1, a^2, \ldots a^M\}$. If the sample set D is partitioned according to attribute A, the information gain is obtained.

$$\text{Gain}(D, A) = E(D) - \sum_{\text{m}=1}^{M} \frac{|D^m|}{|D|} E\left(D^m\right) \tag{9.2}$$

Let M be the number of branch nodes produced after the division of the sample set D. D^m comprises all samples within the set D which have the value A^m for attribute A. The larger the gain in information, the more critical is the better purity.

9.1.5.2 Gain Rate

A large number of values tend to be favored by the information gain ratio (Gain ratio (D, A)). This preference, when comparing predictive models, could provide hazards; thus, it should be tested to confirm the MC-RF model's performance.

NN techniques comprise BP and ELM. The number of nodes for the hidden layer affects both BP and ELM's performance. Therefore, having too many or too few hidden-layer nodes could affect the performance of the technique. In general, two key guidelines are to be applied when one selects the node count for the hidden layer for a NN:

1. Let the number of hidden layer nodes be L and the number of training samples be D, then $L < D - 1$.
2. Let the neural network have n inputs and m outputs, then $L = \sqrt{m + n} + \partial, \partial \in [1, 10]$. Therefore, $L < m + n + \partial$.

In this section, in the sterilized milk data, input and output numbers are 9 and 1, respectively, so the number of hidden layer nodes is less than 20. For the small interval of the number of hidden layers, taking the enumeration method to get the optimal number of hidden layer nodes of the BP and the ELM. Figure 9.4 depicts the optimal selection of hidden nodes for BP and ELM.

It is obvious that the BP effect is the best when the number of nodes of the hidden layer is 10, and the effect of the ELM is the best when the number of hidden layers is 14. Thus, the two algorithms have optimal numbers of nodes of the hidden layer: 10 and 14, respectively.

This article performs a comparative experiment to verify the reliability and validity of MC-RF in predicting food safety risk by separately calculating the accuracy, the macro-F1, and the micro-F1 values of the performance evaluation indexes for MC-RF, MC-SVM, MC-BP, MC-ELM, MC-CNN, RF, SVM, BP, ELM, and CNN. In comparative tests, all the nodes in the hidden layers of BP and ELM algorithms are optimal.

The MC-RF, MC-SVM, MC-BP, MC-ELM, and MCCNN express the accuracy of the food safety risk prediction model integrated with the MC technique. In order to decrease the production of negative effects, a gain rate is suggested. The phrase "gain rate" has been used when choosing an ideal partition property.

$$\text{Gain\_ratio}(D, A) = \frac{\text{Gain}(D, A)}{IV(A)}$$
$$IV(A) = -\sum_{m=1}^{M} \frac{|D^m|}{|D|} \log_2\left(\frac{|D^m|}{|D|}\right) \tag{9.3}$$

9.1.5.3 The Gini Coefficient

The Gini index Gini(D) represents the probability that a randomly selected sample in sample set D is wrongly divided. The Gini(D) formula is shown in Eq. (9.4):

$$\text{Gini}(D) = \sum_{k=1}^{b1} \sum_{k} \sum_{k} \text{p}_k P_k = 1 - \sum_{k=1}^{|y|} \text{Pk}^2 \tag{9.4}$$

Gini_index(D, A) of the attribute A is obtained in Eq. (9.3):

$$\text{Gini\_index}(D, A) = \sum_{m=1}^{M} \frac{|D^m|}{|D|} \text{Gini}(D^m) \tag{9.5}$$

Higher purity is achieved due to a lower chance of sample misclassification within the set, which is associated with a smaller Gini index. The best partition attribute is usually selected as the one with the smallest Gini index after partitioning. Although there are many DTs in the RF, overfitting does not occur as the number of trees increases. This is mainly due to the use of the law of big numbers in RF. The main procedure is as follows:

Suppose a series of classifiers $h(x, \theta_1), h(x, \theta_2), \ldots, h(x, \theta_k)$ are given. Randomly select training samples, and set X as the sample vector and y as the classification label vector for the correct classification. Then, the marginal function is defined in Eq. (11).

$$m_g(X, Y) = av_k I(h_k(X) = Y) - \max_{\zeta Y} a_k l(h_k(X) = j) \tag{9.6}$$

Here, $I(\cdot)$ is an indicator function and av($\cdot$) represents the average. The marginal function serves to give a measure of how many more votes X might get right over wrong as opposed to wrong, as opposed to any other wrong classifications. The $\mathrm{av_k}\mathrm{l}(h_k(X) = Y)$ is the chance that the classification is correct and $\max_{j'\mathrm{Yav_k}} I(h_k(X) = j)$ is the highest probability that the classification is wrong. The bigger the marginal function value, the greater the classification confidence, and vice versa.

During the construction of the RF, when each DT is executed, the original dataset is split into two subsets: the initial dataset and the subset that has not been extracted (denoted as $\mathrm{O}_k(x)$). Additionally, $\mathrm{O}_k\left(\mathrm{x}, \mathrm{y}_j\right)$ represents the proportion of the classification result x in $\mathrm{O}_k(\mathrm{x})$ for the input random vector x, as shown in Eq. (9.7).

$$Q_k\left(x, y_j\right) = \frac{\sum_k I\left(h_k(x) = y_j(x, y) \in O_k(x)\right)}{\sum_k I(h_k(x), (x, y) \in O_k(x))} \tag{9.7}$$

In this sense, the numerator is the sum of samples that the k th decision tree correctly classifies with its corresponding unselected dataset, whereas the denominator is the sum of data samples in all unselected datasets. The more base classifiers are, the more θ_k and the generalization error PE^* for all sequences are close to the following:

$$\begin{aligned} &P_{X,Y}\left(P_\theta(h(X,\theta) = y) = y - \max_{\zeta YY} P_\theta(b(x,\theta) = j) < 0\right) \\ &PE^* = P_{X,Y}(mg(X,Y) < 0) \end{aligned} \tag{9.8}$$

As the number of classifiers increases, the generalization error will not increase infinitely, but rather become finite. With the growth in decision trees, overfitting will not be observed in the random forest.

9.1.6 The Structure of the MC-RF Model

The following steps may explain the structure of the MC-RF model:

Step 1: Data augmentation
The MC presents information about food safety.

Step 2: The data augmentation process was okay
The hypothesis test has proven that the data extension gained from the MC is good, since it is within the reliable limits.

Step 3: Training the MC-RF model
The expanded MC dataset was divided into predetermined ratios to train the RF model.

Step 4: Testing of the model
With performance measures—macro-F1 and micro-F1—the suggested model was correctly established in relation to its reliability. Its performances are compared and ranked against these risk prediction models: BP, ELM, RF, SVM, MC-SVM, MC-BP, and MC-ELM.

The MC-RF model conducts a risk prediction on the risk based on data taken from empirically sterilized milk from China. The performance of this MC-RF model risks is tested; its outcome indicates that it yields an early risk warning for potential sterilized milk hazards. Figure 9.5 depicts the flow chart of the random forest classifier.

Data Analysis and Processing: This chapter uses the food safety data of a brand of pasteurized milk sampled by a domestic institution in January 2016 for the risk early warning analysis. Generally, food safety data possesses high-dimensional and nonlinear characteristics. There are multiple indexes in detection, with high timeliness of processing. Primary indexes for detection are nine, including protein, fat, sm non (non-fat milk solid), acidity, mercury, lead, arsenic, chromium, and aflatoxin m1 in sterilized milk. If some aspects lack data, the accuracy of data-driven models toward training data is reduced. The diluted expansion of sterilized

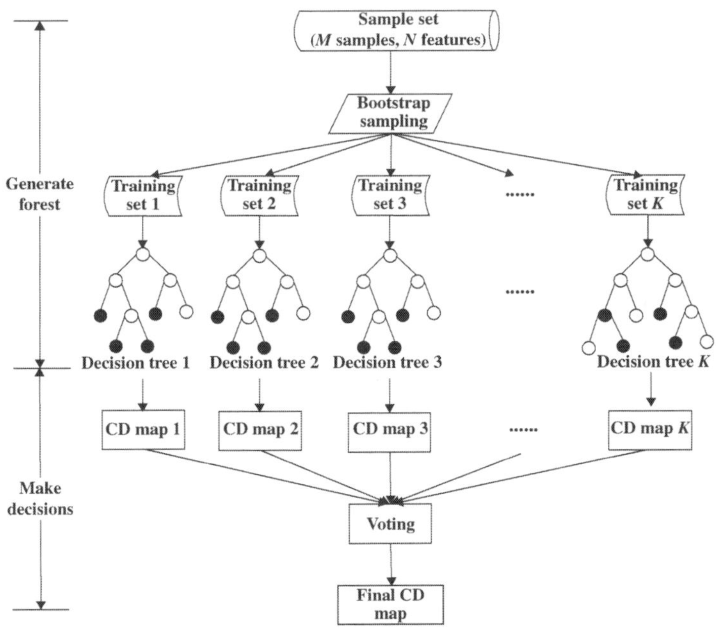

Figure 9.5 The flowchart of the RF classifier (Feng et al. [2018]/Taylor & Francis).

milk data was done through MC. In the spot-checking process, different heterogeneous data might be involved with detection data in spot checking that results in data inconsistency, abnormal data, and also inconsistent detection index units, meaning that this needs to preprocess detection data in the expansion for convenient use later in experiments. In a special situation of missing data 0, digit or modes are used to fill and replace the missing data. Data in the abnormal range is first understood and then corrected according to the three-sigma principle. Then, existed in the work of data such as blank space, special symbols such as "—", "?" and other nonnumeric types; they are deleted and 0 is used to replace them. Then, risk values of sterilized milk have to be worked out and the divided risk levels calculated. A technique called GRA has been employed to determine weight values of risk indices within an enlarged dataset. With regard to the formula for the difference in the sterilized milk index proposed, the values of the enlarged data index were calculated based on the national standard through the application of the difference formula and were multiplied by their respective index weight in order to add up in the acquisition of the last value for food safety risk. At last, under the five scale method, comes the risk value on average, it is divided into five grades of I–V with (0.0626, 0.1708), (0.1708, 0.2789), (0.2789, 0.3871), (0.3871, 0.4952), and (0.4952, 0.6034). The higher the level is, the lower the risk. The virtual samples and food safety risk levels are taken as inputs and the output of the RF to build up the risk prediction model of the sterilized milk. In general, 70% of the virtual samples are selected as the training set to train the risk prediction mode. In addition to the MC-RF model, the MC-SVM, MC-BP, MC-ELM, RF, SVM, BP, and ELM risk prediction models are constructed and evaluated against the MC-RF model to demonstrate the efficacy of the MC-RF model (MC-RF model: Multiclass Random Forest model; MC-SVM: Multiclass Support Vector Machine; MC-BP: Multiclass Backpropagation Neural Network; MC-ELM: Multiclass Extreme Learning Machine; RF: Random Forest; SVM: Support Vector Machine; BP: Backpropagation Neural Network; ELM: Extreme Learning Machine). The U-test is a nonparametric test used to evaluate assumptions regarding whether two independent sequential data samples originate from the same distribution. If the nihilistic hypothesis is not refuted, it can be asserted that the two samples do not differ in the dependent variables to calculate the U statistical value. In a group design, when comparing the means of two samples, it is necessary to determine the standard error of the difference between the two sample means. Consequently, the representation of the statistic U in the calculation is as follows:

$$u = \frac{|\overline{X}_1 - \overline{X}_2|}{S_{\overline{X}_1 + \overline{X}_2}} = \frac{|\overline{X}_1 - \overline{X}_2|}{\sqrt{s_1^2/n_1 + s_1^2/n_2}} \tag{9.9}$$

where $\bar{X}_*$ is the mean of the sample, the n_* is the number of samples, and the S_* is an estimate of the standard deviation. Therefore, assuming that when HO: significant >0.05, there is no significant difference in the distribution of each indicator in the two samples.

As shown, the significance of each indicator is greater than 0.05. Therefore, it is suggested to retain the original hypothesis, that is, the original data and the extended data belong to the same distribution, and the extended data are valid. The relationship between the real sample data and the extended virtual data.

9.1.7 The Evaluation Index of the Food Safety Risk Prediction Model

The evaluation index for generalization performance is able to quantitatively determine the generalization ability of the food safety risk prediction model. The multiclassification problem under discussion becomes a binary problem and follows the construction of the confusion matrix. For TP, it refers to the number of samples correctly indicated to be positive examples. For FN, it is defined as the number of samples wrongly classified as a counterexample. The false positive (FP) is the number of samples in which negative examples are predicted as positive examples. And the true negative (TN) is the number of samples to predict the negative example. In this section, three model evaluation indexes, namely, accuracy, macro-F1, and micro-F1 are shown in the equations as follows. Testing the validity and reliability of the predictive model for food safety risks.

$$\begin{aligned}\text{Accuracy} &= \frac{\sum_{i=1}^{n} TP_i + TN_i}{\sum_{i=1}^{n} TP_i + FP_i + FN_i + TN_i} \\ \text{macro-F1} &= \frac{1}{n} * \sum_{i=1}^{n} F1_i \\ \text{F1}_i &= \frac{2 * P_i * R_i}{P_i + R_i}\end{aligned} \tag{9.10}$$

$$R_i = \frac{TP_i}{TP_i + FN_i}, P_i = \frac{TP_i}{TP_i + FP_i} \tag{9.11}$$

Accuracy is the proportion of samples that were classified correctly among all samples or the classifier's accuracy in classification. The closer it gets to 1, the higher the accuracy in classification. Macro-F1 average value of F1 score over all classes. The value ranges between [0,1] and the closer, the more accurate will be the classifier's performance. P = Precision Ratio denotes the proportion of actual positive samples against all those samples predicted as positive examples.

R = Recall Rate denotes the proportion of the number of samples which have been predicted as positive examples over the total number of samples which are in fact positive examples.

$$\text{micro-F1} = \frac{2^{*}\text{micro-P}^{*}\text{micro-R}}{\text{micro-P}+\text{micro-R}} \tag{9.12}$$

The value of the micro-F1 is between [0,1]. When 1 is taken, the effect is the best. The formulas for micro-P and micro-R are shown in Eq. (9.13).

$$\text{micro-P} = \frac{\sum_{i=1}^{n} TP_i}{\sum_{i=1}^{n} TP_i + \sum_{i=1}^{n} FP_i} \quad \text{micro-R} = \frac{\sum_{i=1}^{n} TP_i}{\sum_{i=1}^{n} TP_i + \sum_{i=1}^{n} FN_i} \tag{9.13}$$

The risk prediction models are compared as a comparative test to verify the performance of the MC-RF model.

Both the BP and the ELM are NN methods, and their performance is related to hidden layer nodes. Therefore, too many or too few hidden layer nodes of the BP and the ELM will have a certain impact on their performance. Generally, the selection of the number of hidden layer nodes of a NN follows two rules: (i) Let the number of hidden layer nodes be L and the number of training samples be D, then $L < D - 1$. (ii) Let the neural network have n inputs and m outputs, then $L = \sqrt{m+n} + \partial, \partial \in [1, 10]$. Therefore, $L < m + n + \partial$.

In this section, the sterilized milk data has 9 inputs and 1 output. So, the number of hidden layer nodes is less than 20. Due to the small range of the number of hidden layers, we use the enumeration method to obtain the optimal number of hidden layer nodes of the BP and the ELM. The BP algorithm performs best when the number of nodes in hidden layers is 10 and, when the number of hidden layers is 14, the effect of the ELM algorithm performs best. Thus, the optimal numbers of nodes in hidden layers of the two algorithms are 10 and 14, respectively.

According to the identified indicator equations and the mentioned problem, calculate the values of the respective accuracy, macro-F1, and micro-F1 performance evaluation indicators of the MC-RF, MC-SVM, MC-BP, MC-ELM, MC-CNN, RF, SVM, BP, ELM, and CNN risk prediction models in this paper to form a comparative experiment to verify the accuracy and reliability of the MC-RF food safety risk prediction model (Geng et al., 2022).

Then, the MC-RF, MC-SVM, MC-BP, MC-ELM, and MCCNN are used to stand for the accuracy of the food safety risk prediction model combined with MC method. The experiments demonstrate that, compared with the accuracy of the MC-RF, MC-SVM, MC-BP, MC-ELM, and MC-CNN, the accuracy of MC-RF is the highest, which is up to 91.63%.

MC-ELM and MC-SVM in AI: MC-ELM and MC-SVM methods are variants of the traditional algorithms for ML that are fortified with Monte Carlo techniques. This technique aims at enhancing performance based on stochastic sampling, randomness, or probabilistic computations.

We can see that the macro-F1 value of the MC-RF is 80.8% higher than other models. It shows that the proposed model has high accuracy in predicting each class in the multiclassification problem, which shows that the model is reliable.

To sum up, the three performance indicators of the MC-RF risk prediction model are higher than those of other risk prediction models mentioned above and the correct attributes of every category are able to be predicted better on multiclassification problems. It can prove that the proposed MC-RF method in the paper has good reliability and accuracy, thus effectively improving the generalization ability of the model of risk early warning and the precision prediction of all kinds of categories when the number of samples is inadequate, as well as providing a more solid theoretical basis and basis for risk early warning analysis.

9.1.8 Early Warning Analysis

For instance, taking 148 sterilized milk original datasets, the MC-RF method-based risk prediction model constructs the ability to predict and analyze potential quality risks in sterilized milk products. Products ranked at risk level I and risk level II are categorized under high-risk products. Items classified as risk level III are termed as moderate-risk items. Products classified with risk ratings IV and V are termed low-risk. High-risk items may be disqualified for many reasons. Results of food safety risk prediction for sterilized milk products.

The prediction accuracy for each risk category in this sample batch is highly notable, as 47 out of 52 low-risk samples were accurately predicted, only 2 medium-risk samples were misclassified, and 25 out of 28 high-risk samples were correctly predicted, with the result that the prediction accuracy exceeds 89% for high-risk samples. Therefore, experimental findings demonstrated that the proposed model revealed the possible dangers associated with sterilized milk. Food products have a specified shelf life, implying that there is an intrinsic linkage between food quality, safety, and time-related factors. This allows for the forecasting and assessment of the hazards that may be associated with the sterilized milk batch test data in combination with time-related attributes. The data are divided into five classes based on the time sequence of the production dates of the tested sterilized milk batches. Based on the temporal factors, as shown in Figure 9.4, changes in the quality and safety of sterilized milk are predicted by the proposed modeling approach. The first three groups have greater potential risks, and the fourth and fifth groups have nearly no potential risks, that is, the sterilized milk tested in the same batch, and the sterilized milk with the earlier date of production have greater potential risks. The reason may be that when the storage time increases, it results in the increase of the proportion of microbial content and contaminants such as aflatoxins present in the sterilized milk after long-term storage. The relevant departments should increase the number and frequency of quality inspections of the sterilized milk to avoid high-risk and substandard products entering the market and affecting the lives and health of consumers, according to this phenomenon.

If the training samples of the sterilized milk are not sufficient, the potential risk of the quality of the sterilized milk can be predicted and warned using the model proposed. At the same time, with the involvement of its time characteristics, high-risk samples together with the time point of the sample can be identified immediately if the high-risk samples appear often. Detection of high-risk samples shall have priority. In addition, by identifying the sources of risk for high-risk samples and tracing them back, food safety risks and unknown dangers can be fundamentally eliminated or reduced as much as possible to achieve all-round management of food safety risks. In this respect, the supervision of dairy production and processing enterprises should be strengthened in such a manner that enterprises should be able to control microorganisms tightly in raw milk and other substances, which can easily lead to risk-generating factors while producing and processing. Companies and merchants engaged in the transportation and storage shall pay attention to the temperature and hygiene of the storage environment and strengthen the control of food safety risks. The departments of sampling inspection relating to food should carry out regular sampling inspection and trace its source promptly according to the results of the sampling to decrease the probability of the occurrence of risks.

And the products with risk levels IV and V are low-risk products. High-risk products may become unqualified products for some reason. The food safety risk prediction results of sterilized milk products are shown in Figure 9.4.

It can be seen that the success rate of prediction at every level of risk in this test sample is significantly high. In the case of the samples that are low-risk, 47 out of 52 samples have been predicted with only 2 cases that failed to be predicted with precision. The number of high-risk samples is at 28, and 25 of these have been predicted, and for the high-risk samples, the accuracy of prediction has been above 89%. The results thus indicate that the proposed model is a better predictor of the potential risk of sterilized milk. The shelf life of the food products implies that, indirectly, some indication can be drawn that there exists some relationship between food quality and safety characteristics with respect to time. For that reason, the risk of the batch of the test data for sterilized milk combined with time characteristics is predicted and analyzed. Divided the data in accordance with the sequence of the production date of the batch of tested sterilized milk into five groups. It is visible from the Figure 9.4, but the changes in quality and safety of sterilized milk under the time factor influence are foreseen by the proposed model.

The first three groups have higher potential risks, and the fourth and fifth groups have almost no potential risks, that is, the sterilized milk tested in the same batch, the sterilized milk with an earlier production date has a greater potential risk. This may be attributed to an increase in storing time, leading to an incomplete sterilization as a result of processing raw milk. The proportion of microbial content and contaminants, such as aflatoxins, in the sterilized milk increases with increasing time of storage of milk, thus increasing the risk levels of the sterilized milk. Relevant departments ought to raise the quantity and frequency of good quality inspections on sterilized milk, so that high-risk and low-quality products should not hit the market and influence the lives and health of consumers caused by this phenomenon.

With fewer training samples of sterilized milk, it can predict and warn the potential risk of quality of sterilized milk by using this model. At the same time with time characteristics, with time period wherein high-risk samples occur, it can be easily detected when high-risk samples happen frequently. Stress should be given in the detection of high-risk samples. By analyzing the causes of high-risk samples and tracing their risk sources, food safety risks and hidden dangers can be fundamentally avoided or reduced as much as possible to achieve all-round management of food safety risks. The government should strengthen the dairy production and processing enterprises so that they may have close supervision of their enterprises, thereby requiring the firms to control microorganisms in raw milk and other substances prone to risks during the production and processing. In transportation and storage, concerned enterprises and merchants must pay close attention to temperature and hygiene conditions in the storage environment, thus strengthening food safety risk control to avoid food safety risks. Food-related sampling inspection departments shall also conduct regular sampling inspection and trace the source in time according to the sampling results to minimize the probability of risk occurrence.

9.2 Sensors for Automated Food Process Control

Tables 9.1, 9.2, and 9.3 represent various studies on fruits, grains, oils, and vegetables using electronic nose.

9.2.1 Methods Adopted

Sensor Construction and Mounting: In-house developed and contracted for fabrication capacitive electrodes were used. A polyamide flexible substrate with interdigitated gold-plated copper electrodes (spacing approximately 0.5 cm) was the sensor and ground electrodes for capacitance measurement in this study. A metal plate placed at the bottom of the capacitive electrodes acted as a Faraday shield that ensured the electric field was directed only upward to the powder, thus

Table 9.1 Various electronic nose devices in fruits, grains, oils, and vegetables.

References	The objective of the study	Data processing algorithm	Main findings
Bayberry (Wu et al., 2019)	Investigate changes in the physicochemical and sensory characteristics of bayberry during growth stages	PCA	The GC-MS and e-nose techniques combined with PCA allowed differentiation of all maturation phases of the bayberry clearly
Grape (Longobardi et al., 2019)	Characterization of the geographical origin or agronomic practice in grape samples	PCA, LDA	The e-nose with the LDA enabled discrimination of grapes based on cultivation method and geographical origin
Mulberry (Aghilinategh et al., 2020)	Detection of maturation degrees of whiteberry and blackberry by e-nose	PCA, ANN, LDA	The e-nose together with ANN allowed evaluation of the degree of maturation in mulberry

Table 9.2 Various electronic nose devices in grains.

Sample (Reference)	The objective of the study	Data processing algorithm	Main findings
Cocoa (Barbosa-Pereira et al., 2021)	Evaluation of volatile fingerprint by GC-MS and e-nose for the classification of cocoa bean shells	PCA, SCA	The GC-MS and e-nose techniques combined with PCA allowed clear separation of some cultivars and geographical origins
Rapeseed (Rusinek et al., 2020)	Evaluate the quality of rapeseed during storage after harvest	PCA	The proposed method discriminated between rapeseed samples infected with field and storage microflora
Rice (Du et al., 2020)	Discrimination of the geographical origin of rice based on mineral elements and volatile components	PCA, LDA	The LDA showed better discrimination of rice samples when ICP-MS and e-nose data had data fusion

eliminating any electrical interference from outside sources that may interfere with the readings. A plastic container of standard size placed above the electrodes in a slot ensured uniform placement of the container throughout the experiment. The container with several powders weighing 10 g each was checked noninvasively for mixing and caking.

Table 9.3 Various electronic nose devices in vegetables.

Sample (Reference)	The objective of the study	Data processing algorithm	Main findings
Carrot (Gaggiotti et al., 2019)	Check the changes in the carrot aroma profiles stored for different periods and temperatures	PCA	The array of sensors developed proposed results capable of tracking the relevant changes in the aroma of carrots
Maca (Li and Sun, 2020)	Identification of maca origin using electronic nose coupled to GC-MS	BP-ANN, OPLS-DA	The e-nose and the BP-ANN algorithm made it possible to trace the Chinese Maca samples' origin accurately
Spinach (Galvan et al., 2022)	Evaluate the application of the e-nose to detect the shelf life of fresh food	SVM, BP-ANN	Based on the results, the e-nose can predict the freshness degree of spinach

9.2.2 Capacitance Measurements

Images of the sensing device used in this experiment for powder mixing and caking analysis mohan2022novel A Protocentral FDC1004 (Frequency to Digital Converter) breakout chip from India is used to produce an AC excitation signal with a frequency (f) of 25 kHz and a peak-to-peak voltage ($V_{p\cdot p}$) amplitude of 2.4 V for the real-time measurement of powder capacitance. The FDC1004 is a high-resolution, 4-channel capacitance-to-digital converter designed for capacitive sensing applications that measure the capacitance of an electrode relative to ground. This breakout board can be connected to Arduino's Inter-Integrated Circuit (I^2C) or a comparable interface via four connection pins: ground (GND), serial clock (SCL), serial data (SDA), and voltage common collector (VCC) (5 V to VCC, GND to GND, A4 to SDA, and A5 to SCL). The capacitance measurement range is $\pm$15pF with an offset capacitance of maximum 100 pF and resolution 0.0005 pF. FDC1004 was programmed using the Arduino IDE through Arduino Pro Mini, 3.3 V, 8 MHz, Sparkfun, USA, to acquire the data. To connect the FDC1004 breakout chip with Arduino, an FTDI basic breakout (3.3 V, 8MHz, Sparkfun, USA) including the FT232RL was utilized for the USB to UART interface, facilitating asynchronous data transmission. When interfaced with a microcontroller such as Arduino, the FDC1004 exhibits minimal noise of $\pm$0.001 pF. The baud rate was set at 9600, and data acquisition intervals of 50 ms were used for the immediate readings while 5 minutes were used for prolonged readings. RealTerm was an open-source serial terminal application, which was applied to log data. The sensor, DHT11 (Adafruit Industries, USA), was utilized in measuring temperature and humidity coupled to an Arduino Pro Mini. In a parallel finger configuration, the two electrodes of the flexible patch-acting sensor and ground-were utilized to produce a strong fringing electric field whose penetration depth was adjustable in order to scan the base and the contents of the container. The different powders used in this experiment were Nestle MILOB (M) that consisted of malt barley, milk powder, sugar, cocoa, and eight vitamins and minerals; Nestle Coffee Mate powdered creamer (C) that contained corn syrup solids, hydrogenated vegetable oil (coconut and/or palm kernel and/or soybean), sodium caseinate, and trace amounts of dipotassium phosphate, sodium aluminosilicate, mono- and diglycerides, artificial flavor, and annatto color; MILO plus creamer (MC); and sweet whey (SW). The ingredients were mixed in the following weight ratios (w/w): 100:0 (M), 0:100 (C), 75:25 (M-75), 50:50 (M-50), and 25:75 (M-25). To demonstrate measurement reproducibility and repeatability for the immediate evaluation of powder mixing and caking, each vial was observed for at least 5 s, and at least 5 samples per powder type were assessed unless otherwise stated. In long-term powder-caking monitoring tests, records were taken at 5 minutes intervals for a period of 2 hours for each type, with at least 5 repeated trials.

Data was taken under room temperature at approximately 23–25 °C and a level of relative humidity of about 70%, except otherwise stated. The objective way to measure sensor performance was obtained by measuring powder capacitance, before to, and following mixing and caking over time. Relative capacitance is given by $\Delta C/C_0$ that measures the difference in capacitance ($\Delta C = C - C_0$) of electrodes contacting the container. Here, C_0 and C refer to capacitance of an empty container and one filled with powder, either well, partly or completely stuck together. The sensitivity of the proposed system is defined as the slope of the relative capacitance ($\Delta C/C$) in terms of the mixing/caking percentage. The limit of detection (LOD) is defined as $\frac{3\sigma}{s}$, where σ is the standard deviation of the blank container readings. A linear fit $y = a + bx$ (where b denotes the slope s and a signifies the y-intercept) to the calibration curve yields the coefficient of determination R^2.

9.2.3 Sensor Characterization

The capacitance of the electrodes in an empty container was measured for at least 4 hours ($n = 109$ data points). The mean ± standard deviation ($\mu \pm \sigma$) of the bare electrode was ±2.160.

The measurements for the blank container were 7.485 ± 0.011pF, with a mean standard error of ±0.001pF and a relative standard error of ±0.013%. The noise of the FDC1004 with an electrode connected (±0.011 pF) increased tenfold as compared to the FDC1004 interfaced only with Arduino and without an electrode (2.160 ± 0.001 pF). This noise originates from external long wiring and parasitic capacitance, which can be mitigated through an integrated design approach that incorporates both the FDC1004 and electrodes on a single PCB substrate to minimize routing, while employing the activated shield driver in the FDC1004 to protect against cable parasitic capacitance from the routing. Notwithstanding the tenfold increase in noise, the capacitive signal response to the empty container is significantly higher (over 200-fold) than the increase in noise levels. The minimal inaccuracy confirms the persistent stability of the capacitance readings throughout the recordings. The introduction of a new dielectric material (powder) in the electrode sensing area significantly increases capacitance, due to the powders' superior dielectric constant compared to air. The capacitance trend correlates with the dielectric constants of the powder constituents, clearly differentiating across powder types: exposed < M < C < SW. The estimated residual standard deviation, σ_{res} ($\pm 0.3\%$), for various blank containers is at least 20 times lower than the multiple powder-induced responses ($15 \pm 5\%$), underscoring the robust capacitive signal for detecting powder presence. The minimal variation among samples justifies the reproducibility and repeatability of the measurements. The two t-tests applied to the hypothesis testing confirm that the results are statistically significant at a level of $p < 0.05$ for the existence of powder (see Figure S1a–c in SI). The sensor showed excellent stability and minimal variability, coupled with a rapid data acquisition rate of 50 ms, which is advantageous for the inline screening of powder kinds poured into bottles from hoppers.

9.2.3.1 Sensing Mechanism and Sensor Response

The capacitance (C) of the system is related to the relative permittivity or dielectric constant ($k = \varepsilon_o \varepsilon_p$) of the material being tested, where ε_o denotes the permittivity of free space and ε_p represents the dielectric permittivity of the powder. In the absence of any powder, ε_p is replaced with ε_a (dielectric permittivity of air, about 1), which has the lowest known dielectric permittivity. Since ε_{r} is always less than that of the material intercepting the capacitive field, the output capacitance in the presence of air is always inferior to that with powder. Figure 9.6 is a schematic diagram of powder mixing working principle, where C, C_{lp}, C_{hl}, and C_{hp} are the capacitances of empty container, filled with low permittivity, ϵ powder (lp), powder mixture (hl), and the container filled with high permittivity powder (hp). d is the distance between the sensor and the powder, which is the bottom wall thickness of the container (approximately 0.2 mm in this experiment). The addition of powder increases the dielectric material's ability to capture the electrostatic field, thus increasing the output capacitance ($C_{lp}, C_{hl}, C_{hp} >> C$). The mixing of two powders leads to a partial dielectric contribution $\left(\varepsilon_{lp} < \varepsilon_{hp} < \varepsilon_{hp}\right)$. From both powders, the following trend is observed: $C_{lp} < C_{hl} < C_{hp}$. For caked powders, the food powders are hygroscopic, absorb moisture that causes an additional dielectric contribution from highly polar water molecules with a relative permittivity of around 80, where ε_r refers to the relative permittivity of water compared to a relatively drier fine powder (ε_p). The net dielectric constant is therefore given by: $\varepsilon_{cp} = \varepsilon_p + \varepsilon_r$. Therefore, the caked powder will have a larger dielectric constant ε_{cp}

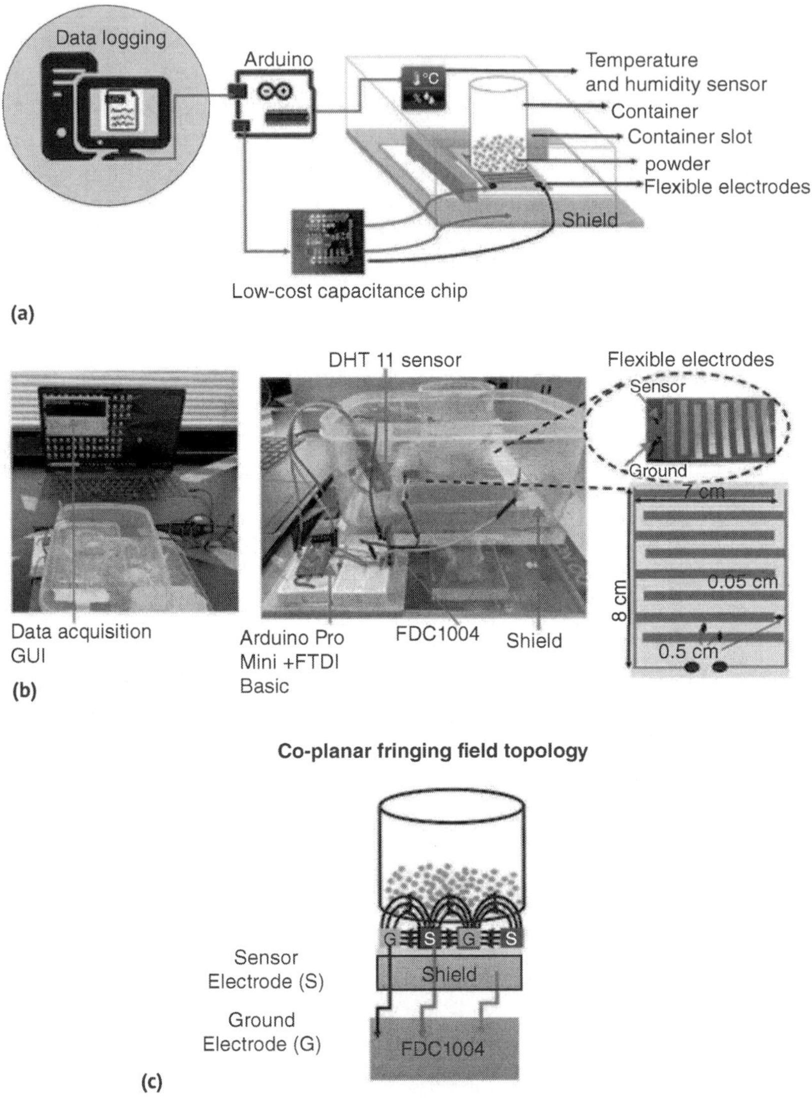

Figure 9.6 (a) Schematic representation of the setup for analyzing powder mixing and caking through capacitive sensing. (b) Image of the capacitive measurement configuration, highlighting the FDC1004, bendable electrodes, holding vessel, powder sample, sensors for temperature and humidity, along with the data collection software utilized in this research. (c) Diagram showcasing the fringing field sensing arrangement (Mohan and Malcolm (2022)/Elsevier).

and, since $C \propto k$, caking yields a capacitive signal that is much larger $(C_{cp} >> C_p)$. The capacitances C_{cp} and C_p are the values for excellent powder and caked powder, respectively. The powder cakes due to agglomeration and reduced airspace between particles, resulting in a greater packing density. The restriction of more particles inside the same electrostatic field improves the capacitive response. Anti-caking agents in the powders make them less hygroscopic; the contribution of ε_r to the total permittivity is minimized: ($\varepsilon'_{cp} < \varepsilon_{cp}$) where ε'_{cp} represents the powder's permittivity with an anti-caking agent). Experiments were carried out to demonstrate the capability of capacitance-based detection to qualitatively and quantitatively assess powder mixing and powder caking. Figure 9.6 represents the setup for analyzing powder mixing and caking through capacitive sensing.

In another study, it was explored how food behaves within vacuum packaging by employing a unique electronic setup, which we detail below. Our methodology leverages wireless sensor networks (WSNs) to facilitate communication between the data acquisition board and the computer, with a keen focus on minimizing power usage, a known limitation within WSN frameworks. The design of our proposed system is bifurcated into two main, cost-effective, and energy-efficient components.

The initial component is dedicated to data acquisition, while the secondary component is responsible for the user interface, storing data, analyzing it, and enabling the remote transmission of findings. The data collected through this system sheds light on the food's condition, its degradation level, and the environment it's stored in. A schematic representation of our system's two components is illustrated in Figure 9.7.

The core of the data acquisition unit is an Arduino board that carries temperature and humidity sensors. These sensors are intended primarily to monitor the status of the electronics within the system, which fits into the case designed with thermal insulation. A separate compartment with gas, pressure, temperature, and humidity sensors is provided for interaction with the product under test. Main elements of this module are:

1. The Bosch Sensortec integrated environmental sensor BME680: This is an all-in-one unit for mobile applications, integrating gas, temperature, pressure, and humidity sensing capabilities all in a single digital package.

   The DHT22 is a digital temperature and humidity sensor and uses a capacitive humidity sensor and a thermistor to sense the ambient humidity and temperature with the data being provided in a digital format.

   This is one in the series of MQ sensors. It has been designed with an electrochemical sensing mechanism in combination with a heating element to detect a large number of gases.

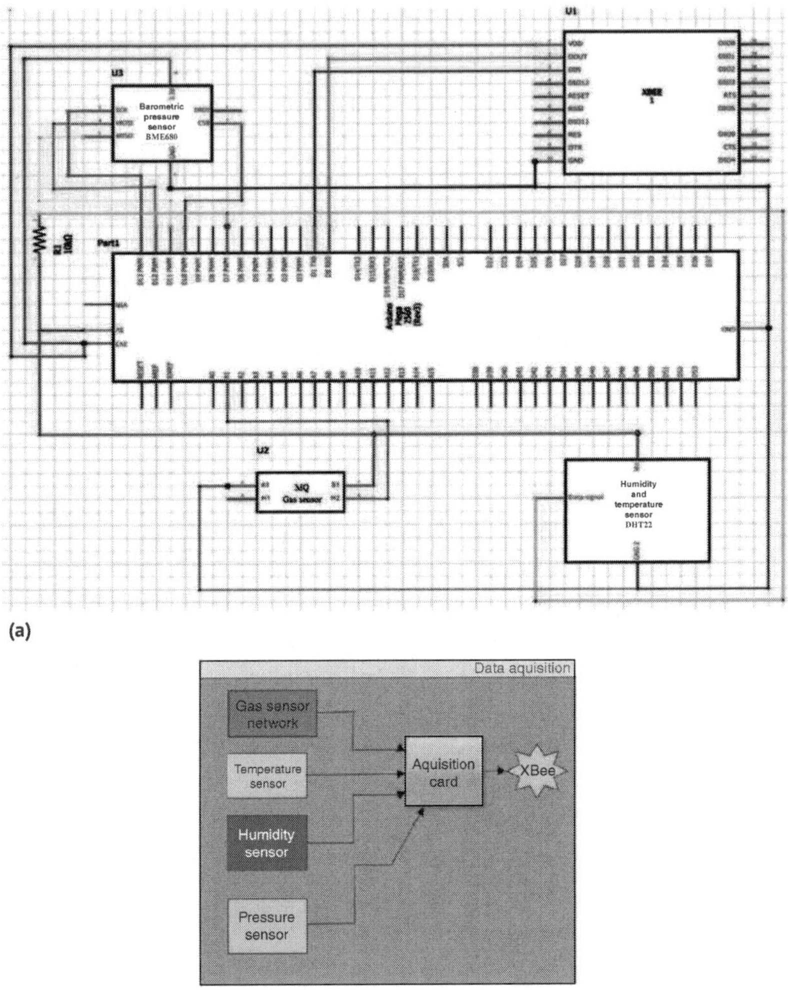

Figure 9.7 (a) Sensor circuit wiring diagram, (b) Block diagram of the vacuum-packed product monitoring system consisting of the modules: the data acquisition module and the man-machine interface.

An Xbee shield adapter is used to ease the integration of a wireless Xbee module along with the Arduino board. With Xbee technology, the Arduino can communicate information to some peripheral device located at a distance extending hundreds of meters, perfectly suited for applications that require fast and trustworthy communication, like Digi XBee 802.15.4 RF modules facilitate.

2. The Arduino Mega 2560 sensor data collection card is renowned for its "ease of programming in an open source framework." The system design consists of two dosing boxes, one of which encases the acquisition board in a thermally insulating shell. This is done to shield the microcontroller from chilliness in what will eventually become the refrigerator. Apart from that, this multiparameter monitoring device can be set up to transmit the output directly into smartphones or tablet PCs. DHT22 sensors keep a track of the surrounding temperature and humidity of the protective case of the microcontroller; they inform users if anything is wrong with this microcontroller. The whole setup runs on a 9V DC battery pack.

The whole process of information gathering is carried out in two phases: calibration of sensors followed by data collection itself. The calibration phase included developing a mathematical model for each of the MQ gas sensors since each interacts differently with the gases, as described. For experimental purposes, a rapid prototype for a data collection interface was created using LabVIEW 2013 on a PC. Its adaptation for mobile phones and tablet PCs, whose extensive use is already well-established plans, is part of future plans. Calibration adjusts the sensors so they can accurately account for readings. There's a drawback to using MQ series sensors, too: they are built to detect multiple gases simultaneously, which tends to give less-than-definitive results if a sample emits more than one detectable gas at once. To allow fresh air to reach the sensor and considering that the sensor supplies voltage Vs, alongside the load resistance $R_L = 20\ k\omega$ and measurement of the voltage Vout across this resistance, let's calculate the clean air sensor resistance parameter R_{SCA} from a specified formula.

$$R_{\text{SCA}} = R_L \frac{V_s - V_{\text{out}}}{V_{\text{out}}}, \tag{9.14}$$

To find the resistance value, R0, of the reference sensor, which corresponds to a hydrogen (H2) concentration of 1000 ppm in fresh air, the following equation can be used:

$$R_0 = \frac{R_{\text{SCA}}}{\text{RatioAir}}, \tag{9.15}$$

where RatioAir is a parameter of each MQ sensor. For MQ5, the RatioAir parameter is shown in Eq. (9.16)

$$\text{RatioAir} = \frac{R_{sCA}}{R_0} = 9.56 \tag{9.16}$$

The sensor output equation is solved using the nonlinear regression

$$ppm = a\left(\frac{R_s}{R_0}\right)^b, \tag{9.17}$$

$$\log(ppm) = \log a + b \log\left(\frac{R_s}{R_0}\right) \tag{9.18}$$

$$ppm = 10^{\log a + b \log\left(\frac{R_S}{R_0}\right)} \tag{9.19}$$

where R_s is the measured internal sensor resistance and ppm is the gas concentration in parts-per-million. R_s is computed from the measured output sensor voltage using Eq. (9.20)

$$R_S = R_L \frac{V_s - V_{\text{out}}}{V_{\text{out}}}, \tag{9.20}$$

The table represented below (The formula obtained for the MQ5 sensor for measuring different gases) presents the formula obtained for the MQ5 sensor for measuring different gases.

Gas	$\log(\text{ ppm }) = f\left(\frac{R_s}{R_0}\right)$
LPG	$\log(ppm) = -2.5279 \log\left(\frac{R_s}{R_0}\right) + 1.8771$
CH_4	$\log(ppm) = -2.5474 \log\left(\frac{R_s}{R_0}\right) + 2.2636$
H_2	$\log(ppm) = -4.2302 \log\left(\frac{R_s}{R_0}\right) + 3.0935$
Alcohol	$\log(ppm) = -4.5008 \log\left(\frac{R_s}{R_0}\right) + 4.8216$
CO	$\log(ppm) = -7.358 \log\left(\frac{R_s}{R_0}\right) + 6.4758$

9.2.4 Applications of Sensors in Automated Food Process Control

A sensor is, fundamentally, a device that senses a range of inputs that could be measured in any aspect, feature, or change that it can sense within its surrounding environment and convert them into a quantifiable electronic signal. Sensed inputs by sensors are indeed a very wide array of environmental factors, including pressure, force, flow, light, heat, motion, and moisture, among many others. The signal produced in response typically takes the form of voltage, current, capacitance, resistance, or frequency. The signal can then be displayed in a form that can easily be read or electronically transmitted across a network for further processing or analysis. Figure 9.8 depicts various sensors for Industry 4.0.

9.2.4.1 Different Types of Sensors

These sensors are divided based on their applications: detecting changes in pressure, temperature, flow, proximity, smoke, gas, alcohol, and many more. Output from these sensors may take many forms, such as voltage, current, capacitance, resistance, frequency, or any other electrical parameter that varies with time. Sensors can be broadly categorized into various groups based on criteria such as physical quantity measured, application fields, principle of conversion, energy domain of the measured quantity, and thermodynamic factors, among others. A classification of sensors by their industrial applications is given in Table 9.4.

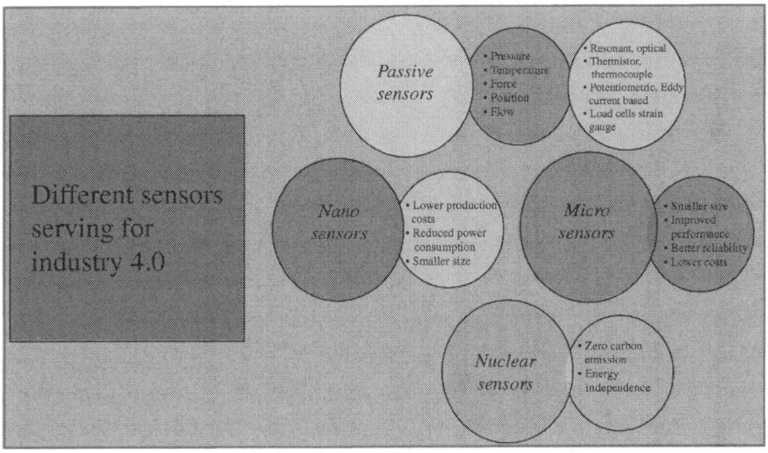

Figure 9.8 Different sensor serving Industry 4.0.

Table 9.4 Classification of sensors in the view of their applications in industries/manufacturing.

S. No.	Types of sensors	Description	References
1	Pressure sensor	Pressure sensors are usually electromechanical devices that measure the pressure of gases or liquids and provide signals to the inputs of control and display devices. Usually, pressure sensors can vary significantly in their technology, design, application suitability, performance, and cost. These sensors are used for control and monitoring in a vast array of everyday applications.	Patel and Doddamani (2019)

(*Continued*)

Table 9.4 (Continued)

S. No.	Types of sensors	Description	References
2	Temperature sensors	A temperature sensor is an electronic device that detects thermal parameters of its environment and converts the input data into electronic data to record, monitor, or signal temperature changes. A temperature sensor typically relies on resistance temperature detector (RTD) or thermocouples to detect temperature changes. A thermometer is the most common temperature sensor used to measure the degree of hotness or coldness. These sensors are frequently used in the automobile sector, aviation sector, medical sector, computers, cooking appliances, and other daily life applications.	Balasubramaniam and Sastry (1995)
3	Proximity sensors	Proximity sensors are electronic devices used to detect the presence of nearby objects without any physical contact. These sensors generally emit an electromagnetic field and look for changes in the field or back return signal. These sensors are used in numerous manufacturing operations to detect the presence of parts.	Sharma et al. (2024)
4	Humidity sensors	Humidity sensors measure the amount of water in the environment and convert these measurements into signals that can be used as input stimulus. The applications of these ranges far and wide and are generally used in the meteorology, medical, automobile, HVAC, and manufacturing industries.	Farahani et al. (2014)
5	Force sensors	Force sensors or load cells are the devices that translate applied mechanical forces, such as tensile and compressive force, and give output digital signals, whose values reflect the magnitude of the force applied. These sensors are used in a wide range of applications such as consumer products, computer devices, musical instruments, medical devices, automotives, sports, to name a few.	Pedersen (1991)

S. No.	Types of sensors	Description	References
6	Flow sensors	Flow sensors are electronic or electromechanical devices used to measure the flow of a fluid such as a gas or liquid and provide output signals to the controller. Flow measurement is vital for controlling many industrial processes, so these sensors have numerous applications in automobiles and medical industries.	Pedersen (1991)
7	Gas sensors	Gas sensors are fixed or portable electronic devices used to sense the presence and properties of a range of gases and relay signals to the controller in the form of an output signal. These sensors act as vital devices to detect harmful gases and provide a vital way to monitor gas concentration and environmental information with many safety concerns.	Wawrzyniak (2023)
8	Flaw sensors	Flaw sensors/detectors are devices used in a diversity of manufacturing processes to show inconsistencies in surfaces or various underlying materials. These sensors/detectors generally use ultrasonic, acoustic, or other means to identify defects in materials. These devices are considered the most crucial aid in quality control during the manufacturing process and inspect components during nondestructive testing and analysis of materials.	Xiaobo et al. (2016)
9	Color sensor	The color sensors are a type of photoelectric sensor, which emits light from a transmitter, and then detects the light reflected from the detection object with a receiver. These sensors help build color sensing applications in image processing, color identification, industrial object tracking, medical diagnosis systems, health fitness systems, industrial process control, etc.	Kryuk et al. (2022)
10	Light sensor	A light sensor is also known as a photoelectric device that converts light energy (photons) detected to electrical energy (electrons). These are invaluable tools in numerous industries. However, different light sensors, such as photo-resistors, photo-diodes, photo-transistors, etc. operate slightly differently and have unique applications.	Ohashi and Karube (1993)

More importantly, sensors are indispensable to increasing the overall market and mitigating a rising labor shortage. Individually, such devices are quite indispensable in industrial settings where they help analyze compressed air and monitor equipment reliant on compressed air. Advanced sensors and predictive maintenance technologies can more precisely detect pressure changes in these systems, which may indicate a looming problem. In addition, these sensors are sensitive to tensile strength and can sense vibrations. This feature makes them very important in automation processes. The application of smart sensors is quite extensive; they can be fitted in ovens and furnaces in order to optimize heating, or fitted in cooling rooms to sense the pre-packaged food temperature and process of packaging. Classification of sensors in Industry 4.0 is expressed in Table 9.4.

Figure 9.8 illustrates the array of sensors that are pivotal in implementing the Industry 4.0 vision within the primary industrial sectors aimed at enhancing community services. The array includes nanosensors, microsensors, nuclear sensors, and the passive variety. Further categorization of the sensors occurs due to their functionalities, which include pressure, temperature, force and positioning, flow control sensors, and more. Significantly, nanosensors, microsensors, and nuclear sensors are the most advanced in this broad area of sensor technology developed so far.

In the field of apparatus for smart factory, manufacturers are manufacturing smart connectors that come with standardized interface connectors along with integrated modules. Such modules have a variety of mounting, blending, and shielding options. In addition, they are designed to integrate power and data in a single module box, optimizing the use of space without diminishing performance. Such integration is very valuable for devices tasked with transmitting data from a sensor or node to a processor, then relaying instructions back to the node or an actuator. For design engineers, the saving in power usage and conserving of space are important factors in their decision-making process. Accurate target location identification, pickup position monitoring, and the strength of the end effector are some detections that are done using sensors with an important role in a distributional force. Some touch sensors can even sense thermal changes. Many industrial applications are based on the use of chemical sensors for fluidic change monitoring or for detecting the changes in air chemistry. These sensors are essential in massive urban areas in order to capture transitions and protect the population.

These gas sensors are applied in the determination of changes that take place in air quality and the identification of various gases. Gas sensors can be applied in various fields, such as control of air pollution, detection of radioactive or fuel gases, monitoring hazardous gas mines in coal mines, and usage in the oil and gas industry, chemical laboratories, packaging, chemicals, rubber, pharmaceuticals, and even pet care, among others.

The concept reveals the existence from cameras to lasers, which is made to slow down the movement of a robot till an employee can reach some area, thereby stopping its activities in small areas, as an organization needs to gear up for the competitive market and must generate profit as well. Therefore, manufacturers need to cut down every cost, leading to operational and maintenance costs. This promotes early detection of problems within the organization, hence minimizing both machine downtime and early replacement of parts with reduced unexpected repairs, thus minimizing downtime and financial impact through process failure by intensive monitoring of the production machinery.

Sensors are applied in industrial applications, such as cooling towers, fermentation tanks, storage containers for fuels, and mixing vessels, to measure solvent levels. They are applied very broadly in different industries. Industrial automation brings a lot of opportunities to deploy wireless sensor networks in modular sensors-based applications requiring scalability, where traditional wiring is quite expensive. The field is further gathering pace in proving the efficacy of wireless sensors in networks. In this regard, the recent trend toward the adoption of wireless technology, which the manufacturing and process control sectors are increasingly adopting, is reflected in this efficacy. Very promising for deployment in tough environments—too, where nuclear facilities, oil and gas installations, and high-temperature furnaces are said to benefit from it—wireless sensor networks are capable of providing sensor data from difficult-to-access locations.

There is a growing need in the manufacturing world for advanced sensor technologies that top sensor manufacturers have been trying to fill. Software developers have also been working on advanced sensor fusion software which merges multiple types of sensor information toward better precision and overall depiction of sensor data. This sophisticated approach allows the manufacturers to create a more comprehensive and clear view of the operations and, by implication, optimize the standards of efficiency, safety, and health. Modern factory equipment is fitted with smart sensors that monitor operational parameters like temperature, which are then reported to the control center. This meant that these intelligent sensors, which contained on-board processors and memory, were brilliant at recording and analyzing data to provide insight.

9.2.5 Robotics and Automation for Food Packaging and Handling

The food industry covers various industrial activities, such as the production, processing, packaging, distribution, preparation, preservation, and service of food. For a long time, the food industry was perceived to be low-technology. In recent times, however, it has experienced an enormous spurt of technological integration, testified by the rising share of research and development expenditure in its sales. As a result, the growing aged population and workforce shortages in countries like Japan have further put pressure on the industry to seek more and more automation to sustain profit margins. The requirement for further automation has been further compounded by the pandemic from COVID-19, where the message of automation at every stage of food production has been boosted.

There are some operations in the food manufacturing industry requiring direct contact with the final products. For the production and processing of such operations, specific machine equipment is desired. Rice cookers, dumpling machines, and automatic chocolate molding equipment are some examples. Such machines operate continuously over a period without change, so adjustments or changes are not required. In general, manipulation of food material is not required for this type of equipment unless it involves packaging. Since packaging often involves pick-and-place operations as well as significant transport of food products through various processing stages, manipulation end-effector-equipped robots with sensors such as cameras to handle pick-and-place operations, are very common in foodstuffs manipulation for efficiency and flexibility.

Robotic end Effectors: To counter the diversity and variability in food product characteristics, there is also a growing need for developing a variety of robotic end-effectors. As Chua et al. (2003) stated, lying at the heart of the barriers that hinder the swift integration of robots in the food industry is the lack of efficient robotic end-effectors. It is challenging to design appropriate grippers for robots to grab food products with both hands of the gripper because food objects are generally soft, have irregular surfaces, various shapes, and must be hygiene-compliant. In addition, the robotic end-effectors should be able to address wetness and stickiness on surfaces of the food. These end-effectors should, therefore, have simple motion, operate at a rapid pace for efficient cycle time, and be cost-effective but hygienically designed per EHEDG 2018.

Food Recognition: Challenges concerning the identification of food items, in reality, are pretty different depending on the context in which one is dealt with. In the food manufacturing industry, situations for identifying food come in two major kinds: (i) situations where food items are correctly positioned or distributed along a conveyor belt with no overlap, and (ii) situations in which food items are randomly scattered about inside a container sometimes overlapping, sometimes tangling with each other. The latter situation is referred to as the RBP scenario. In the first scenario, food items or materials can be spread over a flat surface so that the required identification is possible based on traditional 2D imaging techniques exploiting color or a distinctive pattern. Only the location and orientation of the food item on a 2D plane need to be determined. In food production sites, automated robotic systems used for such applications include robots that can pick pizzas or package powdered products based on the pattern recognition feature. In the case of RBP, recognition of food is difficult because the objects overlap and lie in three-dimensional space, which requires the definition of its position as well as orientation in three dimensions. The identification of multiple food items in either

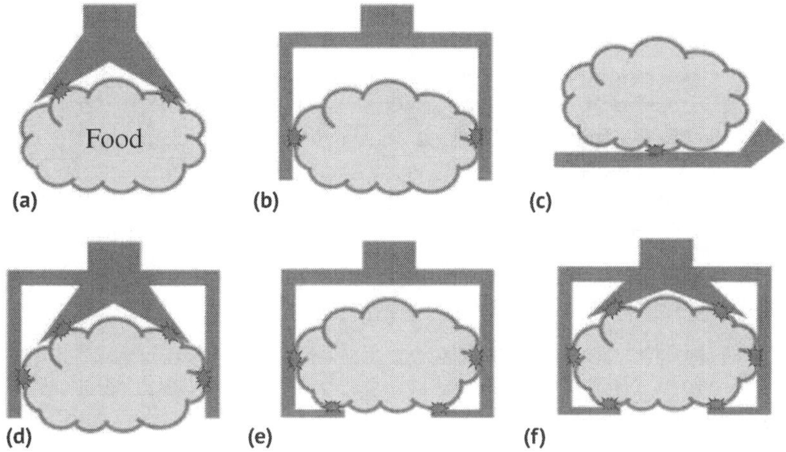

Figure 9.9 Various robotic end-effectors and where they are handled, surfaces that are (a) top, (b) side, (c) bottom, (d) top and side, (e) side and bottom, and (f) top, side, and bottom. Contact positions are indicated by star markings.

a 3D or an RBP scenario remains a challenging problem. In figure, the above-mentioned different robotic end effectors have been represented with their handling positions represented in Figure 9.9.

Grasping at the Top Surface: As illustrated when referring to the action of picking up objects consisting of foodstuffs from their topside as shown in Figure 9.9, conventional gripping systems, in general, can be categorized into four basic types: (i) suction cups, (ii) Bernoulli principle-based grippers or Coanda effect-based grippers, (iii) adhesive force-based tools, which either act with a rolling mechanism or which immobilize the moist outer layer of the object, and (iv) needle-insertion type grippers, which have been studied. Suction cup grippers, which are vacuum pressure-operated, have advantages in that they arc designed relatively simply and function very reliably. However, they are best suited for products with flat, even, and dry surfaces and may mark or bruise raw items. These disadvantages led to the evolution of grippers using the Bernoulli principle or Coanda effect. Though noncontact designs that don't touch the food are designed, they can cause instability in the grip and unwanted spinning because of fast air flow. Adhesively gripped products are normally those with intrinsic stickiness, such as dough, or those with moist surfaces, like fresh fish. Grippers that work by freezing the product in place also need a means to release the product, either by cutting or heating.

Grasping at the Side Surface: This category focuses much attention on grippers and is used the most in food industries. In order to successfully grip an object's lateral surfaces, the dimensions of the object must be known, and the gripper should have an adequate range of motion. The friction force basically plays a determining role in the effectiveness of the grip, and therefore, there should be enough friction to preserve a good hold. This particular grasping technique has an advantage in being more accurate, since the object is constrained within the gripper, and it is easy to align the object. Some of the classic two-fingered and multifingered grippers that are used within manufacturing automation were directly transferred to food processing when the changes ensure that they are food-grade and the products cannot get bruised. One of the best examples is the food gripper provided by SCHUNK, which offers customized compliant components and gripping mechanisms for the requirements of the food industry (SCHUNK, 2021).

Grasping at the Bottom Surface: Because of the problems with the instability or sliding of the food when it is subjected to fast motion, this technique of touching the food only from the bottom, which is normally done with a spatula in most recipes, is rarely applied by automated systems. The only known robotic gripper that applies this type of design is the FURUKAWA KIKO SWITL hand. The device itself is composed of a stainless steel plate and Teflon film. The Teflon is wrapped around the plate like a conveyor belt so that it slides beneath the food item. According to a 2011 study by Kusuda, creative hands can even handle a wide range of foods, including slices of ham and cheese, semi-liquid gels, mousses, and sauces like mayonnaise or melted cheese.

Advances in Food Recognition: 3D Recognition: As illustrated in Figure 9.9 the necessity for 3D perception often arises in the context of RBP. This problem has been extensively researched, especially for picking up rigid objects with predefined CAD models, particularly in the industrial domain. Commercial solutions abound to overcome the challenge of RBP. Numerous robot manufacturers add a vision system to their robotic arms, such as FANUC's 3D Vision Sensor, as explained by Connolly in 2007 and FANUC in 2021, Mitsubishi Electric's MELFA-3D Vision (Domae et al., 2024), and KUKA's PerceptionTech (KUKA., 2021). The companies consist of Pickit (Pickit, 2021) and Kyoto Robotics with their "TVS" 3D vision sensor (KYOTO ROBOTICS, 2021) that only offer 3D vision solutions for RBP applications. Those methods tend to be based on known, rigid objects and, thus, apply less to food items as pointed out by Castaman and others in 2020. So far, different ML techniques have been applied to address the RBP task of food product recognition. For instance, Joffe et al. (2019) adapted a Faster R-CNN standard framework with Resnet 101 feature extractor for chicken detection and used two approaches for pose estimation, namely the augmented autoencoder along with the direct regression approach that involves a suction cup gripper in order to grasp chicken. In 2020, Nishina and Hasegawa developed a method for optimal grip points via a deep neural network implemented successfully both on the robotic type two-fingered hand and on a suction cup gripper. In 2021,

Low et al. (2021) proposed a YOLOv3-based object detection algorithm for detecting sausage, potato, broccoli, and tomato items in RBP scenarios, which achieved an object classification accuracy of 67.06% with a speed of pose estimation being 92.7 ms.

9.2.6 Trends in Packaging and Materials, Active Packaging, Material Handling Systems

The paper discusses the design and development of a material handling system integrated with autonomous intelligent vehicles (AIVs) aimed at enhancing flexible manufacturing systems (FMS) in an increasingly competitive industrial environment. The integration of these technologies is positioned as a solution to reduce manual handling operations and improve efficiency in manufacturing processes. The developed prototype for autonomous material handling by Cronin et al. (2020) is represented in Figure 9.10.

AIV and Material Handling System Integration: Kostal Ireland GmbH employs a linear manufacturing flow line made of traditional conveyor systems

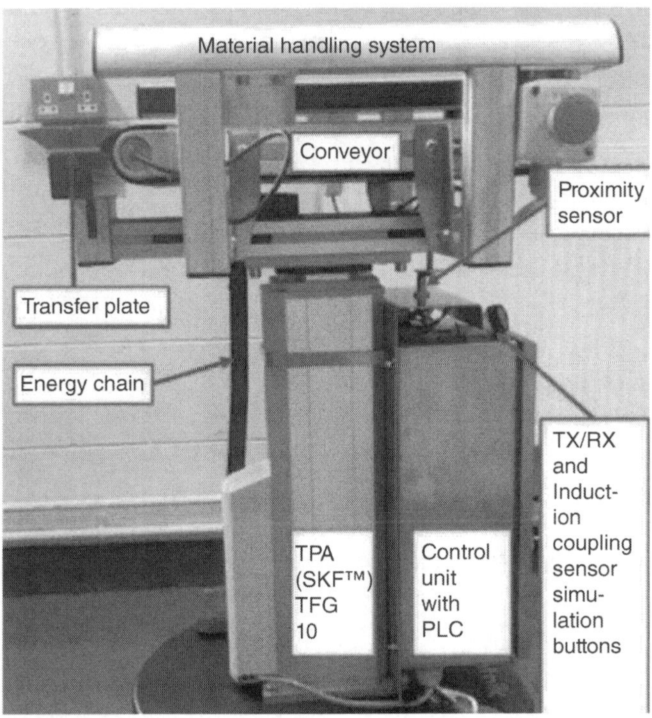

Figure 9.10 Autonomous material handling system prototype (Cronin et al., (2020)/Elsevier).

that do not allow flexibility and consume highly valuable floor space. In answer to the need for flexibility in manufacturing, new industry-related technologies recognized as Industry 4.0 were a result of that discovery. The project involved designing an affordable material handling system coupled with an AIV, seeking to gain increased manufacturing flexibility in the direction of minimizing the time delays experienced with fixed conveyor systems.

9.2.6.1 Design and Development

Material Handling System: A custom-designed conveyor system operating on the x-y plane was designed since off-the-shelf solutions were not available. Kostal Ireland GmbH was engaged in the project so that it would be verified to be suitable for industrial usage.

Electromechanical Integration: The integration of the electromechanical hardware was what made the material handling system work. The design was tested extensively at the individual component levels prior to full integration.

9.2.6.2 Testing and Challenges

Integration Testing: During the testing process, it became apparent that a red LED error signal was on the Interroll™ roller drive, which indicated that the drive had problems at the power supply when the AIV was in motion. This was corrected by the addition of a capacitor to stabilize the power supply.

Signal Handling: Ability to send and receive signals from the AIV for loading and unloading was crucial in this. It is noted that errors do occur, and thus signal isolation would have to be used for proper diagnosis.

Successful Model: The project developed a model that brought cost savings but was operationally effective. As realized, the integration of the material handling system and AIV proved the viable solution to many of the modern manufacturing challenges.

Future Implications: This case study highlights the importance of flexible manufacturing systems and the use of smart technologies to increase production capacity. The results serve as a validation of the current trend toward Industry 4.0 in manufacturing contexts.

Flexibility: Embedding AIVs with material handling systems can significantly increase the flexibility in the manufacturing process and efficiency.

Testing Importance: Testing must be intensive, and the components isolated from the system to guarantee the integration and operation end correctly.

Cost-effective: The project was able to show an economical alternative to conventionally used fixed conveyor systems, thus opening up wider industry applications.

9.2.7 Controls, Supervisory Control, Online Instrumentation, and Data Acquisition

SCADA refers to a group of industrial control protocols developed over many years to enhance the management of intricate operations. SCADA originally built on top of the PLC or programmable logic controllers but the systems over time have developed enormously to become integral parts of various operations, ranging from manufacturing plants to large-scale utilities, water treatment plants, and nuclear energy facilities. Although advanced technologies exist in wide distribution in many food processing plants, the new technologies generally have not received much acceptance. The new systems are complex, and integration into existing structures is complicated, thus it will be difficult to make anyone believe in positive cost-benefit analysis.

Need for SCADA: This brings a higher dependency on control technologies in industrial systems and calls for better efficiency, meaning that the mechanisms need to be made much more complex than what a human is able to track. Consider an intricate operation where a high volume uses a recipe that has been perfected; decisions need to be made every second as flow rates and temperatures are changing. This is where SCADA comes into play, bringing to the scene a master terminal unit that doesn't only oversees but also controls various lower-level processes through remote terminal units. In fact, these RTUs are run at such high speed of data within their own local subsystems, which send process information back to the MTU for adjustments; it makes the process optimized through complex algorithms executed at the MTU level.

Consider, for example, the mass production of bread. It's a multistep process: flour and water mixing, kneading the dough, fermentation, baking, slicing, then packaging (Matz, 1991). This involves a lot of equipment in production, including conveyors, mixers, slicers, ovens, and packaging machines. However, the factors affecting the production process are endless; the temperature of water and dough, time spent in baking, the color of the baked goods, environmental temperature, energy costs, the need for storage, and materials used for packaging, as well as the slicing speed are some examples. This combination of these variables is so tough that even a little modification can drastically alter the quality of the final product. Being timely in decision-making is important; for example, if bread loaves become too dark in the oven, then it may cause wastage of thousands of units. An SCADA model of a baking operation indicates how a high-level controller can ensure optimal performance and quality in such a process.

9.2.7.1 SCADA Systems Components

Field Devices: Include the sensors that acquire the physical information and transmit it to digital form, and the actuators that carry out the instructions for control and are found in the field to be studied.

Remote Telemetric Units (RTUs): High-technology electronic devices, which have the function of transferring the sensed data to the MTU. They can also, depending on cases whereby PLCs may not be present, send instructions directly to the actuators.

PLCs: These are smart electronic devices that pass on collected data and control instructions to the MTU. PLCs may further have direct control of field devices at the central unit by following their own logic.

Master Terminal Unit (MTU): MTU is the central unit that collects data from each of the control level units and sends them to the supervisory control tier. It also receives the control commands to be transmitted to the field devices.

They are interfaces through which the operators can access SCADA. From them, they allow a single view of the schematic of the plant. From the analysis of information that relates to MTU, it is pretty simple to send control commands to the physical world.

Servers: High-performance servers that would maintain data from field devices in databases. OPC servers can be defined as specialized software serving as a gateway between industrial hardware and windows-based software.

Communication Network: It represents all the different kinds of communication technologies that facilitate the connection between the SCADA control room, PLCs, RTUs, among other appliances in the field. They include Ethernet, fiber optics, Wi-Fi, and satellite, among others, including switches, routers, and modems. Figure 9.11 represents the SCADA architecture for the baking process.

Advantages:

1. Mass data can be stored
2. Data presented personalized
3. Different kinds of sensors are integrated
4. Advanced algorithms in control
5. Data of various types are monitored
6. Access to data is allowed from remote locations
7. Improved product uniformity
8. Automated data capture that will eliminate the need for personnel in other things
9. Automatic notification mechanism

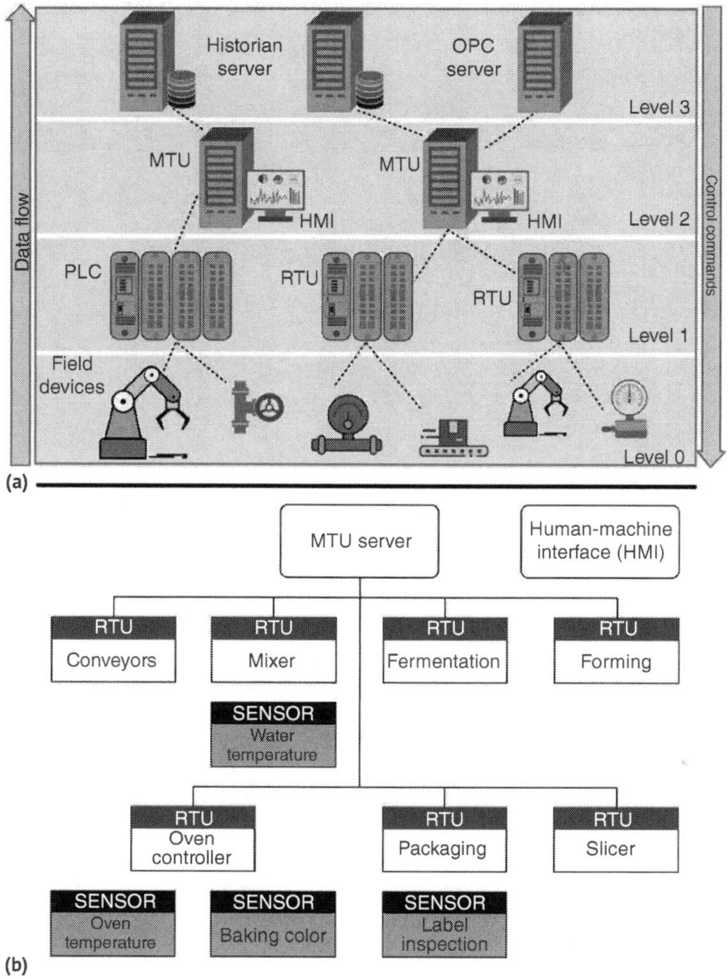

Figure 9.11 (a) SCADA architecture. (b) Notional SCADA network for baking process.

Disadvantages:

1. Complexity of software and hardware
2. New skills needed in the workforce
3. New security threats that may compromise process safety or private data
4. Implementation of the system is costly

9.2.7.2 Infographic Case Study: Low-cost Data Acquisition System for Milk Powder Production

9.2.7.2.1 Overview

Project Title: Development of a Low-cost Data Acquisition System for Milk Powder Production Line Monitoring represented in Figures 9.12 and 9.13 by Cleary et al.

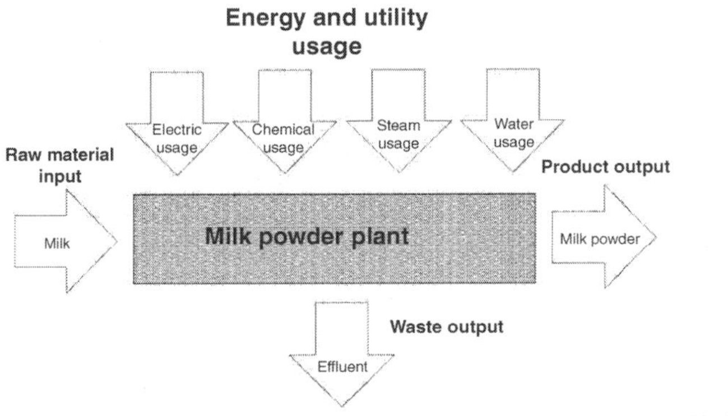

(a)

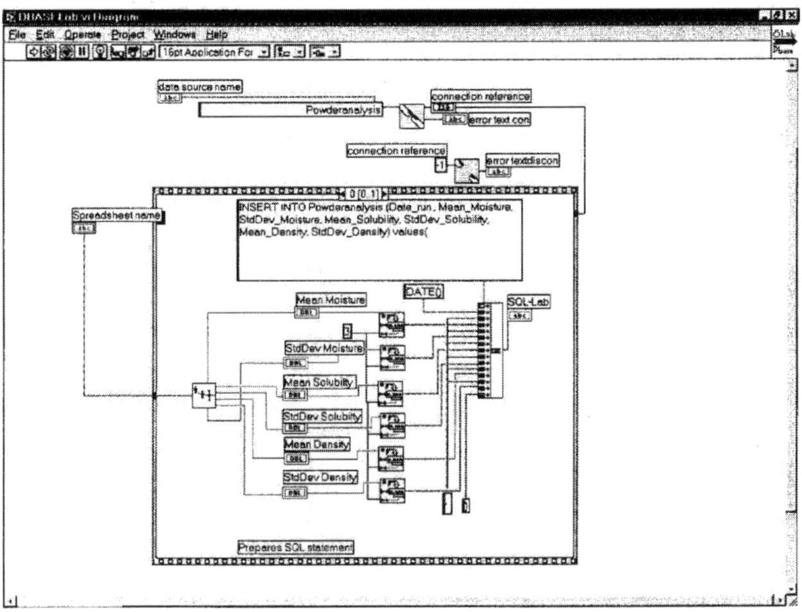

(b)

Figure 9.12 (a) A block diagram illustrating the inputs and outputs of the milk powder production process, which was monitored by the data acquisition and supervisory and data acquisition (SCADA) system developed in Carbery. (b) An example of a sub-virtual instrument (VI) written in LabVIEW.

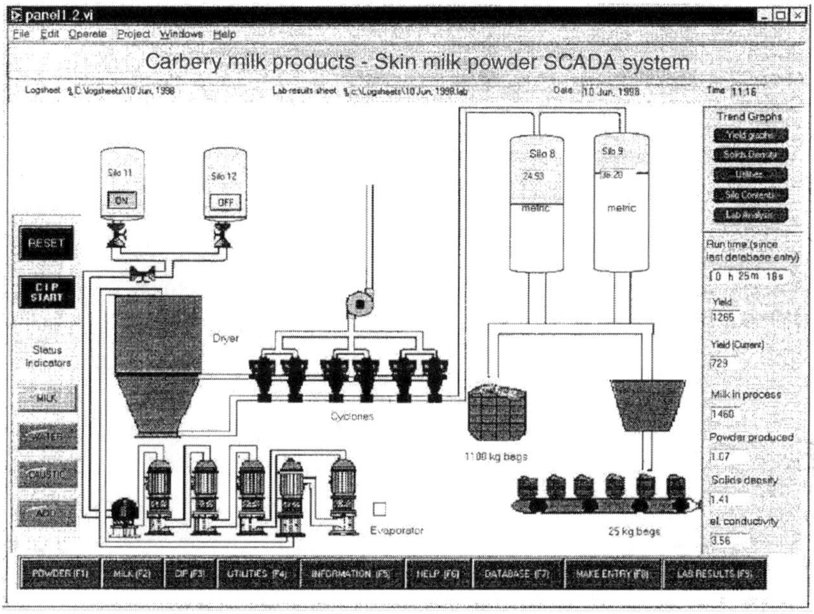

Figure 9.13 The HMI of the data acquisition system. The interface is designed to maximize ease of use by the operators. CIP = cleaning in place; SCADA = supervisory control and data acquisition.

(1999) Location: Carbery Milk Products Ltd., Ballineen, County Cork, Republic of Ireland

Industry: Dairy Production

Objective: A cheap data acquisition monitor system is offered which was to be used in acquiring in-line data handled in the production process of milk powder.

9.2.7.2.2 Critical Features of the System

Data Acquisition: The commercially available, inexpensive hardware and software tools are used in monitoring the production process.

Yield Calculation: Determines tonnes of powder produced per gallon of milk used; calculated every 15 minutes.

Data Storage: Schemed in relational database that will allow for historical trending and analyzing the efficiency of the process.

9.2.7.2.3 Benefits

Efficiency is increased; data collection is automated, therefore less dependent on manual recording methodologies

Potential cost accounting, with rates for utilities, chemicals, and energies, thus allows for realistic calculation for cost per tonne.

Meeting regulatory demands intensifying requirements for traceability and accountabilities in food processing

9.2.7.2.4 Software Development

Development Environment: Developed in LabVIEW™ which makes programming less difficult with graphic interfaces.

Human-Machine Interface (HMI): Compliant design friendly to the users with virtual instrument displays, plant mimics, and operator trending graphs.

9.2.7.2.5 Future Developments

Advanced Control Systems: The future stages are going to be installed with NNs and fuzzy logic for improved control processes in the evaporators.

9.2.7.2.6 Conclusion

It represents an advancement in the development of the small- to medium-scale manufacturers of dairy products, as it steps up their productivity, ensures regulatory compliance, and strengthens economic evaluations of the production processes.

9.2.8 Nondestructive Inspection

As the kernel matures, pistachios crack open, indicating that they are ready to be harvested. One of two methods—hand examination or pin-selection devices—is used to sort out the ones that do not crack open from the rest. Pin-picker devices have the disadvantage of sometimes cracking the pistachio kernel when it pops out. The pistachios were cracked by inserting a needle into their crack. This will change the color of pistachios, and it would appear that the customers know the kernel has wormholes. Further, a bad kernel is said to be less oily and likely to deteriorate. The disadvantage of visual inspection is that man is not created to work for long hours on repetitive work. Therefore, the quality of the inspection will vary. Additionally, human contact with consumables can contaminate it. In fact, in 1943, the first artificial neural network was developed. McCulloch and Pitts (1943) (El-Mesery et al., 2019; Abasi et al., 2018)

9.2.8.1 Nondestructive Robotic Sorting of Cracked Pistachio

9.2.8.1.1 Overview

Title: Nondestructive Robotic Sorting of Cracked Pistachio Using Deep Learning
Problem Statement

There was a long time ago when the separation between open-shelled pistachios and closed-shelled pistachios was only done through the use of manual labor. It is a very laborious process, and also an inefficient one.

But today, there are automated ways for this, such as pin-picker machines, but these machines can damage the kernels of pistachios, and hence the nonusage of these in processing. A deep-learning-based algorithm to classify pistachios without breaking the kernels has been designed.

The system uses a double-camera configuration that can accurately identify the one-sided split on a pistachio, which cannot be easily identified with one camera.

9.2.8.1.2 Methodology

Data Preparation: Images of pistachios captured in various lighting for strength of detection. 377 images and images of over 3700 pistachios are annotated to train the algorithm.

Dataset is labeled into the Pascal VOC format and converted to TFRecord to use in tensorflow.

Camera Calibration: Images of chessboard pattern are captured.

Intrinsic and extrinsic parameters of the cameras were determined for the calculation of the real-world coordinates of the pistachios. Using these parameters, the real-world coordinates of the pistachios are calculated.

9.2.8.1.3 Results

The accuracy of detection was 98% of open-shelled pistachios and 85% of closed-shelled pistachios.

The system is developed in a step-by-step procedure and thus can be applied in industrial environments that are not subjected to extremely high speeds. Figure 9.14 represents the non-destructive robotic sorting of Pistachio.

Future Implications: The approach can potentially be adapted for classifying other agricultural products with different characteristics, broadening its application in the agricultural sector.

Improving workspace conditions, such as lighting, could further enhance detection performance.

This innovative method provides a nondestructive solution for sorting pistachios, addressing the limitations of manual labor and existing automated methods. The integration of deep learning and robotics in agricultural processing represents a significant advancement in the field.

(a)

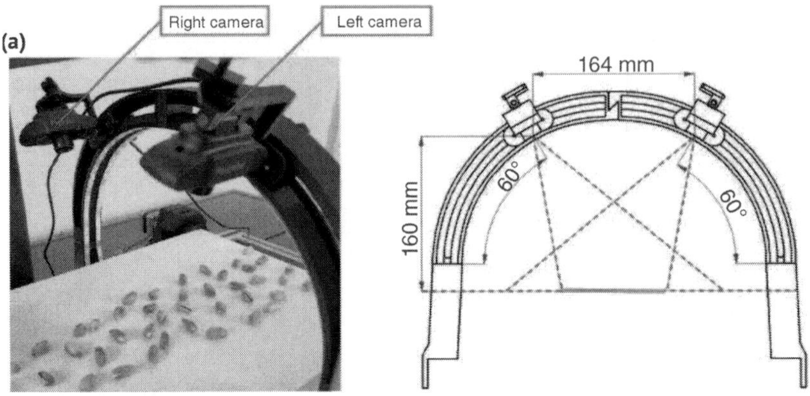

(b)

```
Import libraries
Start serial communication to robotic system
Import the model; zero the cartesian manipulator; Start the
conveyor system

while TRUE do                                                    ▵ Infinite loop
    Take two consecutive images on one camera
    Separate objects from the background in the images
    Subtract first image from the second image
    if there is difference then
        Stop the conveyor system; Take an image with the right camera
        Run the detection on the image
        if there is object in the image then
            Take an image with the left camera
            Run the detection on the image
            Find the centers of the objects in both images
            Delete duplicate detection if there is any
            Find the real-world coordinates of the objects in both images
            Calculate the Euclidian distances of every objects in the images
take with the both cameras
            Pair one object in the image taken with the right camera with
another object in the image taken with the left camera if it has similar real-
world coordinates in both images. Delete the farther one if two objects in
one image are paired with one object in the other. Pair that object with
an unpaired one if it exists
            Calculate the objects relative distance to manipulator's origin
            if the first object will be out of reach when the conveyor moves
then
                Separate objects in the first detection
            end if
            Move the conveyor belt until the last object is out of the vision
of the right camera
            Separate all the object on the conveyor system
        end if
    end if
end while
```

Figure 9.14 (a) Experimental setup and position of cameras. (b) Pseudocode for nondestructive robotic sorting of cracked pistachio using deep learning (Karadağ et al., (2023)/Elsevier).

9.2.9 Multispectral, Hyperspectral, and X-ray-based Systems

It presents an overview of advancements in imaging techniques for robotic grading systems of fruits and vegetables. Human eyes are sensitive to the primary colors: red (R), green (G), and blue (B), which are fundamental to computer vision, producing monochrome images via three filters centered at 700 nm using 546 nm (G) and 435 nm (B) wavelengths to replicate human eyesight. Computer vision is the discipline dedicated to the investigation and implementation of techniques that facilitate computers to comprehend the information of images, and this analysis entails the extraction of specific traits that are crucial for a specified objective. Computer vision is an evolving engineering discipline. Integrating mechanics, electromagnetic sensing, optical instrumentation, AI, and image processing technologies. Computer vision systems possess the ability to deliver exceptional spatial information encompassing significant data to assess several quality-related characteristics of fruits and veggies. As economical, swift, consistent, dependable, and objective, computer vision technologies have been extensively included as inspection tools. Utilized in automated sorting machines for the assessment of fruit quality, systems for computer vision have been identified as an effective instrument for ascertaining not just hue, form, dimensions, texture, as well as surface imperfections. Computer vision systems can capture analogous color images seen by human vision, and the extraction of quality attributes related to subtle differences in color images (initial deterioration, minor). The identification of bruises and other imperceptible flaws continues to pose significant challenges. In numerous applications, photographs, including geographical information alone, are inadequate to derive quality-related attributes. One method to augment the efficacy of conventional computer vision systems. It aims to integrate spatial and spectral data through the utilization of hyperspectral technology and multispectral imaging systems (Lorente et al., 2012; Teena et al., 2013). Hyperspectral imaging technology amalgamates imaging and spectroscopic methodologies into a singular system, enabling the acquisition of a series of monochromatic images at nearly continuous hundreds of thousands of wavelengths. Hyperspectral imaging devices can deliver spatial information, identical to traditional imaging systems, for every wavelength in the complete spectrum and spectral information, analogous to a spectroscopic apparatus. A multispectral imaging system differs from a hyperspectral imaging system's method in the quantity of monochromatic images obtained in the spectral domain (the distinction between hyperspectral imaging).

Hyperspectral imaging systems are frequently employed for basic study, while the multispectral imaging employs selected wavelengths and is consistently utilized in engineering applications. Alternative imaging modalities (ultrasound imaging, magnetic resonance imaging, thermal) imaging, X-ray imaging, olfactory imaging, computed tomography, 3D imaging and spectroscopic approaches for nondestructive inspection of fruit quality have also been documented in the literature.

9.2.10 Optical Sensors and Online Spectroscopy for Automated Quality and Safety Inspection of Food Products

9.2.10.1 Optical-based Nondestructive Quality Inspection in Robotic Fruit and Vegetable Grading Systems

Optical Techniques: Since the last few decades, advances in computer vision, spectral imaging, and near-infrared technology have greatly improved the quality inspection of fruits and vegetables in a nondestructive manner.

Challenges: However, with all these advances automatic detection of quality is challenging because several factors make it difficult to commercialize.

9.2.10.2 Key Challenges

Physical and Biological Variability: High variability in texture and color complicates image analysis, making defect detection difficult. Colors usually depend on varieties and thus often result in low contrast with detection of defects.

Integrated Surface Detection: The whole surface of fruits and vegetables is supposed to be captured as well for reliable inspection. Techniques that use several mirrors or cameras achieve this.

Replicated sampling and shape distortion might degrade the quality of images, but they are tolerable since the objective is qualitative analysis.

Defect Discrimination: Effective grading is possible only by differentiating the real defects from natural traits such as stems and calyxes.

Advanced imaging techniques like NIR-coded structured lighting have emerged to enhance the possibility of defect detection.

Robustness of Features and Algorithms: One of the major issues is the reliability of the algorithms under various conditions. Popular feature selection methods include CARS and RF to enhance the algorithm's performance.

Quick Development of Rapid Optical Detection System: Most importantly, the need for faster and more efficient optical detection systems is critical to the enhancement of the automation of grading processes.

Solutions and Innovations Feature Selection Techniques: Methods such as successive projections algorithm (SPA), genetic algorithm (GA), and principal component analysis (PCA) are popular for selecting relevant features in quality inspection. Figure 9.15 presents optical-based nondestructive quality inspection in robotic fruit and vegetable grading systems.

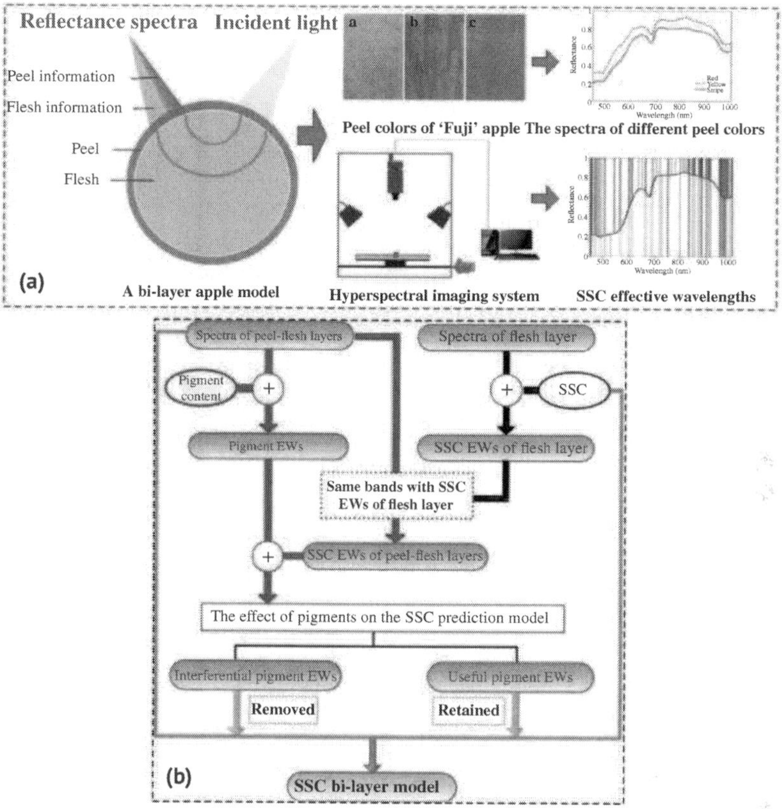

Figure 9.15 Bi-layers model for SSC prediction. (a) Description of the bi-layer model. (b) The experimental scheme of the bi-layers model (Tian et al., (2018)/Elsevier).

Hyperspectral Imaging: This technique combines spatial and spectral information, allowing for better quality-related character extraction.

Engineering Approaches: The review emphasizes the importance of engineering solutions to address the challenges and improve the automation level of grading systems.

Conclusion: The review highlights the ongoing challenges in optical-based nondestructive quality inspection for robotic grading systems and discusses potential solutions. Future trends and remaining engineering challenges are also explored, emphasizing the need for continued innovation in this field.

9.2.11 Optical Sensing and Spectroscopic Techniques

Recently, a large number of environmentally friendly analytical techniques have been developed—spectroscopic and imaging methods—in large numbers, primarily in the agricultural, food, chemical, and medical sectors. They have emerged acknowledged as highly rapid, noninvasive, low-cost methods that require little to no sample preparation or cause no environmental damage (An et al., 2023). For all these reasons, they have become a staple in many researchers' toolboxes.

In food technology, IR spectroscopy is highly promising, which, in turn, incorporates NIR and MIR spectroscopy. This technology provides detailed chemical analysis of food and agricultural products and delivers information about some key components, such as moisture, fat, and protein, at a molecular level. The result of combining imagery with IR spectroscopy forms the latter as a potent nondestructive analysis tool called hyperspectral imaging (HSI). HSI is a combination of the strengths of spectroscopy and imaging in order that both spatial and spectral information can be obtained collectively about a sample. Such an integration conquers the individual weaknesses of spectroscopy and imaging reported by researchers like Kandpal et al. (2016) and Pu et al. (2023). In this regard, HSI is becoming increasingly adopted these days as a highly effective and sought-after method for the nondestructive assessment of quality in food and agricultural products.

Optical-based Nondestructive Quality Inspection in Robotic Fruit and Vegetable Grading Systems: Advances in IRS and HSI for cereal quality detection introduction cereals are an important source of human nutrition, and the evaluation of their quality has become extremely critical for health and safety evaluation. IRS and HSI are being widely developed as promising, nondestructive techniques for cereal quality evaluation, especially when combined with AI.

9.2.11.1 Key Applications

9.2.11.1.1 Qualitative Analysis

These agencies can distinguish cereal varieties, detect adulteration, identify unsound kernels, trace geographical origins, and determine the years of harvest as well as maturity stage.

For instance, this research differentiated Korean native rice from imported Chinese rice using PLS-DA to point out that spectral preprocessing is essential to results accuracy.

9.2.11.1.2 Quantitative Analysis

All of these techniques will predict nutrient content, hardness index, and toxin levels in cereals. These will enhance food safety and quality control.

Technological Integration: The integration of IRS and HSI with AI, notably DL, boasts superior performance in analyzing the complexities of spectral data compared to the traditional techniques. DL algorithms automatically learn features during training. Such learning is better suited to high-dimensional and nonlinear data typical of spectral analysis.

Challenges and Future Directions: Though much has been achieved in this direction, much remains to be worked out, especially if one is taking into account variations in spectra due to geographical origins and storage conditions. Further research studies are recommended for improving model accuracies and the study of new applications of IRS and HSI in cereal quality assessment.

Conclusion: The integration of IRS, HSI, and AI represents a great step forward in cereal quality detection by offering quick, reliable analysis methods. Continued research and development of these technologies will be applied in the agricultural sector, improving control quality and consumer safety. Application of a blend of advanced spectroscopic techniques with AI to bring forth innovative ways of enhancing cereal quality assessment will bring much-needed change in food quality management and safety, and that is as depicted in Figure 9.16 (An et al., 2023).

9.2.12 Spectroscopic Techniques, IR Spectroscopy, Raman Spectroscopy, and NIR Spectroscopy

Spectroscopic Techniques in Food Quality Evaluation Overview of Spectroscopic Techniques: VisIR Spectroscopy: 400–5000 nm: It ranges from visible light down to infrared and is most commonly used for the evaluation of food quality because it is able to interact through electromagnetic radiation with the food components involved (He and Sun, 2015).

Raman Spectroscopy: It uses laser light for the provision of more detailed vibrational and rotational spectra of food molecules. It is particularly effective in the study of complex food materials since it is relatively less influenced by water.

Fluorescence Spectroscopy: It is a rapid technique without damaging the samples. Spectrum signatures of food can provide quick insights into its quality.

Applications in Food Microbial Analysis Total Viable Counts (TVC): NIR and FTIR spectroscopy can predict TVC in poultry and meat products. For instance, a PLS model that has R^2 of 0.804 with an RMSE of 0.799 for sliced chicken breast.

Detection of Specific Microorganisms: FTIR is utilized as a means to detect and estimate the amount of specific bacteria like *Salmonella enterica* in chicken breast, along with very promising metrics of up to 0.95 in R^2.

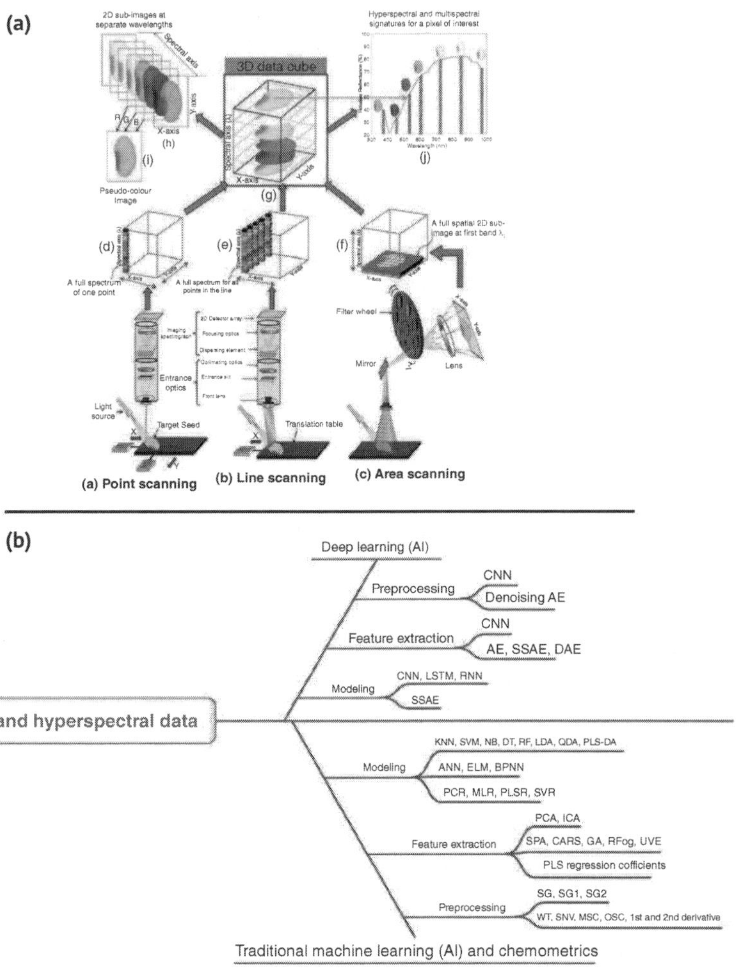

Figure 9.16 (a) Three scanning methods for collecting hyperspectral cube data. (b) Various ways to process IR and hyperspectral data.

Case Studies Demonstrating Utility Chicken Breasts Spoilage: The research determined that FTIR and NIR spectroscopy can classify spoiled chicken breasts at a 100% accuracy using PLSDA. Thus, the researches have demonstrated potential in the application of these techniques for the real-time detection of spoilage.

Cheese Quality Assessment: FTIR has been highly useful in predicting cheese quality as applied in cheese research. It is known to identify several bacteria strains, including *Listeria monocytogenes* with high sensitivity, which explains its safety application in dairy products.

The spectroscopic methods are a successful way to integrate into food quality evaluation, promising speedy analysis in a nondestructive manner. Spectroscopic methods enhance the detection of spoilage and, importantly, ensure food safety as individual pathogens are effectively identified. Further research and development is the need of the hour for improvement in food quality assurance practice.

9.2.13 Applications in Food Industry

9.2.13.1 Robotics in Dairy and Food Industries

9.2.13.1.1 Introduction

The dairy and food processing industries are very labor-intensive. At times, they have even reached as high as 50% of the product costs. This opens a significant avenue for process automation, enhancing productivity while lowering costs.

9.2.13.1.2 Current Applications of Robotics

Packaging and Palletizing: At present, these are the most dominant applications of robots. Their applications are expanding (Prasad, 2017).

Grading and Sorting: Food products can be graded through robotics to obtain consistency and quality.

Pick and Place Operations: Robots will top the list in placing products, which is very critical for high-volume production environments.

Meat Processing: Robotics will make the process of processing meat products safer and more efficient.

Milk Production and Processing: Techniques and processes of dairy production can be streamlined by automation.

9.2.13.1.3 Challenges in Implementation

Dimension, shape, and weight are very variable in food products, for which it is very difficult to design effective manipulators.

Many food items are fragile and are covered with slippery substances, which require high sensing and handling capabilities.

Highly High Initial Costs : Investments in robotic systems might be highly significant, and skilled engineers are needed to operate and maintain these systems. Future Prospects: Potential for robotics in the dairy and food industry is enormous with continuous research and development going on to tackle some of the current challenges.

Industry-specific Expectation: Of course, with the passage of time, it is expected that the use of robots in the industry will rise significantly and attain levels like those in the automotive sector.

Specifications for Food Sector Robots Hygiene Design: The robot has to be hygienically designed. This is especially true in processing sensitive products like poultry and fish.

Material Requirements: The stainless steel, AISI-304, shall be used in the construction of the robots. They will be made in such a way that they can be easily cleaned and inspected.

Ingress Protection: The ingress protection rating of the robots shall be IP67 so that they can be operated in a food-processing environment.

The integration of robotics in the dairy and food industries has presented both opportunities and challenges. With advancement in technology and increased research, the future of robotics in these sectors will undoubtedly shine with a potential impact on efficiency, safety, and profitability.

9.2.14 Applications of Image Sensing Technology

9.2.14.1 Efficient Extraction of Deep Image Features Using CNN for Food Analysis

9.2.14.1.1 Background

Quality and safety of products of food have been hard to detect and analyze in the food industry.

Most traditional algorithms in ML fail because they cannot display multifaceted attributes of food well (Liu et al., 2021).

CNN: CNNs are the most effective feature extraction tools applied in the analysis of food commodities.

They are now used widely in detecting and analyzing complex food matrices in quality control operations as well compared to traditional methods.

9.2.14.1.2 Key Features of CNN Network Design

CNNs employ an array of architectures, 1D, 2D, and 3D, for feature extraction from food images.

Feature Pooling: Feature combinations that had been impossible have been made possible using the help of batch normalization and auxiliary classifiers. Thus, performance improves.

9.2.14.1.3 Applications in Food Examination

Applications of CNNs are made for the following:

Meat and Aquatic Products: The classification accuracy was up to 98.21% with VGG-16 for the freshness detection.

Nutritional Content Evaluation: The proposed model employs multitask 2D CNNs to predict the calorie values based on food images.

General Food Detection: Models were able to classify the type of food and also assess quality for all major categories such as fruits, vegetables, and cereals, owing to the inherent ability of CNNs to identify food items.

9.2.14.1.4 Challenges and Future Directions

Data fusion remains an open problem; other than this, the classic approaches might sometimes produce redundant information and decrease detection accuracy.

Future work will be put on the improvement of feature extraction methods for better extraction of the informative features while reducing noises.

CNNs are the result of great progress toward food detection and analysis, giving better performances than traditional ones.

The possibilities of real-time application on mobile devices are promising, opening the doors to future developments toward food quality assessment.

This case study specifically highlights the impact of CNNs in changing the food industry through their efficacy in enhancing food safety and quality assessments.

9.2.15 Case Studies

9.2.15.1 Case Study: Advances in Raman Spectroscopy for Fish Quality and Safety

Overview: Fish is the most commercially in-demand aquatic product, and its quality and safety matter most to public health. Traditional methods of quality assessment of fish tend to be very time-consuming and hence costly. Hence, there is a call for effective and rapid, as well as noninvasive detection techniques.

9.2.15.2 Key Applications of RS and SERS

Figures 9.18 and 9.19 present molecular fingerprinting molecular information which is rich in Raman spectroscopy (RS) and surface-enhanced Raman scattering (SERS), and it is possible to identify the hazardous substances in the fish. Figure 9.17 represents the various feature extraction process such as 1D, 2D, and 3D CNN.

Microplastics Detection: RS is used for the identification of microplastics in stomachs of fish through the development of methods that suppress fluorescence

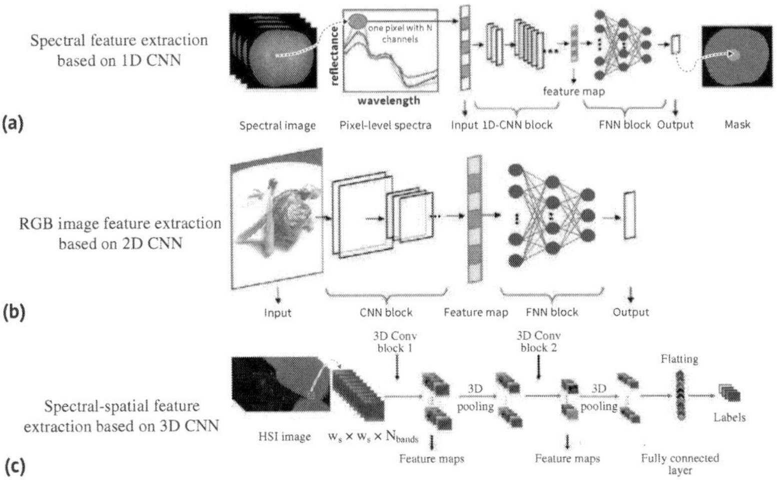

Figure 9.17 Various feature extraction processes. (a) 1D CNN (Adapted from Zhou et al. [2019]), (b) 2D CNN (Liu et al. [2021]/Elsevier), (c) 3D CNN (Sarayreh et al. [2020]/Elsevier).

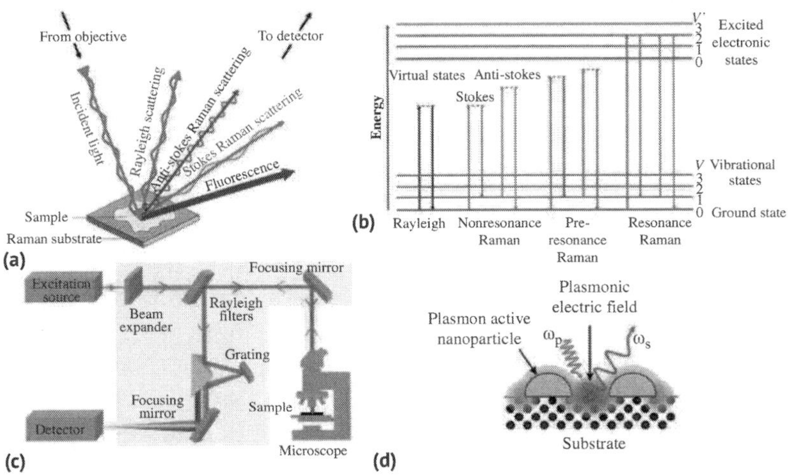

Figure 9.18 SERS and RS. (a) Illustration of light scattering at a sample surface after laser irradiation (Butler et al., 2016/Springer Nature). (b) Rayleigh scattering, Raman scattering, and fluorescence energy level diagram (Auner et al., 2018/Springer Nature/CC BY 4.0). (c) Butler et al. (2016) outline the general instrumentation choices available in a typical spontaneous Raman spectroscopic microscopy system, SERS (d). The incident pump light induces surface plasmon resonance. According to Jones et al. (2019), this results in a subsequent increase in the strength of the oscillating electro-magnetic field on the surface, which boosts the light-matter interaction to further increase the intensity of the Raman scattered light (Jones et al., 2019/Springer Nature/CC BY 4.0).

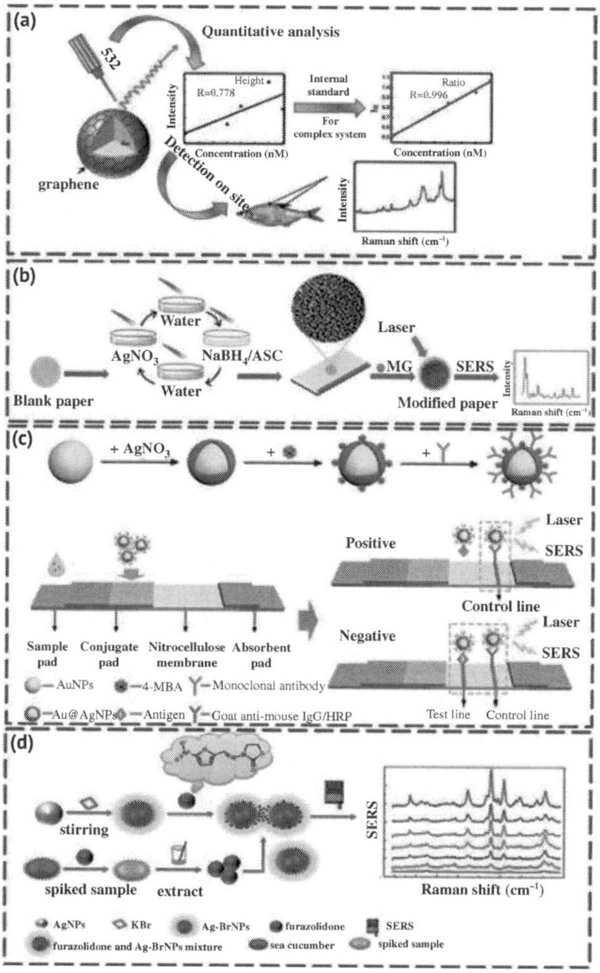

Figure 9.19 An example of the four major veterinary drug residues found by SERS. (a) Graphene-isolated-Au-nanocrystal (GIAN) platform schematic diagram of SERS measurement for crystal violet (CV) determination on fish muscle and scale (Zhang et al., 2020). (b) Malachite green (MG) detection scheme and substrate preparation using glass fiber paper-based SERS technology (Deng et al., 2019/Elsevier). (c) Schematic representation of the competition-based LFA platform connected to gold-silver core–shell nanoparticles (Au-Ag NPs) and SERS for the detection of chloramphenicol (CAP) (Pan et al., 2021/Springer Nature). (d) A schematic representation of the preparation and application of Ag-Br NPs for the detection of furazolidone (FUZD) (W. L. Fan et al., 2021/Elsevier).

interference. PCA-assisted Raman imaging enhances the visualization and identification of microplastics.

Formaldehyde Quantification: SHINERS can be applied to the trace quantitation of formaldehyde in fish for food safety.

Changes in Structural Proteins of Fish: RS can detect the protein structural changes that occur due to storage in the fish, and hence it is crucial for their freshness determination.

9.2.15.3 Challenges and Future Prospects

Data Analysis: Chemometric approach is needed for the complex data interpretation resulting from RS analysis, which allows differentiation and quantification of several compounds.

Problem of Sensitivity: Determination of multiple analytes at one time may compromise the sensitivity and precision parameters, which require high-grade algorithms and ML for the effective analysis.

Detection of Adultery: Combination of RS with chemometric methods is rapid as well as sensitive for differentiation between fresh and frozen-thawed fish for customers to have confidence in quality.

Sample analysis of fish quality and safety, when integrated with RS and SERS, allows the comparison of fish quality and safety tests with at least several critical advantages over traditional methods in terms of speed and sensitivity combined with the nondestructive testing capability offered by these technologies. Thus, with further development of the technology, they can be used to good advantage in routine food safety inspections, improving public health from there.

9.3 Review Questions

Q1 Define infrared spectroscopy.

Q2 Explain nuclear magnetic resonance (NMR) spectroscopy.

Q3 Briefly explain the absorption spectrum.

Q4 Cost Analysis: If a handheld near-infrared (NIR) spectrometer costs $5,000 and is used for quality assessment in a food processing plant that produces 100 tons of product per day, how much would the cost per ton be if the equipment has a lifespan of 5 years?

Q5 In a study, 200 spectral readings of a food sample are taken, and 150 readings are classified as "safe" based on established thresholds. What percentage of the readings indicates that the food is safe?

Bibliography

S Abasi, S Minaei, B Jamshidi, and D Fathi. Dedicated non-destructive devices for food quality measurement: a review. *Trends in Food Science & Technology*, 78:197–205, 2018.

N Aghilinategh, MJ Dalvand, and A Anvar. Detection of ripeness grades of berries using an electronic nose. *Food Science & Nutrition*, 8(9):4919–4928, 2020.

M Al-Sarayreh, MM Reis, WQ Yan, and R Klette. Potential of deep learning and snapshot hyperspectral imaging for classification of species in meat. *Food Control*, 117:107332, 2020.

D An, L Zhang, Z Liu, J Liu, and Y Wei. Advances in infrared spectroscopy and hyperspectral imaging combined with artificial intelligence for the detection of cereals quality. *Critical Reviews in Food Science and Nutrition*, 63(29):9766–9796, 2023.

VM Balasubramaniam, and SK Sastry. Use of liquid crystals as temperature sensors in food processing research. *Journal of food engineering*, 26(2):219–230, 1995.

L Barbosa-Pereira, S Belviso, I Ferrocino, O Rojo-Poveda, and G Zeppa. Characterization and classification of cocoa bean shells from different regions of Venezuela using hplc-pda-ms/ms and spectrophotometric techniques coupled to chemometric analysis. *Foods*, 10(8):1791, 2021.

F Cleary, MJ McGrath, M Gaiser, JF O'Connor, M Everard, and J Holland. Development of a low cost data acquisition system for milk powder production line monitoring. *Journal of Dairy Science*, 82(9):2039–2048, 1999.

C Cronin, A Awasthi, AC, D O'Riordan, and J Walsh. Design and development of a material handling system for an autonomous intelligent vehicle for flexible manufacturing. *Procedia Manufacturing*, 51:493–500, 2020.

F Du, P Liu, K Wang, Z Yang, and L Wang. Ionomic responses of rice plants to the stresses of different arsenic species in hydroponics. *Chemosphere*, 243:125398, 2020.

HS El-Mesery, H Mao, and AE-F Abomohra. Applications of non-destructive technologies for agricultural and food products quality inspection. *Sensors*, 19(4):846, 2019.

H Farahani, R Wagiran, and MN Hamidon. Humidity sensors principle, mechanism, and fabrication technologies: a comprehensive review. *Sensors*, 14(5):7881–7939, 2014.

W Feng, H Sui, J Tu, W Huang, and K Sun. A novel change detection approach based on visual saliency and random forest from multi-temporal high-resolution remote-sensing images. *International Journal of Remote Sensing*, 39(22):7998–8021, 2018.

S Gaggiotti, F Della Pelle, V Masciulli, C Di Natale, and D Compagnone. Monitoring shelf life of carrots with a peptides based electronic nose. In: *Sensors: Proceedings of the Fourth National Conference on Sensors, February 21-23, 2018, Catania, Italy 4*, 69–74, 2019. Springer International Publishing.

D Galvan, A Aquino, L Effting, ACG Mantovani, E Bona, and C-JC Adam. E-sensing and nanoscale-sensing devices associated with data processing algorithms applied to food quality control: a systematic review. *Critical Reviews in Food Science and Nutrition*, 62(24):6605–6645, 2022.

Z Geng, X Duan, J Li. C Chu, and Y Han. Risk prediction model for food safety based on improved random forest integrating virtual sample. *Engineering Applications of Artificial Intelligence*, 116:105352, 2022.

H-J He and D-W Sun. Microbial evaluation of raw and processed food products by visible/infrared, Raman and fluorescence spectroscopy. *Trends in Food Science & Technology*, 46(2):199–210, 2015.

Y He, W Xu, M Qu, C Zhang. W Wang, and F Cheng. Recent advances in the application of Raman spectroscopy for fish quality and safety analysis. *Comprehensive Reviews in Food Science and Food Safety*, 21(4):3647–3672, 2022.

JF Holmes, G Russell, and JK Allen. Supervisory control and data acquisition (SCADA) and related systems for automated process control in the food industry: an introduction. In: *Robotics and Automation in the Food Industry*. Elsevier, 130–142, 2013.

J Wawrzyniak. Advancements in improving selectivity of metal oxide semiconductor gas sensors opening new perspectives for their application in food industry. *Sensors*, 23(23):9548, 2023.

M Javaid, A Haleem, RP Singh, S Rab, and R Suman. Significance of sensors for industry 4.0: Roles, capabilities, and applications. *Sensors and International*, 2:100110, 2021.

AE Karadağ and A Kılıç. Non-destructive robotic sorting of cracked pistachio using deep learning. *Postharvest Biology and Technology*, 198:112229, 2023.

RV Kryuk, MG Kurbanova, AY Kolbina, KB Plotnikov, IB Plotnikov, AN Petrov, and MEA Khelef. Color sensors in smart food packaging, 2022.

J Li and L Sun. Forecasting of wood moisture content based on modified ant colony algorithm to optimize lssvm parameters. *IEEE Access*, 8:85116–85127, 2020.

Y Liu, H Pu, and D-W Sun. Efficient extraction of deep image features using convolutional neural network (CNN) for applications in detecting and analysing complex food matrices. *Trends in Food Science & Technology*, 113:193–204, 2021.

F Longobardi, G Casiello, V Centonze, L Catucci, and A Agostiano. Electronic nose in combination with chemometrics for characterization of geographical origin and agronomic practices of table grape. *Food Analytical Methods*, 12:1229–1237, 2019.

HKSV Mohan and AA Malcolm. A novel low-cost capacitance-based approach for non-destructive evaluation of powder quality for food processing industries. *Measurement: Sensors*, 23:100395, 2022.

E Ohashi and I Karube. Sensors for the food industry. *Food Control*, 4(4):183–188, 1993.

P Patel and A Doddamani. Role of sensor in the food processing industries. *International Archive of Applied Sciences & Technology*, 10:10–18, 2019.

LD Pedersen. Assessment of sensors used in the food industry. *Food Control*, 2(2):87–98, 1991.

A Popa, M Hnatiuc, M Paun, OGeman, DJ Hemanth, D Dorcea, LH Son, and S Ghita. An intelligent IoT-based food quality monitoring approach using low-cost sensors. *Symmetry*, 11(3):374, 2019.

R Rusinek, H Jeleń, U Malaga-Toboła, M Molenda, and M Gancarz. Influence of changes in the level of volatile compounds emitted during rapeseed quality degradation on the reaction of MOS type sensor-array. *Sensors*, 20(11):3135, 2020.

S Prasad. Application of robotics in dairy and food industries: a review. *International Journal of Science, Environment and Technology*, 6(3):1856–1864, 2017.

N Sharma, MA Choudhari, DN Dabhade, and SJ Bari. Sensors in food industry: a review. *Sensors*, 4(4), 2024.

X Tian, J Li, Q Wang, S Fan, and W Huang. A bi-layer model for nondestructive prediction of soluble solids content in apple based on reflectance spectra and peel pigments. *Food Chemistry*, 239:1055–1063, 2018.

X Wang, Y Bouzembrak, AGJMO Lansink, and HJ Van Der Fels-Klerx. Application of machine learning to the monitoring and prediction of food safety: a review. *Comprehensive Reviews in Food Science and Food Safety*, 21(1):416–434, 2022a.

Z Wang, S Hirai, and S Kawamura. Challenges and opportunities in robotic food handling: a review. *Frontiers in Robotics and AI*, 8:789107, 2022b.

D Wu, H Cheng, J Chen, X Ye, and Y Liu. Characteristics changes of chinese bayberry (*Myrica rubra*) during different growth stages. *Journal of Food Science and Technology*, 56:654–662, 2019.

Z Xiaobo, H Xiaowei, and P Malcolm. Non-invasive sensing for food reassurance. *Analyst*, 141(5):1587–1610, 2016.

10

Case Studies

After reading this chapter, you should be able to:

- Practical robotics applications in food manufacturing focus on real-life application examples, from modeling and simulation robotic applications to human-following and agricultural support robots for useful insights into automation in food processing.
- State-of-the-art techniques: This work will give readers knowledge of advanced techniques, which include deep reinforcement learning for training of the task, OpenPose for robotic cooking, and compliant grippers to estimate the cooking time, essentials for fine-tuning robotic precision, and adaptability in food processing tasks.
- This chapter focuses on optimizing food service robotics, but particularly through case studies with humanoid parameter tuning and object detection in additive manufacturing, thus emphasizing innovations that actually make the efficiency and performance of service robots better geared toward the dynamic needs of the food industry.

This chapter discusses the application of robot simulation tools to an industrial manufacturing cell in order to analyze, model, and improve it. The chapter draws on the potential applications of such developments as contributing to the enhancement of the digital factory initiative. The general discussion conducted in the text also pertains to packaging and palletization operations within the food industry and other operations in the food industry, which serve as a case study.

Food Engineering Automation with Robotics and AI, First Edition. Abir Chakravorty.

10.1 Modeling Simulation and Practical Application of Robotics in Food Manufacturing Different Simulation Scenario

This chapter discusses the use of robot simulation tools in the analysis, modeling, and improvement of industrial manufacturing cells by synthesizing approaches toward upgrading the digital factory initiative. This report bases its discussion on a case study of the aerosol can packaging and palletizing operations of the metal can industry catering to food and products. In this paper, consideration will be given to the case in which the study of workers' manual packaging and palletizing and their time and motion analysis will be taken for the assessment of the processes applied. The paper focuses on the transition from manual operations to an advanced automated packaging and palletizing system after cycle time analyses. This proposed system is based on putting together a robotic manipulator and a production system with automated equipment and devices which were developed and tested for performance. From simulation and cycle time analysis, the feasibility of this new system was determined. The software of MATLAB, which is fuzzy logic software, can be used to quantify the efficiency of the system in terms of productivity and use existing facilities. A holistic design phase of the manufacturing cell can also be achieved through the use of the 3D simulation software DELMIA V6 (Rösen 2024). Implications of the simulation results indicate that a system with a level of productivity equivalent to approximately 4.3 manual cells could be possible using one robotized manipulator and an automated packaging and palletizing cell. These results underscore the need to shift to automation for the aerosol can packaging and palletizing operations.

10.1.1 Case Study: Automation in Aerosol Can Packaging

Introduction Industry Application: Metal Canning of Food and Aerosol Products: In the status quo, aerosol cans are packed and palletized manually with high time and manpower involvement (Srasrisom et al., 2020). Problem Statement Inefficiencies in Manual Operations: Cycle times increased, production rates lower with the attained productions. Workers get tired of the repetitive operations, and levels of productivity decline.

Needs Automation: It explains that automated systems are only in line for better efficiency and also lower the cost of labor.

Proposed Solution

Concept of Digital Factory: Robot simulation tools to be implemented for modeling and analysis for enhanced output through manufacturing cells.

Robotic Manipulator: A robotic system to be implemented on packaging and palletizing lines; it would be the step wherein the production rate is expected to shoot multifold times.

Economic Analysis: The economics of such a new automated system in comparison to the current manual work-cells has been studied by using such performance metrics as net present value and payback period. Production Rate Improvement: With the availability of automation, the production rate improves from 130 to 521 pas/hr, which implies absolute amounts of increased productivity.

Simulation and Results: Delmia V5 and MATLAB were used for the in-depth manufacturing cell design and analysis.

Cycle Time Analysis: The simulation results revealed that the output of the automatic process was able to produce 4.3 manual work-cells, thus manifesting efficiency.

Demand for Change: Based on the findings, this transformation would favor the automation of aerosol can packaging and palletizing with increased productivity and lower operational costs.

Implications for the Future: This report says that automation would bring improvements in manufacturing efficiency that is suitable for the needs of the new markets as well as for technological progress.

Current Work-cell Analysis

A worker who has received a live workout undergoes a cycle time analysis, that is, human cycle time (*CT*) by using one of the work measurement techniques called Method Time Measurement (MTM). Equation (10.1) has human cycle time for packaging and palletizing aerosol cans as cited in:

$$CT = \sum_{i=1}^{n} t_i \tag{10.1}$$

where CT = cycle time, sec /pa; t_i = material handling/assembly operation time at workstation i, sec; i = workstation.

The production rate (R_p) of a human work-cell can be illustrated as in Eq. (10.2):

$$R_p = \frac{60}{T_p} \tag{10.2}$$

R_p is in work units per hour, pa/hr; T_p is the production time per work unit on the workstation that takes the longest, minute. the matching production time measurement for a single pack on each workstation. The cycle time of a human work cell that packages aerosol cans in a single pack before palletizing them can be determined using Eq. (10.1). As a result,

$$CT = 15 + 20 + 10 + 10 + 30 + 15 = 100 \text{ sec} \tag{10.3}$$

It is also seen that aerosol cans' packaging at workstation 5 takes the longest, that is,

$$T_p = 30 \text{ sec} = 30/60 = 0.5 \text{ min} \tag{10.4}$$

Calculate the production rate from Eq. (10.2) will be: ["pas" packets]

$$R_p = \frac{60}{0.5} = 120 \text{ pas/hr} \tag{10.5}$$

The only alternative would be to reduce the cycle time (CT) to keep up with higher-paced job demands. This approach, however, can result in some levels of physical stress and entails muscle soreness and stiffness particular to neck and shoulder regions common with repetitive tasks when cycle time is reduced. Thus, increasing the production rates and meeting the demand might be done by increasing the work-cells or shifts. However, this will also increase the operating costs. So, for those areas of packaging and palletizing, automation is suggested as a remedy.

Robot Cycle Time Analysis

Cycle time analysis in robot operation helps facilitate the process of finding the amount of time needed for particular tasks before the workstation and robot programming. Using the basic cycle time equation derived in Eq. (10.6), the cycle time of the proposed robotic applications involves some form of material handling in palletizing/de-palletizing and machine loading/unloading applications (Srasrisom et al., 2020). The robot is used to support an automatic can transferring machine and an automatic carton sealer:

$$T_c = T_o + T_h + T_w \tag{10.6}$$

where T_c = cycle time, min/pa; T_o = time of the actual palletizing/de-palletizing, machine loading/unloading operations, min/pa; T_h = work part handling, min/pa; and T_w = average waiting time, min/pa.

The period that the robot manipulator will be engaged in all activities for packaging and palletizing aerosol cans. The robot cycle time, T_c can be calculated from Eq. (10.6), which is the total time of all activities:

$$\begin{aligned} T_c &= 10 + 20 + 171.19 + 95 + 258.78 = 554.95 \text{ sec} \\ &= 9.25 \text{ min}/19 \text{ pas} \end{aligned} \tag{10.7}$$

The production rate of a robotic cell is based on the average production time, which must include the time to set up the cell.

$$T_p = \frac{T_{\text{su}} + Q_{\text{Tc}}}{Q} \tag{10.8}$$

where Q = the number of work units produced throughout the production run; T_p = the average production time per work unit, min; and T_{su} = setup time, min. The reciprocal of the average production time is the production rate:

$$R_p = \frac{60}{T_p} \tag{10.9}$$

where R_p is expressed in work units per hour, pa/hr. For long-running jobs, R_p approaches the cycle rate R_c, which is the reciprocal of T_c. That is, as Q becomes very large $(T_{\mathrm{sin}}/Q) \rightarrow 0$ and

$$R_p \rightarrow R_c = \frac{60}{T_c} \tag{10.10}$$

where R_c = operation cycle rate of the machine, pa/hr; and T_c = operation cycle time which takes the longest, min /pa, from Eq. (10.3).

From Eq. (10.7), it can be seen that the T_c that takes the longest is 27.63 s, or

$$T_c = 27.63/60 = 0.46 \text{ min} \tag{10.11}$$

The production rate from Eq. (10.9) is obtained as:

$$R_p = \frac{60}{0.46} = 130.43 \approx 130 \, \text{pas/hr} \tag{10.12}$$

However, in the actual implementation of the system, the results of the production rate, which is obtained from the fuzzy logic model, will be enhanced by approximately 4 times.

$$R_p = 130.43 \times 4 = 521 \, \text{pas/hr} \tag{10.13}$$

Economic Analysis

Modern industrial work-cells should be compared to the economics of a new automated cans' packaging cell. One popular statistic for assessing the value of an investment is net present value. The following formulas are used to determine it: the capital recovery factor is denoted as A/P, C_t is the net cash flow at time t, n is the investment's economic lifetime, and EUAC is the equivalent uniform annual cash. Another useful number is the payback period, which is determined by solving. Here, the payback period is represented as T.

$$\begin{gathered} \text{EUAC} = -\text{ Investment cost } (A/P, 30\%, n) + \text{Annual income} - \text{Operation cost} \\ \sum_{t=0}^{T} C_t = 0 \end{gathered} \tag{10.14}$$

Using Eq. (10.14), where investment cost = 7,000,000 Baht, operation cost = 660,000 Baht/year, annual income = 5,745,600 Baht, and A/P which MARR = 30%, and $n = 5$ year, will be 0.41058 (note: 30 Bath = 1 US dollar). Thus,

$$\begin{aligned} \text{EUAC} &= -7{,}000{,}000(0.41058) + 5{,}745{,}600 - 660{,}000 \\ &= +2{,}211{,}540 \text{ Baht} \end{aligned} \tag{10.15}$$

Note: When the result of EUAC is positive, this robot project should be good for investment. Payback period can be calculated from Eq. (10.16). Thus,

$$\text{Payback period} = \frac{7{,}000{,}000}{5{,}745{,}600 - 660{,}000} = 1.38 \text{ years} \tag{10.16}$$

10.2 Robotic Cooking

10.2.1 Case Study: Gustoso the Intelligent Project

10.2.1.1 Project Overview

Objective: Development of an efficient and convenient system for preparing meals using intelligence (AlNaibari et al., 2019).

Time Efficiency: It cooks faster than humans; hence, users can order meals from anywhere, thus quite effective in saving a lot of time.

Cost Savings: Food wastage is reduced, and operations are saved for restaurants by not having plenty of chefs and waiters.

Design and Analysis

System Architecture: The project will feature an exhaustive system architecture that details the subsystems and their connectivity. This would enable the user to understand how the system works.

User Interface: The user interface with the system will be through a PC tablet, which controls cooking options and processes.

Major Learnings from the Survey

User Preferences: 80% of the respondents expressed interest in having a smart cooking device in their home. This therefore represents an excellent market opportunity.

Restaurant Service Expectations: 80% of the respondents said that quick service in a restaurant is vital; therefore, effective food service is important.

10.3 Robotic Arm Control and Task Training Through Deep Reinforcement Learning

10.3.1 Case Study: Robotic Arm Control and Task Training Through Deep Reinforcement Learning Overview

Objective: The objective of the study is to explore the deep reinforcement learning approach to managing control in robotic arms, with special attention to reward functions and hyperparameters in the learning of policy (Franceschetti et al., 2021).

Tasks: We consider two tasks: random target reaching and a pick and place application to be addressed using different reward functions. Methodology Algorithms Compared: The paper compares four DRL algorithms in order to estimate which can learn continuous torque control policies more effectively for manipulation tasks.

Performance Measurement: It used average return from each policy update based on test of 10 new task episodes to measure the performance of the agent.

Key Findings Implication on Reward Function: The quality of the learned policy is very much dependent on the reward function chosen. The paper emphasizes the selection of well-defined structures of reward for the purpose of efficient learning.

Algorithm Performance: The paper elaborates on policy architectures in great detail and implementations of formulations with reward in simulated and real-world execution of the best performing algorithms.

Theoretical Insights Algorithms Studied: The research focuses on policy-based algorithms, where VPG as well as TRPO are analyzed as to the strengths and weaknesses of the two algorithms.

Strength and Stability: Both VPG and DQN-NAF show good strength because VPG has a fast convergence; however, stability may be a problem with training; on the contrary, DQN-NAF is more stable and efficient, especially in terms of learning efficiency for continuous action space.

10.3.2 Pick and Place

A cylinder with radius $r_c = 2$ cm and mass $m_c = 0.5$ kg must be grasped by the robotic arm. The robot must transfer the object from its current position on the table, p_{cyl}, to a target position, p_{goal}, about 30 cm above its starting position, once the grip is stable:

$$p_{\text{cyl}} = \begin{bmatrix} 0.7 & 0 & 0.9 \end{bmatrix}^{\mathsf{T}} \quad p_{\text{goal}} = \begin{bmatrix} 0.7 & 0 & 1.2 \end{bmatrix}^{\mathsf{T}}$$

Both cylinder and goal poses are fixed, while the initial pose of the robot end effector p_{ee} is uniformly sampled. The state of the environment is:

$$s = \left[q^{\top}\dot{q}^{\top}p_{\text{ee}}^{\top}p_{\text{goal}}^{\top}p_{\text{cyl}}^{\top}\right]^{\top} \in \mathbb{R}^{45}$$

To finish the task, the robot requires $T \cdot T_s = 4$ s, which is equal to $T = 500$. To motivate the minimization of the following three distances, we set up a payment function based on geometry: The formula $d_1 = ||p_{\text{ee}} - p_{\text{cyl}}||$: the distance between the cylinder and the end effector; $2 = \sum_{i=1}^{3}(||p_{\text{cyl}} - p_{f_i}|| - r_c)$: the distance between the fingers and the cylinder's center of mass (CoM). Since the object is a cylinder on a table, its orientation is irrelevant, and we can find its position using p_{cyl} without loss of generality. Similarly, the second phalanx of finger $i \in \{1, 2, 3\}$ in $\sum_W$ has the 3×1 Cartesian location p_{f_i}. The fingers will never reach the cylinder CoM; because r_c serves as an offset. $d_3 = ||p_{\text{cyl}} - p_{\text{goal}}||$: the distance from the goal to the cylinder. Penalty to make distance weightings equal and opposite, let c_j be some arbitrarily chosen offset. Intuition about reward is:

$$\mathcal{R}_{s,1}^{a} = -\sum_{j=1}^{3} c_j d_j - c_a ||a||$$

An episode is successful if the gripper gets close to the cylinder and the cylinder gets close to the target: t is terminal if $\sum_{j=1}^{3} d_j(t) < 0.1$ m
Encouraging Reward:

The reward function is:

$$\mathcal{R}_{s,2}^{a} = \frac{1}{1 + \sum_{j=1}^{3} c_j d_j} - c_a ||a||$$

while the episode is terminated on the following state check:

$$t \text{ is terminal if } \begin{cases} d_1(t) > 0.035 \text{ m} \\ d_3(t) > 0.035 \text{ m} \end{cases}$$

1. VPG: The SGA algorithm is based on the premise that judgments in the future are independent of past rewards until the policy is altered. Therefore, it uses only N_τ policy rollouts to estimate the policy gradient. Moreover, even when data is generated from a highly stochastic system, it guarantees convergence to the actual gradient at the fastest theoretically attainable error reduction rate of $\mathcal{O}\left(N_\tau^{-1/2}\right)$. The disadvantage is that the gradient is derived by averaging values of a sample of transitions from the existing policy. Therefore, the gradient depends on transitions sampled at random and the variance generated is far from being insignificant.

2. TRPO: The algorithm allows for more control, as there is more precision than in VPG, due to the introduction of the surrogate loss. It also improves the stability of the training by imposing, at each interaction, a K L divergence constraint on the size of the update in the policy. Therefore, there is no update such that one step changes the policy drastically. On each interaction, TRPO solves a constrained optimization problem which maximizes the expectation of an advantage function estimated through the GAE algorithm, at the cost of monotonic increase in performance. It proved very successful when used by high-dimensional robots to solve locomotion tasks in contact-rich environments. Considering that we believe our manipulation tasks are no longer computationally demanding, we expect good results. Among the actor-critic algorithms, we select DDPG and DQN-NAF.

10.4 Sensors for Automated Food Process Control

During operations like harvesting, the farmer cooperates with the landscape; they go around the ridges and crops. In order not to damage either of these features accidentally, it's indispensable that the robot performing these operations is able to detect ridges as well as safe paths between crops with all its onboard sensors. This capability enables the robot to follow ridges using a proprietary coordinate system developed uniquely for the purposes of robotic navigation. Established work established how to locate and track a ridge along a robot using a combination of attached sensors, such as LiDAR. According to previous studies, this paper will present a novel approach for ridge detection based on imagery combined with the PCD provided by the RGB-D camera mounted at the front end of the robot. Moreover, the robot was designed in such a manner that it would optimally have a distance from the farmer so that the movement of the farmer would not be restricted if they need interaction with the robot.

10.4.1 Case Study: Human-following Control in Agricultural Robots

Introduction: The present research was motivated by the agricultural robots crisis in Japan resulting from the shrinkage of the agricultural workforce and the desertion of land meant to be used for agricultural purposes. A small agricultural robot is proposed which can help to move the crops harvested and follow the user in furrowed fields.

Problem Statement: Agricultural robots must identify ridges and furrows to enable appropriate navigation. However, whenever a robot is being followed by a user at large, the issues of occlusion tend to occur more frequently, such that the identification of ridges becomes challenging.

The robot must become acclimated to changing different widths of furrows without hitting ridges or crops.

Methodology

Ridge Detection: The robot employs RGB-D camera to extract features corresponding to the edges of ridges from images. Contours detect in an HSV image, and ridge lines are calculated by the application of Hough transform.

User-following Control: This paper applied fuzzy set theory to track the user while keeping ridges away by creating a map of it. In this, the robot can adjust its path with the state of detected ridges on the map.

Experimental Results: Experiments were performed in a 1.1 m wide furrow, where the robot was successfully able to detect the ridge lines as well as adjust its motion toward avoiding collisions with the right ridge.

The proposed ridge detection and user-following control methods enable the robot to work effectively in the agricultural environment, and therefore ensures that transportation support is safe and smooth. The study highlights the requirement of utilizing onboard sensors to discern navigable furrows and obstacles to detect. This would be highly significant for the real-world implementation of the robot in agriculture. This work contributes to the development of an autonomous agricultural robot that solves labor shortages in agriculture and builds productivity through robotic perception. Future work could be focused on how the robot can detect ridges placed in a more challenging environment while enhancing the user-following algorithms.

10.4.2 Velocity Command Based on Mixed MF

According to the mixed MF μ_{mix} which is obtained through the integration of μ_{user} and μ_{ridge}, the moving direction, θ_{out}, and the translational moving speed, v_{out}, are calculated as:

$$\theta_{\text{out}} = \arg\max\left(\mu_{\text{mix}}(\theta)\right)$$

$$v_{\text{out}} = v_{\text{robot}}^{\max}\,\mu_{\max}\left(\theta_{\text{out}}\right)$$

where $v_{\text{robot}}^{\max}$ is the maximum translational speed of the robot. Then, the velocity command is calculated as follows:

$$v_x = v_{\text{out}} \cos\theta_{\text{out}}$$

$$\omega_z = \begin{cases} K_\omega \theta_{\text{out}} & \text{if } K_\omega \theta_{\text{out}} < \omega_{\max} \\ \omega_{\max}\text{sgn}\left(\theta_{\text{out}}\right) & \text{else} \end{cases}$$

where $\omega_{\text{robot}}^{\max}$ is the maximum angular speed and K_ω is an angular velocity gain.

10.5 Indirect Teaching of the Robot Hand

Another work is on the indirect teaching of controlling a six-axis robotic arm, which is further divided into a process where remote control is utilized in operation. The flaw with such a method is that it introduces a time delay inherent in the system, which degrades the performance of the controlling system. An approach called a Smith predictor has been envisioned for overcoming the problem with an idea that incorporates a model-based structure but remains prone to uncertainties and disturbances. This paper uses a model neural network (MNN) set up to approximate the dynamics of the system closely so that the capabilities of prediction are supported by the Smith predictor. Furthermore, to compensate for the time delay, and in order to make the system more responsive toward human commands, the paper proposes gains in the PID controls of the outer loop be adjusted. This adjustment is achieved with the help of a PID neural network, PIDNN, that also optimizes such control parameters. The proposed PIDNN has a very significant impact on reducing the time of learning taken by the robotic arm during indirect teaching. In this way, this type of control system becomes more efficient and effective. Figure 10.1 shows the indirect teaching of the robotic arm.

10.5.1 Control System Principle and Design

This paper presents a remote-control system divided into three major subsystems: the primary remote controller, the signal transmission mechanism with inherent time delays, and also the secondary controlled subsystem. The notation $s(t)$ is used to represent the command input to the primary remote controller. The main focus of this research work is to reduce the effect generated by these time lags by utilizing a Smith predictor. In this master-slave configuration, the transfer function from the primary system to the robotic arm becomes

$$\Phi(s) = \frac{G_c(s)G_m(s)e^{-\tau s}}{1 + G_c(s)G_m(s)^{-2\tau s}} \tag{10.17}$$

To make the stabilizing effect of this control system more strongly evident, which is undermined by a pure lag link in the equation, a Smith predictor is used in the slave component of the system. This is achieved by the addition of a compensating link to the controller, which is designed to counterbalance the lag associated with the controlled object.

The term $s(t)$ denotes a command for position, $u(t)$ the value of the control for position, $y(t)$ is the desired position (actually, the position to which the end-effector of the manipulator has to move), and $G_p(s)$ is the transfer function for the controlled process. The system introduces a delay, represented as $e^{-\tau s}$, with $G_p(s)$

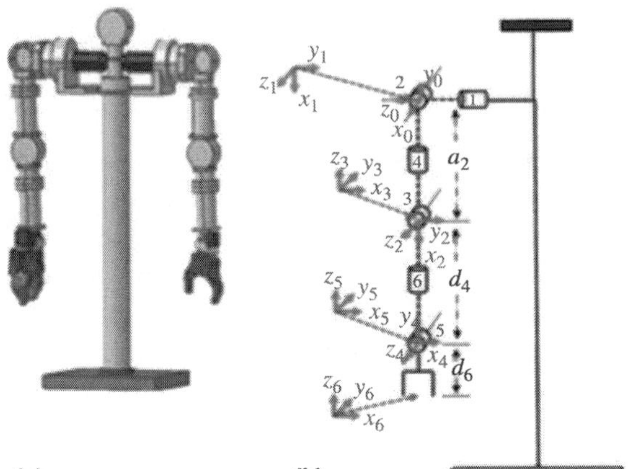

(a) (b) (c)

Link	θ	d	a	a
1	$\theta_1 - 90^\circ$	0	0	-90°
2	θ_2	0	a_2	0
3	$\theta_3 + 90^\circ$	0	0	-90°
4	θ_4	$-d_4$	0	90°
5	θ_5	0	0	-90°
6	θ_6	$-d_6$	0	0

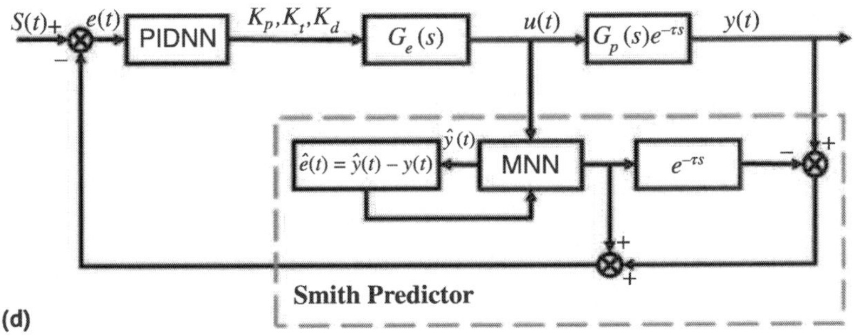

(d)

Figure 10.1 The indirect teaching of the robotic hand. (a) The dual 6 DoF manipulator, (b) coordinate system for various joints, (c) Denavit–Hartenberg parameter table, (d) system structure (Zhang et al. [2021]/Springer Nature).

being the transfer function again for the controlled process, and $G_m(s)$ the model to estimate. The inclusion of the Smith predictor alters the transfer function of the system.

$$\Phi(s) = \frac{G_c(s)G_p(s)e^{-\tau s}}{1 + G_c(s)\left(G_m(s) + G_p(s)e^{-\tau s} - G_m(s)e^{-\tau s}\right)} \tag{10.18}$$

After correction, pure lags do not affect the stability of the system if the model is correct; that is $G_p(s) = G_m(s)$. The premise of the Smith control method is that in order to build a precise prediction model, it is essential to know the precise mathematical model of the controlled item.

10.6 Tuning Humanoid Walking Parameters for Food Service Robots for Better Performance

10.6.1 Case Study: Adaptive PID Control Algorithm for the Two-legged Robot Walking on a Slope

PID Controller: A feedback mechanism for a control loop broadly applied in industrial control systems.

Biped Robots: They are those robots that walk on two legs to mimic human locomotion.

Slope Walking: The capabilities of biped robots to walk on sloping surfaces enhance their utility and application.

10.6.1.1 Research Objectives

To derive the gains of a torque-based PID controller that is trained by nature-inspired optimization algorithms MCIWO and PSO.

To improve the adaptability of the error-based angular displacement PID control.

10.6.1.2 Methodology

Neural Network Training: Optimization of the PID controller is gained by two algorithms:

MCIWO: A newly proposed nature-inspired optimization algorithm.

PSO: A widely known optimization technique.

Dynamic Tuning: The gains of the controller are adjusted online as a function of the robot's performance and changing environmental conditions.

10.6.1.3 Results

The effectiveness of the developed PID controller was tested through simulation and testbed implementations on a biped robot.

Robust Adaptive PID Control Algorithm Improves Walking Stability: The control algorithm contributed a great deal toward improved walking stability along slopes.

10.6.1.4 Challenges Addressed

Research addresses the challenges of biped robots with respect to stability on complex terrain.

Research Topics: Trajectory generation deals with proper walking trajectory generation for stability upon movement.

Improving adaptive PID control algorithms will make biped robots able to walk on slopes and rough grounds. This paper contributes to the development of the robotics area because improvements are made both on adaptability and stability of bipedal locomotion systems.

10.6.1.5 Directions on Future Work

There could be further research into optimizing other types of algorithms and their effects on performance. The performance, in this case, has a tendency to change depending on the different kinds of terrains of biped robots; it even goes so dynamic or uncertain conditions.

This case study captures the innovative approach of research in improving biped robot locomotion in the advanced control algorithms and optimization techniques incorporated.

10.6.2 Modeling and Design of Controller of the Biped Robot

This section deals with the biped robot model and the design of the associated PID controller. Modeling the use of biped robots in both industrial and nonindustrial applications is increasing significantly every day. The robot has a single motor installed on each joint, and all of its joints are serially coupled. On every hand, three degrees of freedom are built to be accommodated by the robot, while every leg of it is allocated with six degrees of freedom. As the balanced walk of a biped robot needs to follow an appropriate polynomial trajectory, this foot is considered in this paper under various types of polynomial trajectories as quadratic, cubic, and fifth-order. They found that, unlike the other trajectories, the cubic polynomial trajectory generated dynamically more balanced gaits. The coefficient of friction (μ) determines the greatest angle of slope at which the robot may walk without slipping. All the joints are allowed to move in unison in order to address the issue properly. The concept of inverse kinematics is applied to move the links in a systematic way while satisfying the kinematic constraints. The foot of the robot should always be parallel to the ground while walking, and the ZMP should lie within the foot support polygon. The following equation gives the formula for calculating the ZMP in the X and Y directions as presented in Eq. (10.19):

$$\begin{aligned} x_{\mathrm{ZMP}} &= \frac{\sum_{i=1}^{n} \left(I_i \dot{\omega}_i - m_i \ddot{x}_i z_i - m_i x_i \left(g + \ddot{z}_i\right)\right)}{\sum_{i=1}^{n} \left(m_i \left(\ddot{z}_i + g\right)\right)} \\ y_{\mathrm{ZMP}} &= \frac{\sum_{i=1}^{n} \left(I_i \dot{\omega}_i - m_i \ddot{y}_i z_i - m_i y_i \left(g + \ddot{z}_i\right)\right)}{\sum_{i=1}^{n} \left(m_i \left(\ddot{z}_i + g\right)\right)} \end{aligned} \tag{10.19}$$

where $\dot{\omega}_i, I_i, g$ and m_i represent the angular acceleration $(\mathrm{rad/s^2})$, mass moment of inertia $(\mathrm{kg \cdot m^2})$ of the ith link, acceleration due to gravity $(\mathrm{m/s^2})$ and mass (kg) of the link i, respectively. Further, (x_i, y_i, z_i) signify the coordinates of the ith lumped mass and $\ddot{z}_i$ and $\ddot{x}_i$ indicate the acceleration of the ith link moving in z- and x-directions $(\mathrm{m/s^2})$, respectively.

Once the ZMP is calculated, the dynamic balance margin of the biped robot in X- and Y-directions is determined by using the mathematical equations as represented in Eq. (10.20)

$$\begin{aligned} x_{\mathrm{DBM}} &= \left(\frac{f_s}{2} \times \cos\alpha - |x_{\mathrm{ZMP}}|\right) \\ y_{\mathrm{DBM}} &= \left(\frac{f_w}{2} \times \cos\alpha - |y_{\mathrm{ZMP}}|\right) \end{aligned} \tag{10.20}$$

where f_w and f_s represent the width and length of the foot, respectively.

With one exception—the acceleration caused by gravity, or "g," operating in the opposite direction of the root's movement—the gait-generating process for the biped while descending the slope is quite similar to that for the ascending scenario.

10.7 Sensorized Compliant Robot Gripper for Estimating the Cooking Time of Boil-cooked Vegetables

Figure 10.1d depicts a system which includes a Smith predictor used to compensate for stability problems generated from the pure lag in the equation. In this figure, $s(t)$ is defined as the command for position, $u(t)$ stands for the value for position control, while $y(t)$ is defined as the desired position or the location of the manipulator's end-effector. The transfer function of the controlled process is represented as $G_p(s)$. Delay is present in this system, and it is represented as $e^{-\tau s}$. In this, $G_p(s)$ refers to the transfer function of the controlled process and $G_m(s)$ is an estimation model. Adding the Smith predictor modifies the entire transfer function of the system.

10.7.1 The Mechanical Vegetable Model

As shown in Figure 10.2, the raw vegetable is considered to be a core, while the surrounding shell, which is made up of a material of a specific stiffness, is considered to be cooked. By Hooke's Law "F = kDelta x," the potato can be considered to be a spring system. The stiffness (k) can be considered to be the function of cross-sectional area (A) and Young's modulus (E).

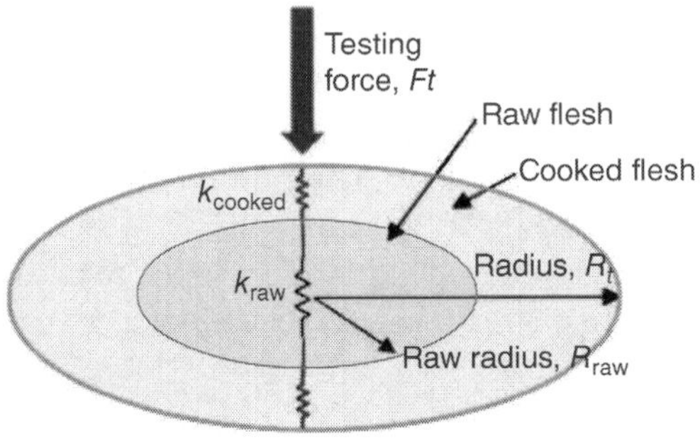

Figure 10.2 The vegetable mechanical model (Sochacki et al. [2021]/Springer Nature).

To give $k = \frac{EA}{L}$. Applied to the potato model, we can define the stiffness of the raw and cooked regions as:

$$k_{raw} = \frac{E_{raw}A_{raw}}{2R_{raw}}, \quad k_{cooked} = \frac{E_{cooked}A_{cooked}}{2(R_t - R_{raw})}$$

such that the stiffness of the total potato system can be given by the sum of the spring system:

$$k_{total} = \frac{k_{cooked}k_{raw}}{k_{cooked} + k_{raw}}$$

We make the simplifying assumption that $A_{raw} \approx A_{cooked}$, which should hold well as the pressure of an implement is usually focused on a small fraction of the vegetable area. The full spring constant becomes:

$$k_{total} = \frac{A}{2} \frac{E_{cooked}E_{raw}}{E_{cooked}R_{raw} + E_{raw}(R_t - R_{raw})}$$

10.7.2 Kinematic Model of Vegetable

Earlier research has demonstrated and examined the physics of boiling potatoes in real experiments. The potato was depicted as a uniform sphere with a constant outside temperature. This leads to a "cooking front" that demarcates the boundary between the cooked and uncooked parts of the potato and varies with time, according to Sochacki et al. (2021). An indicator of the "cookedness" of the potato is the ratio of the inner to outer spheres. A number of 1 indicates that the potato is raw, and a value of 0 indicates that it is fully cooked. The thermodynamics model that outlines this ratio is as follows:

$$\ln\left(\frac{\lambda_1\left(\frac{r_{raw}}{r_{total}}\right)}{\sin\left(\lambda_1\left(\frac{r_{raw}}{r_{total}}\right)\right)}\right) = -\lambda_1^2 \frac{\alpha}{{r_{total}}^2} t + \ln\left(\frac{A_1}{\Theta_f}\right) \tag{10.21}$$

where α is the thermal diffusivity of the potato, which is constant, and Θ_f is the temperature of the cooking front, which is taken to be constant. Thus, the model depends only on three parameters: the size of the potato r_{total}, and the coefficients A_1 and λ_1, which approximate the solution to the one-dimensional heat transfer equation. The last two depend only on the Biot number, which itself is defined using $Bi = h * r_{\text{total}}/k$, where k refers to the potato's thermal conductivity and h for the convection heat transfer coefficient; both parameters are constants. The Biot number is also a function of the potato size, but A_1 and λ_1 do not change much more than 3% for potatoes in sizes that will be likely to be tested. We now take advantage of the facts that $A_1 = 2$ and $\lambda_1 = 3.14$. So, any potato may be regarded as an infinitely large potato. Thus, the cooking function $f_c(t)$ can be taken as a time-dependent straight line. When $f(t) = 0$, the cooking time is the value of the ratio $r_{\text{raw}}/r_{\text{total}}$.

$$f_c(t) = \ln\left(\frac{\lambda_1\left(\frac{r_{\text{raw}}}{r_{\text{total}}}\right)}{\sin\left(\lambda_1\left(\frac{r_{\text{raw}}}{r_{\text{total}}}\right)\right)}\right) = \frac{A}{r_{\text{total}}}t + B \tag{10.22}$$

where A and B are constants, constructed by replacing constants with a single symbol, that must be found through stiffness experiments for potatoes cooked for different amounts of time to provide $\frac{r_{\text{raw}}}{r_{\text{total}}}$ for a known radius of potato.

10.8 Multiple Object Detection and Segmentation for Automated Removal in Additive Manufacturing with Service Robots

10.8.1 Case Study: Service Robot Used to Automate Object Removal in 3D Printing

Problem Background Motivation for Automating the Process: In low-quantity production, particularly with 3D printing, object removal from the build plate after the printing process is a significant bottleneck of the process. The idle periods for printers become economically not viable for the companies over a long period of time.

Current Limitations: Most of the existing 3D printers cannot automatically monitor and remove finished prints, thus requiring human effort.

Proposed Solution

Incorporation of Robotic Arm: This research paper presents a method which makes use of a robotic arm for the automatic monitoring and removal of several objects from a 3D printer. The method consists of two major components:

- It begins by analyzing the G-code in order to hypothesize the number, position, and size of printed objects.
- The robot's depth camera inspects the G-code analysis in order to be able to provide correct results in addition to flagging any errors that have occurred throughout printing.

10.8.2 Methodology

Data Acquisition: The system reads the G-code data and applies algorithms like MeanShift, DBSCAN, and OPTICS to identify the points at which objects should be placed while maintaining an inter-object distance of at least 20 mm.

Point Cloud Analysis: The analysis of the 3D scene using point clouds is performed. Techniques for cropping and segmentation are applied to refit the printing plate and objects.

10.8.3 Results and Test

Stability of G-code Analysis: The G-code analysis is found to have stable performance; thus, confident outputs for placement and object removal.

Functionality of the Robot: There is not only movement of the camera to optimal positions to be verified, but also efficient removal of printed objects by its gripper.

Future Enhancements for Future Work: Improved sensor techniques can be added to generate a point cloud. The width check of the object could be enhanced so that fine parts are not included in the object point cloud.

Effect on 3D Printing: The removal process might be automated. This could minimize downtime to a significant extent, and continuous printing operations may be maintained with higher productivity from the industry. This case study looks at the new trend of merging robotics with 3D printing; and hence, it can give opportunities for more efficient methodologies in manufacturing to be realized in the future.

10.9 Miscellaneous Case Studies

Measurement inaccuracies can greatly compromise the effectiveness of robots, especially in cases where they are supposed to manipulate objects that either happen to be very fragile or have differences in shape and size. This is really a major obstacle in solving the problem of picking and placing in an autonomously operating robot. We use the idea of "embodied" intelligence encapsulated in soft robotic technologies to design an integrated system, which we have termed Soft-Handler. This system addresses some of the critical shortcomings that have been shown to exist within traditional industrial pick-and-place robots. SoftHandler is defined by the incorporation of a novel parallel soft manipulator called SoftDelta, as well as by an innovative soft end effector, Pisa/(Istituto Italiano di Tecnologia) IIT SoftGripper.

10.9.1 Case Study: An Integrated Soft Robotic System for Handling Heterogeneous Objects

10.9.1.1 Overview

A novel robotic system, known as SoftHandler, addresses the challenges of heterogeneous object handling in industrial pick-and-place tasks. This system is provided with the advantages of soft robotic technologies toward making manipulation versatile and efficient.

10.9.1.2 Key Challenges Addressed

Bin-picking objects can be disposed of in random positions and could have more complex geometrical shapes, hence hard to look for pose precisely. This may cause failures in grasp or excessive interaction forces that will break the robot or crush the objects being dealt with.

Variety in Objects Placement: This might be either random or according to given configurations, for instance, specific food handling situations. There ought to be variety that calls for adaptability in handling systems that can adapt to various contexts.

10.9.1.3 System Design and Functionality

Mechanical Design: The system presents a variable-stiffness delta manipulator, which is a soft Delta, and a soft, articulated gripper, known as Pisa/IIT SoftGripper. This system overcomes some major drawbacks of traditional rigid robots, enabling a much more adaptive solution for various handling tasks.

Control Architecture: The control architecture preprocesses visual information captured by an Asus Xtion-Pro camera. The source of this camera is an infopoint for point cloud information about objects. The system segments the information to identify object clusters as well as what the next object is for picking.

Performance Metrics: The measured and approximated forces were compared, and the mean error was 1.24 N, and the peak error was 3.74 N while an object is being picked. This result implies that the system can pick objects with different stiffness and shapes without any difficulties.

Setup: In this experiment, there were two crates for picking and placing objects, which proved the practical applicability of the system in real life.

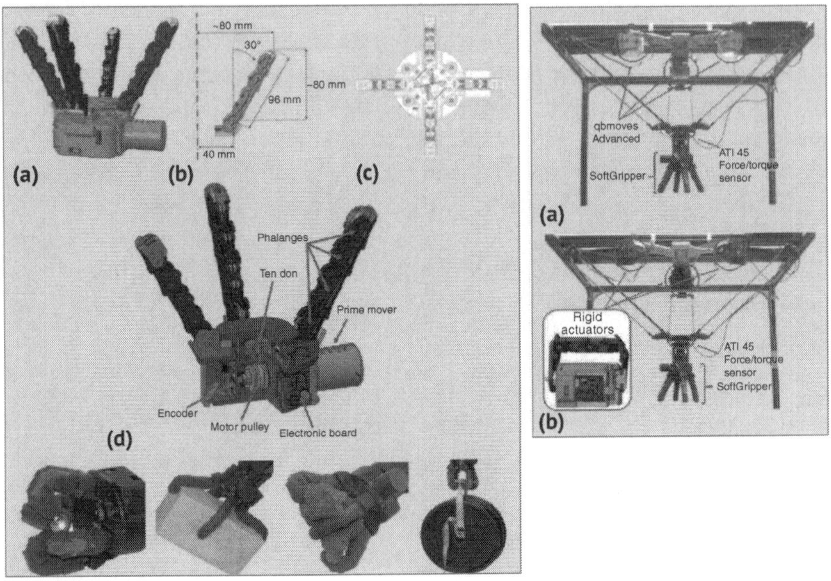

Figure 10.3 Various grippers, manipulators, and delta robots (Angelini et al. [2020]/IEEE/CC BY 4.0).

Figure 10.3 presents the SoftHandler system (Angelini et al., 2020) developed for handling heterogeneous objects. The system is a great breakthrough that can widely address significant problems concerning the handling of such items. Its innovative design, as well as control mechanisms, are powerful steps toward creating more efficient and adaptable robotic solutions for industrial applications.

Since it affects the speed performance of the manipulator, the success rate of the grasp is another important statistic. $T_{\text{tot}} = (n/\gamma)T$, where γ is the success rate for grasping is an estimated value of the total task time to executing a series of n cycles (e.g. to empty a basket of n oranges) if the time needed to complete one grasping cycle is T. This, directly, would mean that the stiff manipulator has to cycle faster in comparison to the time of the SoftHandler.

$$T_{\text{R}} < \frac{\gamma_{\text{R}}}{\gamma_{\text{S}}} T_{\text{S}} \tag{10.23}$$

where γ_{S} and γ_{R} are the probabilities of obtaining a successful grasp with the SoftHandler and the rigid delta, respectively, while T_{S} and T_{R} are the cycling times of the SoftHandler and the rigid delta, respectively. This means that, for example, assuming a maximum displacement error of 30 mm, we get (from Figure 10.3) $\gamma_{\text{S}} = 0.983$ and $\gamma_{\text{R}} = 0.713$; thus, the rigid robot should be 38% faster to recover the loss of performance due to the larger failure rate.

Force at the End Effector: In this section, we analyze the results obtained with the SoftDelta, comparing the experimental force measurements with the force estimated using the Cartesian stiffness model. Given a configuration of the Soft-Delta q and the wrench due to the end-effector weight, it is possible to compute the Cartesian stiffness. We choose $q_i = -0.2443$ rad, $i = 1, \ldots, 3$, that is, the configuration corresponding to the grasping phase. Given K_C, we can estimate the applied force $\tilde{\mathbf{f}}$ as

$$\tilde{\mathbf{f}} = K_C \cdot \Delta x, \tag{10.24}$$

where Δx is the Cartesian displacement from the reference position. Having as reference the grasping phase, the vector of the Cartesian displacement Δx_n at each node n $n(n = 1, \ldots, 81)$ of the grid is equal, in the x-y plane, to the grid displacement itself. Considering this, we compute an estimate for the force $\tilde{\mathbf{f}}_n = \left[\tilde{f}_{x,n}, \tilde{f}_{y,n}, \tilde{f}_{z,n}\right]^T$ in terms of all nodal displacements $n(n = 1, \ldots, 81)$. Since apriori the displacement along the z-axis is not known, we take the value to be 0 and therefore omit the estimation of force $\tilde{f}_{z,n}$ in our further analysis.

The measured values of force experiments $\mathbf{f}_{s,n}(t) = [f_{x,s,n}(t)$ and $f_{y,s,n}(t)$ and $f_{z,s,n}(t)]^T$ where t is the time variable, s is the index of a sphere $(s = 1, \ldots, 6)$, and $n(n = 1, \ldots, 81)$ is the nodal position on the grid, we mean the following force vector $f^*_{s,n} = [f^*_{x,s,n}$ and $f^*_{y,s,n}$ and $f^*_{z,s,n}]^T$, which corresponds to the maximum value of norm force during approaching to grasp.

$$f^*_{s,n} \triangleq f_{s,n}\left(t^*_{s,n}\right) = \begin{bmatrix} f_{x,s,n}\left(t^*_{s,n}\right) \\ f_{y,s,n}\left(t^*_{s,n}\right) \\ f_{z,s,n}\left(t^*_{s,n}\right) \end{bmatrix}, \forall s = 1, \ldots, 6,\ n = 1, \ldots, 81 \tag{10.25}$$

where $t^*_{s,n} = \operatorname{argmax}_t\left(\left\|\mathbf{f}_{s,n}(t)\right\|\right)$. Given (8), we then average the results of the successful grasps among the six spheres.

Thus, we define the vector of the average measured force as

$$\bar{\mathbf{f}}_n \triangleq \frac{1}{N_n} \sum_{s \in S_n}^{N_n} \mathbf{f}^*_{s,n}, \forall n = 1, \ldots, 81 \tag{10.26}$$

where N_n is the cardinality of S_n and S_n is the set containing the indexes of the spheres successfully grasped at the nth displacement. We introduce the performance statistic that corresponds to the nth grid displacement given. It measures the average error between the force measurements and the force vector estimated by the Cartesian stiffness model.

$$P_n \triangleq \left\| \begin{bmatrix} \tilde{f}_{x,n} - \bar{f}_{x,n} \\ \tilde{f}_{y,n} - \bar{f}_{y,n} \end{bmatrix} \right\|, \forall n = 1, \ldots, 81 \tag{10.27}$$

10.10 Review Questions

Q1 What are the latest advancements in robotic cooking technologies, particularly using pose extraction methods like OpenPose?

Q2 How can deep reinforcement learning improve robotic arm control in food manufacturing tasks?

Q3 What role do agricultural support robots play in enhancing food production efficiency through human-following control?

Q4 How does indirect teaching of robot hands facilitate the automation of complex food-handling tasks?

Q5 What parameters need tuning for humanoid robots to perform effectively in food service environments?

Q6 How can sensorized compliant grippers estimate cooking times for various food items, and what implications do this have for food quality?

Q7 What are the challenges and opportunities in automating multiple object detection and segmentation in food manufacturing?

Bibliography

M AlNaibari, T Mariam, M AlQattan, and B Nour. Gustoso the intelligent (cooking robot). 2019.

F Angelini, C Petrocelli, MG Catalano, M Garabini, G Grioli, and A Bicchi. Softhandler: an integrated soft robotic system for handling heterogeneous objects. *IEEE Robotics & Automation Magazine*, 27(3):55–72, 2020.

A Franceschetti, E Tosello, N Castaman, and S Ghidoni. Robotic arm control and task training through deep reinforcement learning. In: *International Conference on Intelligent Autonomous Systems*, 532–550, 2021. Springer.

C Rösen. Analysis of simulation modeling: a study of simulation modeling programs in the context of manufacturing processes. 2024.

G Sochacki, J Hughes, and F Iida. Sensorized compliant robot gripper for estimating the cooking time of boil-cooked vegetables. In: *International Conference on Intelligent Autonomous Systems*, 227–238, 2021. Springer.

K Srasrisom, P Srinoi, S Chaijit, and F Wiwatwongwana. Improvement of an automated can packaging system based on modeling and analysis approach through robot simulation tools. *International Journal of Robotics and Automation*, 9(3):178–189, 2020.

X Zhang, D Kim, J Bang, and J Lee. Time delay compensation of a robotic arm based on multiple sensors for indirect teaching. *International Journal of Precision Engineering and Manufacturing*, 22:1841–1851, 2021.

Appendix

List of Suppliers for Sensors and Automation Equipment

1. Pepperl+Fuchs India Pvt. Ltd.
   Proximity sensors, capacitive sensors, photoelectric sensors, ultrasonic sensors, vision sensors, intrinsic safety barriers, rotary encoders, explosion protection

2. Siemens India
   Temperature, pressure, flow, level sensors, HMIs, SCADA systems, industrial controllers

3. Turck India Automation Pvt. Ltd.
   Inductive, capacitive sensors, magnetic field sensors, fluid sensors, RFID solutions, field I/O systems

4. Honeywell Automation India Ltd.
   Pressure sensors, temperature sensors, humidity sensors, gas sensors
   - SCADA system, Building Automation Solutions

5. General Industrial Controls Pvt. Ltd. (GIC)
   - Sensors: Temperature sensors
   - Products: HMIs, PID controllers, PLCs

6. ABB India Ltd.
   - Sensors: Level, motion, pressure sensors
   - Products: Robotics, drives, smart sensors for condition monitoring

7. Rockwell Automation India Pvt. Ltd.
   - Sensors: Photoelectric, proximity, safety sensors
   - Products: PLCs, SCADA, IIoT Platforms

Food Engineering Automation with Robotics and AI, First Edition. Abir Chakravorty.

8. Omron Automation Pvt. Ltd.
    - Sensors: Photoelectric, proximity, vision sensors, safety light curtains
    - Products: Robotic solutions, industrial controllers

9. SICK India Pvt. Ltd.
    - Sensors: Optical, encoders, LIDAR, distance sensors
    - Products: Vision systems, safety systems

10. IFM Electronic India Pvt. Ltd.
    - Sensors: Proximity, fluid (flow, pressure, temperature), vibration sensors
    - Products: Monitoring condition devices, network interfaces, Industrial Robotics and Automation

11. Fanuc India Pvt. Ltd.
    - Sensors: Encoder sensors, force sensors
    - Products: Industrial robotics and CNC equipment

12. KUKA Robotics India Pvt. Ltd.
    - Sensors: Robotic motion and end-effector sensors
    - Products: Collaborative and industrial robots

13. Yaskawa India Pvt. Ltd.
    - Sensors: Torque, motion sensors
    - Products: Robotic arms, servo motors

14. B&R Industrial Automation Pvt. Ltd.
    - Sensors: Linear motion, temperature sensors
    - Products: Automation controllers, industrial PCs

15. Mitsubishi Electric India Pvt. Ltd.
    - Sensors: Temperature, motion sensors
    - Products: Servo motors, factory automation controllers

16. Delta Electronics India
    - Sensors: Flow, temperature sensors
    - Products: Drives, HMIs, industrial robots

17. Beckhoff Automation Pvt. Ltd.
    - Sensors: Encoder, optical sensors
    - Products: PC-based automation, IIoT solutions

18. Panasonic Industrial Devices Sales India
- Sensors: Motion, optical sensors
- Products: Factory automation systems

19. Lenze Mechatronics Pvt. Ltd.
- Sensors: Torque, motion sensors
- Products: Drives, controllers

20. Inventek India Pvt. Ltd.
- Sensors: Robotic vision sensors
- Products: Robotic systems, assembly automation, specialization in automation solutions

21. Technosys Automation Pvt. Ltd.
- Sensors: Motion and positioning sensors
- Products: Packaging and material handling solutions

22. Advance Industrial Automation Pvt. Ltd.
- Sensors: Industrial process sensors
- Products: Custom manufacturing line automation

23. Prolific Systems and Technologies Pvt. Ltd.
- Sensors: Process automation sensors
- Products: SCADA, process controllers

24. SP Automation and Robotics India Pvt. Ltd.
- Sensors: Vision, robotic sensors
- Products: Food and beverage automation systems

25. Triotech Automation Pvt. Ltd.
- Sensors: Position sensors
- Products: Pick-and-place systems

26. Tetra Pak India Pvt. Ltd.
- Sensors: Flow and liquid level sensors
- Products: Liquid food packaging automation

27. Festo India Pvt. Ltd.
- Sensors: Pneumatic sensors, pressure, flow sensors
- Products: Pneumatic and electromechanical systems

28. Schneider Electric India
- Sensors: Industrial proximity, energy management sensors
- Products: SCADA, control products

29. Phoenix Contact India Pvt. Ltd.
- Sensors: Signal conditioners, temperature sensors
- Products: Industrial connectivity, automation systems

30. Baumer India Pvt. Ltd.
- Sensors: Optical, pressure, capacitive sensors
- Products: Encoders, precision sensors, Interface and Connectivity Solutions

31. Weidmüller Electronics India Pvt. Ltd.
- Sensors: Signal converters
- Products: Interface systems, surge protection

32. Pilz India Pvt. Ltd.
- Sensors: Safety light curtains, pressure sensors
- Products: Safety controllers

33. HMS Industrial Networks India Pvt. Ltd.
- Sensors: IIoT-enabled sensors
- Products: Networking and connectivity solutions

34. TE Connectivity India Pvt. Ltd.
- Sensors: Industrial temperature, flow sensors
- Products: Connectors, relays

35. LAPP India Pvt. Ltd.
- Sensors: Connectivity-related signal sensors
- Products: Industrial cables

36. Balluff India Pvt. Ltd.
- Sensors: Proximity, linear position, photoelectric, RFID sensors
- Physical products: Connectivity solutions

37. Banner Engineering India Pvt. Ltd.
- Physical sensors: Vision, ultrasonic, photoelectric sensors
- Products: Wireless automation systems

38. Leuze Electronic Pvt. Ltd.
- Sensors : Optical, barcode, laser distance sensors
- Products: Identification systems

39. Contrinex Automation Pvt. Ltd.
- Sensors : Inductive, photoelectric sensors
- Products: Industrial safety systems

40. Autonics India
- Sensors : Proximity, motion, rotary encoders
- Products: HMIs, industrial timers
  Rise and Specialized Developers

41. Bosch Rexroth India Pvt. Ltd.
- Sensors : Torque, motion sensors
- Products: Drive and control systems

42. Keyence India Pvt. Ltd.
- Sensors : Vision, laser displacement, barcode readers
- Products: Measurement systems, laser markers

43. Turck Banner Pvt. Ltd.
- Sensors: Inductive, capacitive, ultrasonic sensors
- Products: Joint sensing and automation solutions

44. Hindustan Automation Solutions
- Sensors: Process control sensors
- Products: Turnkey industrial automation

45. Inductive Automation Pvt. Ltd.
- Sensors: Data acquisition sensors
- Products: HMI and SCADA software

46. Pepperl+Fuchs (Process Automation Division)
- Sensors: Intrinsically safe sensors
- Products: Explosion protection systems

47. Bonfiglioli India Pvt. Ltd.
- Sensors: Encoder-based feedback systems
- Products: Gear motors, drives

48. Yokogawa India Ltd.
- Sensors: Temperature, flow, pressure sensors
- Products: Control systems, field instruments

49. SEW Eurodrive India Pvt. Ltd.
- Sensors: Motion control sensors
- Products: Drives, servo motors

50. Eaton Automation India
- Sensor applications: Energy management sensors
- Products: Industrial control systems

Index

Food Engineering Automation with Robotics and AI, First Edition. Abir Chakravorty.

g

h

i

q

r

S

t